普通高等教育“十一五”国家级规划教材

全国高等医药院校药学类[illegible]轮实验双语教材

生物制药工艺学实验与指导

（第2版）

主　　编　何书英

主　　审　高向东

副 主 编　刘　玮

编　　者　（以姓氏笔画为序）

王全逸　刘　玮　何书英

姬晓南　童　玥

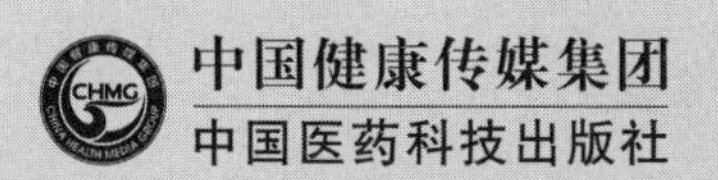

中国健康传媒集团

中国医药科技出版社

内容提要

本教材是“全国高等医药院校药学类专业第二轮实验双语教材”之一，根据生物制药工艺学实验教学大纲的基本要求及课程特点编写而成，内容上覆盖生物制药工艺技术基础实验、生物药物制备综合性实验及设计性实验，总计33个双语实验。本教材不仅对实验方法进行了详尽的描述，而且对相关原理进行了阐述，使实验操作与理论教学紧密关联，有利于学生实验能力和科研素质的培养。

本教材可供生物制药、生物工程及生物技术等专业实验教学使用，同时对生物制药科技人员也具有参考价值。

图书在版编目（CIP）数据

生物制药工艺学实验与指导 / 何书英主编 . —2 版 . —北京：中国医药科技出版社，2019. 12

全国高等医药院校药学类专业第二轮实验双语教材

ISBN 978 -7 -5214 -1391 -5

Ⅰ. ①生…　Ⅱ. ①何…　Ⅲ. ①生物制品 -生产工艺 -实验 -双语教学 -医学院校 -教学参考资料

Ⅳ. ①TQ464 -33

中国版本图书馆 CIP 数据核字（2020）第 000770 号

美术编辑　陈君杞

版式设计　南博文化

出版　**中国健康传媒集团** | 中国医药科技出版社

地址　北京市海淀区文慧园北路甲 22 号

邮编　100082

电话　发行：010 -62227427　邮购：010 -62236938

网址　www. cmstp. com

规格　889 × 1194mm 1/16

印张　16

字数　354 千字

初版　2008 年 3 月第 1 版

版次　2019 年 12 月第 2 版

印次　2022 年 7 月第 2 次印刷

印刷　三河市万龙印装有限公司

经销　全国各地新华书店

书号　ISBN 978 -7 -5214 -1391 -5

定价　45. 00 元

获取新书信息、投稿、为图书纠错，请扫码联系我们。

序

Preface

教学是学校人才培养的中心环节，实验教学是这一环节的重要组成部分。“全国高等医药院校药学类专业实验双语教材”是中国药科大学坚持药学实践教学改革，突出提高学生动手能力、创新思维，通过承担教育部“世行贷款21世纪初高等教育教学改革项目”等多项教改课题，逐步建设完善的一套与药学各专业学科理论课程紧密结合的高水平双语实验教材。

本轮修订，适逢“全国高等医药院校药学类专业第五轮规划教材”及《中国药典》（2020年版）、新版《国家执业药师资格考试大纲》出版，整套教材的修订强调了与新版理论教材知识的结合，与《中国药典》（2020年版）等新颁布的法典法规结合。为更好地服务于新时期高等院校药学教育与人才培养的需要，在上一版的基础上，进一步体现了各门实验课程自身独立性、系统性和科学性，又充分考虑到各门实验课程之间的联系与衔接，主要突出了以下特点。

1. 适应医药行业对人才的要求，体现行业特色，契合新时期药学人才需求的变化，使修订后的教材符合《中国药典》（2020年版）等国家标准及新版《国家执业药师资格考试大纲》等行业最新要求。

2. 更新完善内容，打造教材精品。在上版教材基础上进一步优化、精炼和充实内容。紧密结合“全国高等医药院校药学类专业第五轮规划教材”，强调与实际需求相结合，进一步提高教材质量。

3. 为适应信息化教学的需要，本轮教材全部打造成为书网融合教材，即纸质教材与数字教材、配套教学资源、题库系统、数字化教学服务有机融合，为读者提供全免费增值服务。

4. 坚持双语体系，强调素质培养教材以实践教学为突破口，采用双语体系编写有利于加快药学教育国际接轨，提高学生的科技英语水平，进一步提升学生整体素质。

“全国高等医药院校药学类专业第二轮实验双语教材”历经15年4次建设，在各个时期广大编者的努力下，在广大使用教材师生的支持下日臻完善。本轮教材的出版，必将对推动新时期我国高等药学教育的发展产生积极而深远的影响。希望广大师生在教学实践中对本套教材提出宝贵意见，以便今后进一步修订完善，共同打造精品教材。

吴晓明

全国高等医药院校药学类专业第五轮规划教材常务编委会主任委员

2019年10月

前言

Preface

生物制药工艺学实验是将生物制药的理论和技术融为一体并付诸实践的重要课程。随着生物技术和生物分离工程技术的迅速发展，生物制药工业已成为现代制药工业的重要发展领域，有关生物制药工艺的新理论、新工艺、新技术层出不穷，为此，我们在2008版《生物制药工艺学实验与指导》教材的基础上，结合长期以来在教学实践中所积累的经验以及最新的生物药物进展，编写了这本《生物制药工艺学实验与指导》双语教材。

本教材增加了英文讲义，增加了疫苗、抗体、重组蛋白及反义寡核苷酸等生物技术药物的制备及分析。全书共分为三部分：第一部分为生物制药工艺技术基础实验，包括微生物发酵技术、动物细胞培养技术、固定化酶技术、基因工程技术和生物大分子分离纯化技术等基础实验；第二部分为生物药物制备综合性实验，涉及天然生物药物的制备和生物技术药物的制备；第三部分为设计性实验。全书共包括33个实验，每个实验都按实验目的、实验原理、实验材料、实验方法及思考题的形式进行阐述。本教材还包括电子素材资料，包括实验PPT、实验习题、实验知识点及实验相关视频材料。

生物制药工艺学在生物制药、生物技术、生物工程、海洋药学等专业学生的学习中具有重要地位，是培养高级生物制药技术人才的重要专业课程。本教材除可作为专业教材外，对生物制药科技人员也具有参考价值。

本教材由高向东教授主审，在全书的编写过程中，高教授给予了大量指导性建议，谨此表示衷心的感谢！另外对2008版《生物制药工艺学实验与指导》参与编写的编委一并表示感谢！

由于编者水平所限，对于书中存在不足之处，敬请读者批评指正。

编　者

2019年9月

Contents

第一部分　生物制药工艺技术基础实验

第一节　微生物发酵技术基础实验

微生物发酵（microbiology fermentation）亦称微生物工程，是生物工程的重要组成和基础，它利用微生物的作用并通过近代工程技术来实现有用物质向工业化生产或其他产业过程转化的科学技术体系。它以微生物学、生物化学、遗传学的理论为基础，开发自然界微生物资源和其所有的潜在功能，使之应用于生产实践。其主要包括原料的处理，有用微生物的筛选和诱变，菌种工业应用的最适培养条件的选择，代谢的调节和控制，生物反应等的研究和设计，发酵工艺中各种参数的测试与自控，产物的分离和提取等。微生物发酵技术与基因工程、细胞工程、酶工程相互密切结合、相互渗透和相互促进，是科研成果从实验室向商业化转移的重要课题。总的来说，微生物发酵既是开发生物资源的关键技术，也是生物技术产业化的重要环节。

本节内容主要总结了我校长期开设的生物制药工艺学理论和实验课以及科学研究中的部分工作经验，并在参考国内外有关教材和资料的基础上，兼顾了微生物实验的基本操作和技能训练与专题实验相结合，着重介绍微生物的菌种的分离、纯化、培养、选育和保藏技术等内容。

微生物发酵技术的研究开发内容之一的菌种筛选和诱变技术，为人类社会创造了巨大财富。在这方面，本节介绍了微生物的菌种的分离、纯化、培养、选育。学生学习后可以从各种环境中分离出不同类型微生物，还可以通过诱变（包括物理诱变）和化学诱变和其他手段进行菌种的遗传改造，并从中筛选出性状优良的突变株和重组体。同时还介绍了微生物菌种保藏技术，使得从自然界直接分离的野生型菌株以及经人工方法选育出来的优良变异菌株被保藏后不死亡、不变异、不被杂菌污染，并保持其优良性状，以利于生产和科研使用。

在微生物发酵技术中考察环境因素对微生物生长的影响，测定微生物的生长曲线也很重要，为此介绍了微生物生长曲线测定实验。

微生物发酵技术还包括发酵产物的分离和提取，这方面的内容将在本书第二部分进行介绍。

Part 1　Basic experiments of biopharmaceutical technology

Section 1　Basic experiments of microbial fermentation technology

Microbiology fermentation, also known as microbial engineering, is an important component and foundation of bioengineering. It uses microorganisms and modern engineering technology to achieve useful substances to industrial production or other industrial processes into the scientific and technological system. Based on the knowledge of microbiology, biochemistry, and genetics, along with the development of microbial resources and their potential functions, make it applied to production practice. Microbiology fermentation mainly includes the treatment of raw materials, the selection and mutagenesis of useful microorganisms, the selection of optimal culture conditions for industrial application of strains, the regulation and control of metabolism, the research and design of biological reactions, the test and control of various parameters in the fermentation process, the separation and extraction of products and so on. Microbial fermentation technology and genetic engineering, cell engineering, and enzyme engineering are closely related to each other, mutual penetration and mutual promotion, which is an important topic for the transfer of scientific research results from laboratory to commercialization. In general, microbial fermentation is not only a key technology for the development of biological resources, but also an important part of biotechnology industrialization.

In this section, we mainly summarize the long – term biopharmaceutical technology theory and experimental courses and some work experience in scientific research. Based on the relevant domestic and foreign textbooks and teaching materials, we also designed the basic operations of microbiology experiments. The combination of skill training and special experiments focuses on the separation, purification, cultivation, breeding and preservation techniques of microbial strains.

Strain screening and mutagenesis techniques, one of the research and development of microbial fermentation technology content, has created enormous wealth for the human society. Thus, the isolation, purification, culture, and breeding of microbial strains are introduced. Students could isolate different types of microorganisms from various environments, and carry out genetic modification of strains through mutagenesis, including physical mutagenesis, chemical mutagenesis and other methods, and screen out mutant strains and recombinant strains with good characters. At the same time, the microbial strain preservation technology is introduced, so that wild – type strains directly isolated from nature and excellent mutant strains selected by artificial methods are preserved, not mutated, not contaminated by bacteria, and maintain their excellent traits and keep the excellent characters, so as to facilitate production and scientific research.

In microbial fermentation technology, it is also important to investigate the influence of environmental factors on the growth of microorganisms and to determine the growth curve of microorganisms.

Microbial fermentation technology also includes the separation and extraction of fermentation products, which will be presentedin the second part of the book.

扫码“学一学”

实验一　土壤中细菌、放线菌、酵母菌及霉菌的分离与纯化

【实验目的】

1. 掌握　获得微生物纯培养物的分离方法。

2. 了解　培养细菌、放线菌、酵母菌及霉菌等四大类微生物的培养条件和培养时间。

3. 学习正确使用微生物学实验室中常用的接种工具和各种接种技术。

【实验原理】

自然界的微生物资源十分丰富，广泛分布于土壤、水、空气、人体表面及与外界相通的腔道中。由于微生物通常以混居的群体形式存在，因此首先要分离出各种微生物的纯培养物。纯培养物是来自某种微生物的一个细胞分裂、繁殖而产生的后代。菌种分离与纯化技术是微生物学中重要的基本技术之一。

菌种分离与纯化一般包括采集菌样、富集培养、纯种分离和性能测定四个步骤。采集菌样首先要依据欲筛选的微生物生态及分布情况来选择采集地点。由于在土壤中几乎可以找到任何微生物，所以土壤往往是首选的采集目标。富集培养，又称增殖培养，就是利用选择性培养基的原理，限制不需要的微生物的生长，使所需的微生物大量繁殖。纯种分离的方法有稀释平板分离法、涂布法、划线分离法、单细胞分离法等。通过纯种分离，掌握微生物的接种和分离技术。在整个试验中，无菌操作技术是微生物实验成功的前提，需要严格掌握。

【实验材料】

1. 器材

（1）无菌培养皿　　18 只
（2）无菌移液管　　10 支
（3）涂布棒　　1 支
（4）玻璃珠　　1 瓶
（5）接种针　　1 支
（6）接种环　　1 支
（7）无菌试管　　8 支
（8）培养箱
（9）250ml 锥形瓶　　1 只
（10）酒精灯

2. 试剂

（1）土壤菌样
（2）马丁琼脂培养基

葡萄糖	1.0g
蛋白胨	0.5g
$K_2HPO_4 \cdot 3H_2O$	0.1g
$MgSO_4 \cdot 7H_2O$	0.05g
孟加拉红（1mg/ml）	0.33ml
琼脂	1.5 ~2g
水	100ml

自然 pH

（3）高氏合成 1 号琼脂培养基

可溶性淀粉	2.0g

KNO_3	0.1g
$K_2HPO_4 \cdot 3H_2O$	0.05g
NaCl	0.05g
$MgSO_4 \cdot 7H_2O$	0.05g
$FeSO_4 \cdot 7H_2O$	0.01g
琼脂	1.5~2g
水	100ml
pH	7.2~7.4

（4）马铃薯葡萄糖培养基

马铃薯浸汁（20%）	100ml
葡萄糖	2g
琼脂	1.5~2g

具体配制方法如下：将马铃薯去皮，切成 $2cm^2$ 的小块，放入200ml的烧杯中煮沸30分钟，注意用玻棒搅拌以防糊底，然后用双层纱布过滤，得到的滤液加葡萄糖，补足体积至100ml，自然pH。

（5）50000U/ml的链霉素

标准链霉素制品为10000000U/瓶，先准备20ml无菌水，在无菌条件下反复用无菌水将瓶中的链霉素溶解、转移、再溶解、再转移。最终得到的链霉素溶液为50000U/ml，临用时每1ml培养基加1μl即可。

（6）肉膏蛋白胨培养基

蛋白胨	1.0g
牛肉膏	0.5g
NaCl	0.5g
pH	7.2
水	100ml

固体培养基则加入1.5%~2%琼脂

（7）10%酚液

（8）无菌水

【实验方法】

1. 分离土壤菌样

（1）土壤稀释液的制备

1）土壤的采集　采集离地面5~20cm处的土壤几十克，盛入预先灭过菌的防水纸袋内，置于4℃冰箱中，待分离。

2）制备土壤悬液　称取土壤1g，经无菌操作，迅速倒入一个带玻璃珠并盛有99ml无菌水的锥形瓶中，振荡10~20分钟，这就是 10^{-2} 的土壤悬液。

3）制备土壤稀释液　利用10倍稀释法来制备土壤稀释液：用无菌移液管吸取0.5ml 10^{-2} 的土壤悬液，放入装有4.5ml无菌水的试管中，即得 10^{-3} 的土壤悬液；依此类推，可得系列稀释度的土壤悬液。

（2）分离细菌（稀释平板倾注法）　取四个无菌培养皿，将 10^{-6}、10^{-7} 的土壤悬液各取 1ml，接入平皿中，每个稀释度接两个平皿，做好标记。再将熔化并冷却至约 50℃ 的肉膏蛋白胨琼脂培养基倾入平皿中，立即将平皿轻轻地作旋转晃动，使菌液与培养基充分混匀，平置待凝固。最后将平板倒置于 30℃ 的培养箱中，培养 1 ~ 2 天，观察结果。

（3）分离霉菌（稀释平板倾注法）　取四个无菌培养皿，将 10^{-2}、10^{-3} 的土壤悬液各取 1ml，接入平皿中，每个稀释度接两个平皿，做好标记。再将熔化并冷却至约 50℃ 的马丁琼脂培养基（为了抑制细菌的生长，加入终浓度为 50U/ml 的链霉素）倾入平皿中，立即将平皿轻轻地作旋转晃动，使菌液与培养基充分混匀，平置待凝固。最后将平板倒置于 30℃ 的培养箱中，培养 3 ~ 5 天，观察结果。

（4）分离放线菌（稀释平板倾注法）　取四个无菌培养皿，将 10^{-4}、10^{-5} 的土壤悬液，每管加入 10% 酚液 4 ~ 5 滴（以抑制细菌和霉菌的生长），摇匀后各取 1ml，接入平皿中，每个稀释度接两个平皿，做好标记。再将熔化并冷却至约 50℃ 的高氏合成 1 号琼脂培养基倾入平皿中，立即将平皿轻轻地旋转晃动，使菌液与培养基充分混匀，平置待凝固。最后将平板倒置于 30℃ 的培养箱中，培养 5 ~ 7 天，观察结果。

（5）分离酵母菌（稀释平板涂布法）　取四个无菌培养皿，将熔化并冷却至约 50℃ 的马铃薯葡萄糖培养基倾入平皿中，平置待凝固。将 10^{-4}、10^{-5} 的土壤悬液各取 1ml，接入平皿中，每个稀释度接两个平皿，做好标记。用无菌玻璃涂棒将菌液自平板中央均匀向四周涂布扩散，平置待吸收完全。最后将平板倒置于 30℃ 的培养箱中，培养 2 ~ 3 天，观察结果。

2. 平板划线法分离微生物　混合菌悬液或当菌种不纯时通常用平板划线法进行纯种分离。

（1）制备平板　无菌操作，在火焰旁将熔化并冷却至约 50℃ 的琼脂培养基倾入平皿中，平置待凝固。

（2）划线分离

1）连续划线法　将接种环灭菌后，从待纯化的菌落或菌液蘸取少许菌种，点种在平板边缘处，再将接种环灭菌，以杀死过多的菌体，然后从涂有菌的部位在平板上做往返平行划线。注意划动要利用手腕力量在平板表面轻轻滑动，不要将培养基划破，所划线条平行密集而不重叠。

2）分区划线法　划线时一般将平板分为四个区，故也称四分区划线法。将接种环灭菌后，从待纯化的菌落或菌液蘸取少许菌种，在平板上的第 1 区做往返平行划线。将接种环灭菌后，从 1 区将菌划出至第 2 区，做往返平行划线。接种环再次灭菌，从第 2 区划出至第 3 区。依此类推，从第 3 区划出至第 4 区。

（3）培养　将平板倒置于培养箱中培养适宜时间，观察结果。

3. 接种　将分离纯化菌株进行斜面培养基接种、液体培养基接种和穿刺接种。

（1）斜面培养基接种　将接种环灭菌后，从待纯化的菌落或菌液蘸取少许菌种，然后轻轻在新鲜斜面上以“之”字形从斜面的下部划至上部，注意不要划破培养基。

（2）液体培养基接种　将接种环灭菌后，从待纯化的菌落或菌液蘸取少许菌种，液体培养基管斜放，将菌液涂于液面处管壁上，使得当试管直立以后菌种就在液体中。

（3）穿刺接种　先制备半固体培养基，盛入小试管或带螺口的穿刺培养小瓶内，高度约为总高度的2/3，高压灭菌后备用。可用针形接种针挑取分离良好的单菌落，刺入培养基总高度的 1/2 处，接种针沿原路抽出。盖上塞子或旋紧瓶盖，做好标记，进行培养。

扫码“练一练”

【思考题】

1. 如何从平板菌落的形态、与基质结合的紧密程度等来区分细菌、放线菌、酵母菌及霉菌?
2. 微生物接种有哪些方式?

EXPERIMENT 1 Isolation and purification of Bacteria, Actinomycetes, Yeasts, and Molds from the soil

【Purpose】

1. To master the isolation technology of microbial pure culture.

2. To be familiar with the culture conditions and culture time of bacteria, actinomycetes, yeasts, and molds.

3. To learn how to inoculate the microorganisms with common inoculation tools in microbiology laboratories.

【Principle】

Microbes are common in nature, and they are widely distributed in soil, water, air, the human body surface and the cavities connecting human and the outside world. Since microorganisms are usually mixed, it is necessary to isolate pure cultures of various microorganisms. A microbial pure culture is the result produced by a cell divided and multiplied from a certain microorganism. The technology of isolating and purifying strains is one of the important basic techniques in microbiology.

The isolation and purification of the specimens generally includes collecting the bacteria sample, enriching the cultures, separating the pure species and measuring the performance. The first step in collecting the bacteria sample is to determine the sampling place, according to the microbial ecology and distribution of the microorganisms. Because almost all microorganism can be found in the soil, soil is often the preferred collection target. Enrichment culture, also known as proliferation culture, uses the principle of selective medium to limit the growth of unwanted microorganisms and to multiply the required microorganisms. The methods of isolating pure cultures include: dilution plate separation method, coating method, scribing separation method, single cell separation method, and so on. The techniques of inoculating and separating microorganisms should be mastered by isolating pure cultures. Throughout the test, aseptic technique is a prerequisite for the success of microbial experiments, which needs to be strictly controlled.

【Materials】

1. Apparatus

(1) Sterile culture dish

(2) Sterile pipette

(3) Spreader

(4) Glass beads

(5) Inoculation needle

(6) Inoculating loop

(7) Sterile tube

(8) Incubator

(9) 250ml conical flask

(10) Alcohol Burner/Spirit Lamp

2. Reagents

(1) Soil microorganisms

(2) Martin agar medium

Glucose	1.0g
Peptone	0.5g
$K_2HPO_4 \cdot 3H_2O$	0.1g
$MgSO_4 \cdot 7H_2O$	0.05g
Bengal red (1mg/ml)	0.33ml
Agar	1.5 ~ 2g
Water	100ml
Natural pH	

(3) Gause's Synthetic No. 1 agar medium

Soluble starch	2.0g
KNO_3	0.1g
$K_2HPO_4 \cdot 3H_2O$	0.05g
NaCl	0.05g
$MgSO_4 \cdot 7H_2O$	0.05g
$FeSO_4 \cdot 7H_2O$	0.01g
Agar	1.5 ~ 2g
Water	100ml
pH	7.2 ~ 7.4

(4) Potato Dextrose Agar Medium

Potato dip (20%)	100ml
Glucose	2g
Agar	1.5 ~ 2g

The specific preparation method: peel the potato, cut into small pieces of $2cm^2$, boil in a 200ml beaker for 30 minutes, which needs to pay attention to stir with a glass rod, then filter with double gauze, and add glucose to the filtrate. Make up the volume to 100ml, adjust to natural pH.

(5) 50000U/ml streptomycin

The standard streptomycin preparation is 10000000U/bottle. At first, preparing 20ml sterile water, the streptomycin in the bottle is repeatedly dissolved, transferred, redissolved, and then transferred by using sterile water under aseptic conditions. The finally prepared streptomycin solution is

50000U/ml. Adding 1μl of 50000U/ml streptomycin to per mL of medium when use.

(6) Meat paste peptone medium

Peptone	1.0g
Beef cream	0.5g
NaCl	0.5g
pH	7.2
Water	100ml

1.5% ~2% agar is added to solid medium

(7) 10% phenol solution

(8) Sterile water

【Procedures】

1. Separation of soil bacteria

(1) Preparation of soil diluent

①Soil collection

Collect tens grams soil from 5 to 20cm under the ground, place them in a sterile waterproof bag, and place the bag in a refrigerator at 4℃.

②Preparation of soil suspension

Weigh 1g soil by aseptic operation, quickly pour them into a conical flask with glass beads and 99ml of sterile water, shake for 10 to 20 minutes, which is 10^{-2} soil suspension.

③Preparation of soil dilution

Soil dilution was prepared by using 10 - fold dilution method: using a sterile pipette to draw 0.5ml soil suspension into a tube with 4.5ml sterile water, to obtain 10^{-3} soil suspension and so on, a series of dilutions of soil suspension are available.

(2) Separating bacteria (diluted pour plate technique)

Take four sterile culture dishes, take 1ml of each of 10^{-6} and 10^{-7} soil suspensions, insert them into two plates for per dilution and remember to mark them. Then, pour the meat paste peptone agar medium which has been melted and cooled to 50℃ into a dish, and vortex plate gently, which mixes the bacterial solution and the medium thoroughly. Then place flat to be solidified. Finally, place the plate in an incubator at 30℃, culture them for 1 to 2 days, and the results could be observed.

(3) Isolation of molds (diluted pour plate technique)

Take four sterile Petri dishes, take 1ml of each of 10^{-6} and 10^{-7} soil suspensions to two plates for per dilution and mark them. Then, pour the Martin agar medium (to suppress the growth of the bacteria, add streptomycin at a final concentration of 50U/ml) which has been melted and cooled to about 50℃ into the dish, and vortex the plate gently to mix the bacterial solution and the medium thoroughly. Then place plates to be solidified. Finally, place the plates in an incubator at 30℃, culture them for 3 to 5 days, and the results could be observed.

(4) Separation of actinomycetes (diluted pour plate technique)

Take four sterile culture dishes, add 4 or 5 drops of 10% phenol solution to each tube of 10^{-4} or

10^{-5} soil suspension (to inhibit the growth of bacteria and molds), shake well and take 1ml into the plate. Prepare two plates for per dilution and remember to mark. Then, pour Gause's Synthetic No. 1 agar medium which has been melted and cooled to about 50℃ into the dish, and vortex the plate gently to mix the bacterial solution and the medium thoroughly. Then place plates to be solidified. Finally, place the plates in an incubator at 30℃, culture for 5 to 7 days, and the results could be observed.

(5) Isolation of yeasts (diluted spread plate technique)

Take four sterile culture dishes, and pour the potato dextrose medium which has been melted and cooled to about 50℃ into the dish. Draw 1ml of each 10^{-4} or 10^{-5} soil suspension, insert them into the plates. Prepare two plates for per dilution and remember to mark. Spread the bacterial liquid uniformly to the periphery from the center of the plate by a spreader, and place plates to be completely absorbed. Then place plates to be solidified. Finally, place the plates in an incubator at 30℃, culture for 2 to 3 days, and the results could be observed.

2. Isolation of pure culture – streak plate technique

Mixed bacterial suspensions or impure strains are usually isolated by the spread plate method.

(1) Preparation of the plates

Aseptically, pour the agar medium, which has been melted and cooled to about 50℃, into the dish near the flame, and lay plates to be solidified.

(2) Separation by scraping line method

①Continuous plate streaking

After sterilizing the inoculation loop, take a little strain from the colony or bacterial solution which needs to be purified, spot at the edge of the plate, and then sterilize the inoculating loop to kill bacteria. Make a round – trip parallel line on the plate from the coated area which needs to the use of wrist strength to gently slide on the surface of the plate. Do not cut the medium, and make the lines dense but not overlapped.

②Streak plate method

In this method, the plate is generally divided into four areas, so it is also called the four – part scribe method. After sterilizing inoculation loop, pick up a little from bacteria or bacteria colonies to be purified, round trip parallel marking in area 1 of the plate. After the inoculation loop was sterilized, the bacteria were drawn from the 1st zone to the 2nd zone, and the round – trip parallel scribing was performed. The inoculation loop was sterilized again and was drawn from the 2nd zone to the 3rd zone. And so on, from the 3rd zone to the 4th zone.

(3) Cultivation

The plate was placed in an incubator for a suitable period and the results could be observed.

3. Inoculating

Isolated and purified strains were inoculated on inclined medium, liquid medium and puncture medium.

(1) Inclined medium inoculation

After the inoculation loop was sterilized, pick up a little bit bacteria or bacteria colonies to be purified by the loop, and then gently draw the "z" shape from the bottom of the slope to the top on the fresh slope. Be careful and do not scratch the culture medium.

(2) Liquid medium inoculation

After sterilizing the inoculating loop, pick up a little bacteria or bacteria colonies to be purified by the loop, and the liquid medium tube is placed obliquely. Apply the bacterial solution to the tube wall at the liquid level, so that bacteria are in the liquid when the tube is erected.

(3) Puncture inoculation

The semi - solid medium is prepared at first, and placed in small tubes or puncture culture flasks with screw orifice, and the height is about 2/3 of the total height. All the medium should be used after autoclaving. A well - separated single colony can be picked up with a needle - shaped inoculation needle, pierced into the 1/2 of the total height of the medium, and the inoculation needle is withdrawn along the original path. Cover with a stopper or screw the cap, mark it, and reserved for the following use.

【Questions】

1. How to distinguish bacteria, actinomycetes, yeasts and molds by the morphology of plate colonies and the tightness of matrix binding?

2. What are the methods of microbial transplanting?

扫码“学一学”

实验二　菌种保藏实验

【实验目的】

1. 掌握　几种常用菌种的简易保藏法及其优缺点。

2. 了解　几种常用菌种的简易保藏法原理。

【实验原理】

微生物菌种的来源很广，大体上说，一类是从各种自然条件中分离的原始菌种，另一类是人工选育出来的优良选育种或用基因重组技术得来的重组菌，这些都是重要的资源。微生物的生长周期短、易于工业生产，但是人为的不断传代容易引起遗传变异，从而带来不必要的损失。菌种保藏（preservation of microorganism）的目的非常明确，在基础研究工作中，同一菌种在不同的时间，都应获得重复的实验结果。对于有经济价值的生产菌株，要求保持其高产的性能；对于重组菌，要求保持菌株本身遗传特性的稳定性。

由于菌种的变异主要发生于微生物旺盛生长繁殖过程，所以菌种保藏的原理是使微生物的新陈代谢处于最低水平或相对静止的状态，从而在一定的时间内菌种不发生变异并保持相应活力。低温、干燥、隔绝空气、无营养和添加保护剂是使微生物代谢能力降低或休眠的重要方法，因此大多数的菌种保藏都是根据上述五种原理设计的。

常用的简易菌种保藏法主要有以下几大类。

1. 斜面低温保藏法　为实验室和工厂菌种室常用的保藏法；主要措施是斜面培养后放置在4℃保存，适用于保藏各大类菌种尤其是需氧菌。保藏期限为3～6个月。优点是操作简便；缺点是保藏时间短，菌种经多次转接之后，容易发生变异。

2. 半固体穿刺保藏法 主要措施是穿刺培养后放置在4℃保存，适用于保藏酵母和细菌，尤其是厌氧菌。保藏期限为6～12个月。优点是操作简单易行；缺点是保藏时间较短，使用范围较窄。

3. 液体石蜡保藏法 在斜面培养物或者穿刺培养物上面覆盖1cm灭菌的液体石蜡后放置4℃直立保存。主要措施是低温和隔绝空气，适用于保藏各大类菌种，保藏期限为1～2年。优点是操作简便，不需特殊设备，也不需经常移种；缺点是保存时必须直立放置，占空间大。

4. 砂土管保藏法 主要措施是干燥和缺乏营养，适用于产孢微生物比如放线菌和霉菌，保藏期限为1～10年。优点是保藏期较长，对于产孢微生物保藏效果好，在抗生素工业生产中应用最广；缺点是对于营养细胞的保藏效果不佳。

5. 含甘油培养物保藏法 主要措施是超低温和利用作为保护剂的甘油渗入细胞后，强烈降低细胞的脱水作用，适用于在基因工程中保存含质粒载体的大肠杆菌，保藏期限为0.5～1年。优点是操作简便；缺点是需要－30℃或－70℃冷冻冰箱。

6. 冷冻真空干燥法 先使微生物在保护剂存在下低温（－70℃左右）快速冷冻，然后在减压条件下利用升华现象去除水分（真空干燥），使微生物处于低温、缺乏营养、干燥的条件下，适用于保藏各大类菌种。保藏期限为5～15年。优点是保藏期限长达数年乃至十几年，并且保藏效果好；缺点是保存操作烦琐，设备昂贵。

7. 低温冷冻保藏法 也称为液氮保藏法（－196℃）。适用于长期保藏各种微生物菌种。通常需要加入甘油、二甲基亚砜等保护剂减少低温结冰对细胞的伤害。优点是菌种形状不变异，保藏时间可达10～30年；缺点是需要特殊设备，成本高。

【实验材料】

1. 器材

（1）无菌试管　12支
（2）无菌移液管　2支
（3）1ml无菌的枪头　1盒
（4）1ml移液枪　1支
（5）接种环、接种针　各1支
（6）40目及100目筛子　各1个
（7）干燥器　1个
（8）安瓿管　2～3个
（9）真空泵
（10）冷冻干燥装置
（11）酒精灯
（12）灭菌锅

2. 试剂

（1）待保藏的菌种
（2）无菌液体石蜡
（3）无菌甘油
（4）五氧化二磷或无水氯化钙
（5）黄土、河沙等

(6) 适于培养待保藏菌种的各种斜面培养基

(7) 适于培养待保藏菌种的各种半固体深层培养基

(8) LB 培养基

胰蛋白胨	1g
酵母提取物	0.5g
NaCl	1g
pH	7.2

(9) 脱脂牛奶

(10) 2% HCl

【实验方法】

下列方法可根据实验室的具体条件与需要选做。

1. 斜面低温保藏法

(1) 接种　将不同菌种无菌操作接种在适宜的固体斜面培养基上。在距试管口 2~3cm 处，试管斜面的正上方贴上标签，注明菌株名称和接种日期。

(2) 培养　在菌株相应适宜的温度下培养，使其充分生长。如果是有芽孢的细菌或生孢子的放线菌及霉菌等，都要等到孢子生成后再行保存。

(3) 保藏　将斜面管口棉塞端用油纸包扎好，移至4℃冰箱中进行保藏。

(4) 移种　保藏时间依微生物的种类不同而不同，到期后需另行转接至新配的斜面培养基上，经适当培养后，再行保藏。不产芽孢的细菌最好一个月移种一次；酵母菌 2 个月移种一次；而有芽孢的细菌、霉菌、放线菌可保存 2~4 个月，然后再进行移种。

2. 半固体穿刺保藏法

(1) 接种　先制备半固体培养基，高度约为试管总高度的 2/3，高压灭菌后备用。用接种环挑取分离良好的单菌落，刺入培养基的 1/2 处。盖上试管塞，做好标记。

(2) 培养　在适宜的温度下培养，培养后的微生物在穿刺处和琼脂表面均可生长。

(3) 保藏　将培养好的菌种直立置于4℃冰箱中保藏。

(4) 移种　一般在保藏半年或一年后，需进行移种。使菌种转接到新鲜的半固体培养基中，依据上述步骤进行保藏。

3. 液体石蜡保藏法

(1) 液体石蜡灭菌　将液体石蜡（亦称石蜡油）分装后，121℃高压蒸汽灭菌 30 分钟。保险起见，可进行二次灭菌。灭菌后要将液体石蜡中的水分除去，通常的办法是在 40℃恒温箱中放置两个星期或置于 105~110℃的烘箱内约 1 小时。

(2) 接种和培养　将需要保藏的菌种接种至适宜的斜面培养基上，使生长良好。

(3) 加液体石蜡　用无菌吸管吸取已灭菌的液体石蜡，注入已长好菌的斜面上，液体石蜡的用量以高出斜面顶端 1cm 左右为准，使菌种与空气隔绝。

(4) 保藏　将已注入液体石蜡的斜面试管管口用牛皮纸包好，直立置于 4℃冰箱保存。在保藏期间如果发现液体石蜡减少应及时补充。

(5) 移种　到保藏期后，需将菌种转接至新的斜面培养基上，依据上述步骤进行保藏。值得注意的是从液体石蜡覆盖层下移种时，接种环在火焰上灼烧时菌体会随着液蜡四溅，如果培养物是病原体时，应予以注意。另外，第一代的培养物会因液蜡的残余而导致生长缓慢且有黏性，通常进行第二次转接才适合于菌种保藏。

4. 沙土管保藏法

（1）制备无菌沙土管

1）处理河沙 取河沙若干，用40目筛子过筛。加10%的稀HCl溶液浸泡（浸没沙面即可），除去有机杂质，浸泡2～4小时，倒去盐酸，用自来水冲洗至中性，烘干。

2）筛土 取非耕作层瘦黄土或红土（不含腐殖质）若干，加自来水浸泡洗涤多次至中性，烘干，磨细，用100目筛子过筛。

3）混合沙和土 取一份土加四份沙混合均匀，装入小试管中（10mm×100mm）。装量约1cm即可，塞上棉塞。

4）灭菌 121℃高压蒸汽灭菌1小时。每天一次，连续3天。

5）抽样进行无菌检查 取灭菌后的沙土少许，接入肉汤培养基中，37℃培养48小时，如有杂菌生长，则需重新灭菌。

（2）制备菌悬液 选择有芽孢的细菌或有孢子的菌种，等产生孢子或芽孢之后，吸取3～5ml无菌水至已培养好待保藏的菌种斜面中，用接种环轻轻搅动培养物，使成菌悬液。

（3）加样和干燥 用无菌吸管吸取菌悬液，在每支沙土管中滴加0.5ml菌悬液（大体上目测沙土刚润湿即可），用接种针拌匀，塞上沙土管棉塞。小试管放入真空干燥器或在干燥器中加五氧化二磷或无水氯化钙用于吸水，然后用真空泵抽干水分，通常不应超过12小时。

（4）抽样检查 每10支抽干的沙土管从中抽取1支进行检查。用接种环取少许沙土，接种到适合于所保藏菌种生长的斜面上培养。观察生长情况和有无杂菌生长。

（5）保藏及复苏 若经检查没有发现问题，可存放于冰箱或室内干燥处进行保藏，也可将管口烧熔再置于冰箱中保存，每半年检查一次活力和杂菌情况。恢复培养时，只需将少量的沙土倾撒在斜面上培养，等生长良好后再移种一次就可以使用。

5. 含甘油培养物保藏法

（1）甘油灭菌 用去离子水把甘油稀释至60%，分装后，塞上棉塞，外包牛皮纸，121℃高压蒸汽灭菌20分钟。

（2）接种与培养 将需要保藏的菌种接种至适宜的液体培养基上，过夜活化；以1%的接种量转接到新鲜的液体培养基中，使生长良好，达到对数生长期，一般需要4～6小时。

（3）培养物与灭菌甘油混合 在1.5ml培养物中加0.5ml灭菌的60%甘油，用涡旋器混合，使培养液与甘油充分混匀。然后将含甘油的培养液置于乙醇－干冰或液氮中速冻。

（4）保藏 将已冰冻含甘油培养物置于－70℃（或－20℃）冰箱中保存。

（5）转接 菌种复苏时，用接种环刮拭冻结的培养物表面，立即将黏附在接种环上的细菌划在含有适当抗生素的LB琼脂平板表面，然后在37℃培养过夜。冻结的培养物放回原处保藏。

6. 冷冻真空干燥法

（1）准备安瓿管和无菌脱脂牛奶 安瓿管一般用中性硬质玻璃制成，为长颈、球形底的小玻璃管。先用2% HCl浸泡过夜，然后用自来水冲洗至中性，分别用蒸馏水和去离子水各冲3次，烘干备用。高压蒸汽灭菌，121℃灭菌30分钟。将脱脂奶粉配成40%的乳液，121℃灭菌30分钟，并作无菌检验。

（2）制备菌悬液

1）培养菌种斜面　作为长期保藏的菌种，须用最适培养基在最适温度下培养菌种斜面，以便获得良好的培养物。培养时间掌握在生长后期，这是由于对数生长期的细菌对冷冻干燥的抵抗力较弱，如能形成芽孢或孢子进行保藏最好。一般来说，细菌可培养 24 ~ 48 小时，酵母菌培养 72 小时左右，放线菌与霉菌则可培养 7 ~ 10 天。

2）制备菌悬液　吸取 1 ~ 2ml 已灭菌的脱脂牛奶至待保藏已培养好的新鲜菌种斜面中，轻轻刮下菌苔或孢子（操作时注意尽量不带入培养基），制成悬液，浓度以 10^9 ~ 10^{10} 个/ml 为宜。

3）分装菌悬液　用无菌长滴管吸取 0. 1 ~ 0. 2ml 的菌悬液，通常加入 3 ~ 4 滴即可，要滴加在安瓿瓶内的底部，注意不要使菌悬液粘在管壁上。

（3）冷冻真空干燥操作步骤

1）预冻　装入菌悬液的安瓿瓶应立即冷冻。直接放在低温冰箱中（ -30℃以下）或放在干冰无水乙醇浴中进行预冻。

2）真空干燥　将装有已冻结菌悬液的安瓿管置于真空干燥箱中，真空干燥。并应在开动真空泵后 15 分钟内，使真空度达到 66. 7Pa。当真空度达到 26. 7 ~ 13. 3Pa 后，维持 6 ~ 8 小时，样品可被干燥，干燥后样品呈白色疏松状态。

3）熔封安瓿管　熔封必须在第二次抽真空情况下，当真空度达到 26. 7Pa 时，继续抽气数分钟，再用火焰在棉塞下部安瓿管细颈处烧熔封口。

4）保藏　将封口带菌安瓿管置于冰箱（2 ~ 8℃）中或室温保存。

5）安瓿管的启封　无菌条件下，于火焰上加热熔封口，然后立即加上 1 ~ 2 滴无菌蒸馏水或用酒精棉花轻擦，玻璃管可产生裂缝，接着轻轻敲击即可断落。加入少量的最适培养液使管内粉末溶解，即可接种在斜面或液体培养基中。通常也是使用经过复壮的第二代菌种。

扫码“练一练”

【思考题】

1. 菌种保藏的一般原理是什么？
2. 实验室中最常用哪一种既简单又方便的方法保藏细菌？
3. 试比较各种菌种保藏方法的优缺点。
4. 产孢子的微生物常用哪一种方法保藏？

EXPERIMENT 2　Preservation of pure culture

【Purpose】

1. To familiar with the simple preservation methods of several commonly used strains with their advantages and disadvantages.

2. To learn about the principles of simple preservation methods for several commonly used strains.

【Principle】

The sources of microbial strains are very broad. In general, one is the original strains isolated from various natural conditions, and the other is the excellent species artificially selected or the recombinant bacteria obtained by genetic recombination technology. These are all important resources. Microorganisms have short growth period and are easy to produce in the industrial form. But repeated sub-culturing is easy to cause genetic variation, resulting in unnecessary losses. The purpose of the preservation of strains is very clear. In the basic research work, the experimental results should be the same when using the strains at different times. For the production strains with economic value, it is required to maintain high-yield performance. For the recombinant strains, it is required to maintain the stability of the genetic characteristics.

Because the variation of strains mainly occurs in the process of growth and reproduction of microorganisms, the principle of strain preservation is to make the metabolism of microorganisms at the lowest level or relatively static state so that the strains do not mutate and maintain corresponding vitality within a certain period. Low temperature, dry air isolation, nutrient-free, and protective agents are important methods for reducing microbial metabolic capacity or dormancy. Therefore, most of the strains are designed according to the above principles.

Common preservation methods mainly have the following major categories.

1. Slant cryopreservation method

The slant cryopreservation method is commonly used in laboratory and factory. The main procedure is to store the tubes at 4℃ after slanting cultivation, which is suitable for preserving all kinds of bacteria, especially for aerobic bacteria. The preservation period is 3-6 months. The advantage is that the operation is simple. The disadvantage is that the storage time is short and the strain is easily mutated after being transferred for many times.

2. Semi-solid puncture preservation method

The main measure is to store the tubes at 4℃ after puncture culture. It is suitable for preservation of yeast and bacteria especially for anaerobic bacteria. The preservation period is 6-12 months. The advantage is that the operation is simple and easy, and the disadvantage is that the storage time is short and the use range is narrow.

3. Liquid paraffin preservation method

It is to cover the slant culture or the puncture culture with 1cm of sterilized liquid paraffin, and then store these cultures at 4℃ in an upright position. The main measure is low temperature and air isolation. It is suitable for preserving all kinds of microorganisms. The preservation period is 1-2 years. The advantage is that it is easy to operate and does not require special equipment or frequent transplanting. The disadvantage is that it must be placed upright when stored and needs large space for storing.

4. Sand and soil preservation method

The main measures are dryness and lack of nutrition, which should be suitable for spore-forming microorganisms such as actinomycetes and molds. The preservation period is 1-10 years. The advantage is that the preservation period is longer and the preservation effect on spore-forming microorganisms is good. It is the most widely used method in the production of antibiotics. The disadvantage is that the preservation effect is not good for nutritive cells.

5. Glycerol-containing culture preservation method

The main measures are ultra-low temperature and glycerine. Glycerine could strongly reduce the dehydration of cells, which is used as a protective agent after infiltration of cells. It is suitable for preserving *Escherichia coli* with the plasmid vector in genetic engineering, and the preservation period is 0.5 – 1 year. The advantage is the operation is easy while the disadvantage is that –30℃ or –70℃ freezer should be prepared.

6. Lyophilization (Free – Drying) method

The microorganisms are quickly frozen at a low temperature (about –70℃) in the presence of a protective agent, and then the water is removed by desublimation under reduced pressure (vacuum drying), thus the microorganisms are in a low temperature, lack of nutrition and dry conditions. It is suitable for the preservation of various major strains. The preservation period is 5 – 15 years. The advantage is that the preservation period is several years or even ten years, and the preservation effect is good. The disadvantage is that the storage operation is complicated and the equipment is expensive.

7. Cryopreservation method

It is also known as liquid nitrogen preservation method (–196℃), which is suitable for long – term preservation of various microbial strains. It is usually necessary to add a protective agent such as glycerin or dimethyl sulfoxide to reduce the damage of cells caused by low – temperature icing. The advantage is that the shape of the strain is not mutated, and the preservation time can reach 10 to 30 years. The disadvantage is that special equipment is required and the cost is high.

【Materials】

1. Apparatus

(1) Sterile test tube
(2) Sterile pipette
(3) 1ml sterile pipette tip
(4) 1ml pipette
(5) Inoculating loop
(6) Sieves (40 mesh and 100 mesh)
(7) Desiccator
(8) Ampoules
(9) Vacuum pump
(10) Alcohol burner
(11) Ultra – cold freezer
(12) Autoclave
(13) Sterilizer

2. Materials and reagents

(1) Strains
(2) Sterile liquid paraffin
(3) Sterile glycerol
(4) Phosphorus pentoxide or anhydrous calcium chloride
(5) Sand and soil

(6) Slant medium

(7) Semi – solid medium

(8) LB medium

peptone	1g
yeast extract	0.5g
NaCl	1g
pH	7.2

(9) Degrease milk

(10) 2% HCl

【Procedures】

The following methods can be selected according to the specific conditions and needs of the laboratory.

1. Slant cryopreservation method

(1) Inoculating

Different strains are sterile inoculated on a suitable slant medium. At a distance of 2 – 3cm from the test tube, a label is placed directly above the bevel of the test tube, indicating the name of the strain and the date of the inoculation.

(2) Cultivating

The strain is cultured at a suitable temperature to allow it to grow sufficiently. If it is a spore – forming bacterium or a spore – forming actinomycete and mold, wait for the formation of spores before preservation.

(3) Preservation

The end of the slant tube is wrapped with paper and transferred to a refrigerator at 4℃ for storage.

(4) Transferring

The storage time varies according to the different types of microorganisms. It needs to be transferred to a new slant medium and then preserved after proper cultivation. The spore – free bacteria are preferably transferred once a month; the yeast is transferred once every 2 months; and the spore – forming bacteria, mold and actinomycetes, can be stored for 2 – 4 months, and then transferred.

2. Semi – solid puncture preservation method

(1) Inoculating

The semi – solid medium is prepared at a height about 2/3 of the total height of the test tube, and was used after autoclaving. A well – separated single colony is picked with an inoculating loop and pierced into 1/2 of the medium. Cover the tube stopper and mark it.

(2) Cultivating

After cultured at a suitable temperature, the microorganisms can grow at both the puncture site and the agar surface.

(3) Preservation

The strains are placed upright in a refrigerator at 4℃ for storage.

(4) Transferring

After half a year or one year of preservation, transferring is required. Transfer the strain to fresh semi - solid medium and store them according to the above steps.

3. Liquid paraffin preservation method

(1) Liquid paraffin Sterilization

The liquid paraffin (also known as paraffin oil) is dispensed and sterilized by autoclaving at 121℃ for 30 minutes. To be on the safe side, secondary sterilization is possible. The water in the liquid paraffin should be removed after sterilization. The usual method is to place it in the incubator at 40℃ for two weeks or in an oven at 105 - 110℃ for about 1 hour.

(2) Inoculating and cultivating

Inoculate the strain onto a suitable slant medium to make it grow well.

(3) Liquid paraffin adding

Absorb the sterilized liquid paraffin with a sterile pipette and inject it onto the inclined surface of the grown bacteria. The amount of liquid paraffin is about 1cm above the top of the slant, so that the strain is isolated from the air.

(4) Preservation

The tube filled with liquid paraffin is wrapped in paper and stored upright in a refrigerator at 4℃. If liquid paraffin is found to decrease during storage, it should be added in time.

(5) Transferring

After the storage period, the strains should be transferred to a new slant medium and stored according to the above steps. It is worth noting that when moving from the reserved medium, the bacteria will splash with the liquid wax when the inoculating loop is burned on the flame. We should be aware of these pathogen cultures. In addition, the first - generation cultures are slow - growing and sticky due to the residual liquid wax, and usually the second transfer is suitable for bacterial preservation.

4. Sand and soil preservation method

(1) Preparation of sterile sand and soil

①Take some river sand and sieve it with a 40 mesh sieve. Soak with 10% dilute HCl solution (immersed in sand surface) to remove organic impurities and dip for about 2 - 4 hours. Pour HCl, rinse with tap water to neutral, and dry it.

②Take soil (excluding humus) and add tap water to wash several times to neutral. Dry, grind, and sieve them with 100 mesh sieve.

③Take a portion of the soil and add four portions of sand and mix them well. Put them into a small test tube (10mm × 100mm). The loading is about 1cm and the tampon should be plugged.

④Sterilizing by autoclave at 121℃ for 1 hour. Once a day for 3 consecutive days.

⑤Take a little bit of sterilized sand, inoculate them to the broth medium and incubate them at 37℃ for 48 hours. If there are bacteria growing, resteriling is needed.

(2) Preparation of bacterial suspension

Select spore - forming bacteria or spore - forming strains. After spores are formed, add 3 - 5ml of sterile water to the cultured medium. Gently stir the culture with the inoculating loop to make the bacteria suspension for using.

(3) Sampling and drying

Suck the bacterial suspension with a sterile pipette, add 0.5ml of the bacterial suspension to each sand tube (generally visually measure the sand just wet), mix them well with the inoculation loop and stuff the sand tube with a cotton plug. Put the small test tube into a vacuum dryer or add phosphorus pentoxide or anhydrous calcium chloride to the water in the desiccator, then use a vacuum pump to drain the water, usually the procedure should be progressed in no more than 12 hours.

(4) Sampling inspection

1 out of every 10 drained sand pipes is taken for inspection. A small amount of sand was taken from the inoculating loop and inoculated to a suitable slant medium. Observe the growth condition and whether there are any miscellaneous bacteria.

(5) Preservation and recovery

If no problems are found after inspection, it could be stored in the refrigerator or indoor dry place for preservation. The nozzle can be melted and stored in the refrigerator. The vitality of the bacteria will be checked every six months. When restoring the strains, just pour a small amount of sand on the slant medium surface. Transplant the strains after good growth and then they can be used.

5. Glycerol – containing culture preservation method

(1) Glycerol sterilization

Dilute the glycerol to 60% with deionized water. After dispensing, plug the tampon, wrap the paper and autoclave at 121℃ for 20 minites.

(2) Inoculating and cultivating

The strain to be preserved is inoculated onto a suitable liquid medium and activated overnight. It is transferred to fresh liquid medium at a 1% inoculation amount to grow well. It generally requires 4 to 6 hours to reach a logarithmic growth phase.

(3) Culture mixed with sterile glycerol

0.5ml of sterilized 60% glycerol is added to 1.5ml of the culture and the mixture is mixed with a vortex to thoroughly mix the broth with glycerol. The glycerol – containing medium is then frozen in ethanol – dry ice or liquid nitrogen.

(4) Preservation

The frozen glycerol – containing culture is stored in a –70℃ (or –20℃) refrigerator.

(5) Transferring

When the strain is resuscitated, the surface of the frozen culture is wiped with an inoculating loop. The bacteria attaching to the inoculating loop are immediately placed on the surface of a LB agar medium containing an appropriate antibiotic and then cultured at 37℃ overnight. The frozen culture is returned to its original location for preservation.

6. Lyophilization (Freeze – Drying)

(1) Prepare ampoule and sterile skim milk

The ampoule tube is generally made of neutral hard glass and is a small glass tube with a long neck and a spherical bottom. Soak them overnight with 2% HCl, then rinse them to neutral with tap

water, wash respectively three times with distilled water and deionized water, and dry for use. Autoclaved sterilize at 121℃ for 30 minites. The skim milk powder is formulated into a 40% emulsion, sterilized at 121℃ for 30 minutes, and subjected to sterility test.

(2) Preparation of bacterial suspension

①As a long – term preservation strain, the bacteria slope should be cultured at the optimum temperature to obtain a good culture. The culture time is in the late growth stage, because the bacteria in the logarithmic growth phase are less resistant to freeze – drying. It is the best preservation if spores can be formed. In general, bacteria can be cultured for 24 to 48 hours, yeast could be cultured for 72 hours, actinomycetes and molds can be cultured for 7 to 10 days.

②Take 1 – 2ml of sterilized skim milk into the fresh slant medium surface and gently scrape off the lawn or spores (precautions should be taken into the medium as much as possible) to make suspension. The concentration is preferably from 10^9 – 10^{10}/ml.

③Pipette 0.1 – 0.2ml of the bacterial suspension with a sterile pipette, and usually add 3 – 4 drops to the bottom of the ampoule. Be careful not to stick the bacterial suspension to the tube wall.

(3) Vacuum freeze – drying

①Ampoules filled with bacterial suspension should be frozen immediately. Pre – freeze directly in a low temperature freezer (below – 30℃) or in ice anhydrous ethanol bath.

②The ampoules containing the frozen bacterial suspension are placed in a vacuum oven and dried under vacuum. The vacuum should be 66.7Pa within 15 min after the vacuum pump is turned on. When the vacuum reaches to 26.7 – 13.3Pa, it can be kept for 6 – 8 hours. The sample can be dried and the sample is white and loose after drying.

③The sealing must be under the second vacuum. When the vacuum reaches to 26.7Pa, continue pumping for a few minutes, and then use a flame to seal the neck of the ampules.

④Store the sealed ampoules in the refrigerator (about 2 – 8℃) or at room temperature.

⑤Under the aseptic conditions, heat the sealing on the flame, then immediately add 1 – 2 drops of sterile distilled water or rub with alcohol cotton. The glass tube can crack, and then tap it to break. A small amount of the optimum medium is added to dissolve the powder in the tube, and it can be inoculated in a slant or liquid medium.

【Questions】

1. What is the principle of strain preservation?
2. Which is the simplest and most convenient way to preserve bacteria in the laboratory?
3. Compare the advantages and disadvantages of various strain preservation methods.
4. Which method is suitable for spore – forming microbes?

第二节　动物细胞培养技术基础实验

在生命科学的许多分支中，细胞培养已成为必不可少的一项技术。细胞培养技术包括动物细胞、植物细胞及微生物细胞培养技术。其中常见动物细胞有三类，即原代细胞、二倍体细胞系及转化细胞系。原代细胞是直接取自动物组织器官、经过分散，消化制得的细

胞悬液。供原代培养的动物细胞常见的有肝细胞、血管内皮细胞、脂肪细胞以及淋巴细胞。原代细胞经过传代、筛选、克隆，从而由多种细胞成分的组织中挑选强化具有一定特性的细胞株，其特点是：①染色体组织仍然是2n的模型；②具有明显贴壁依赖和接触抑制的特性；③只有有限的培殖能力，一般可连续传代培养50代；④无效癌性。转化细胞系是通过某个转化过程形成的，常因染色体的断裂而变成异倍体，从而失去了正常细胞的特点，而获得无限繁殖的能力。另外，从动物肿瘤组织中建立的细胞系也是转化细胞，转化细胞具有无限生命力，倍增时间较短，培养条件要求较低，适于大规模生产培养。

营养、pH及温度是动物细胞生长所要求的重要条件。培养细胞的最适温度为37℃ ± 0.5℃，偏离此温度，细胞的正常生长及代谢将会受到影响，甚至导致死亡。细胞培养的最适pH为7.2～7.4，当pH低于6.0或高于7.6时，细胞的生长会受到影响，甚至导致死亡。在保证细胞渗透压的情况下，培养液里的成分要满足细胞进行糖代谢、脂代谢、蛋白质代谢及核酸代谢所需要的各种组成，如各种必需氨基酸和非必需氨基酸、维生素、碳水化合物及无机盐类等。只有满足了这些基本条件，细胞才能在体外正常存活生长。目前已有人工合成培养基出售，如DMEM、RPMI－1640培养液。

动物细胞培养的方法，一般可根据培养细胞的种类分为原代细胞培养和传代细胞培养；又可根据培养基的不同分为液体培养和固体培养；还可根据培养容器和方式的不同分为静止培养、旋转培养、搅拌培养、微载体培养、中空纤维培养、固定床或流化床培养等。但从生产实际看，动物细胞的大规模培养主要可分为悬浮培养、贴壁培养和贴壁－悬浮培养。

在本节中，主要介绍了动物细胞的原代培养、传代培养、细胞冻存的基本知识和操作及利用动物细胞进行生物药物活性测定的一般实验方法。

Section 2　Basic experiment of animal cell culture technology

In many research areas of life science, cell culture has become an indispensable technology. Cell culture technology includes animal cell, plant cell and microbial cell culture technology. There are three kinds of common animal cells: primary cell, diploid cell line and transformed cell line. Primary cells are cell suspensions which are directly prepared from tissues and organs of animals after digested and dispersed. Liver cells, vascular endothelial cells, adipocytes and lymphocytes are the most common types of primary cultured animal cells. Primary cells are sub－cultured, screened and cloned to select cell lines with certain characteristics from different types of cells. The characteristics are as follows: ①Itschromosome tissue is accord with 2n model; ②It has obvious adherence dependence and contact inhibition characteristics; ③It has limited reproductive capacity and can be cultured for 50 generations continuously; ④noncancerous. Transformed cell lines are formed by a transformation process, and often become aneuploidy due to chromosome breakage, thus losing the characteristics of normal cells and gaining unlimited reproductive capacity. In addition, the cell lines from animal tumors are also transformed cells. The transformed cells have infinite vitality, short multiplication time and low culture conditions, which are suitable for large－scale production and culture.

Nutrition, pH and temperature are important conditions for the culture of animal cells. The optimum temperature of cultured cells is 37℃ ± 0.5℃. Deviating from this temperature, the normal

growth and metabolism of cells will be affected, even leading to death. The optimum pH for cell culture ranges from 7.2 to 7.4. When the pH is lower than 6.0 or higher than 7.6, cell growth will be affected, even leading to death. Under the condition of guaranteeing osmotic pressure, the components in culture medium should meet the requirements of cell for glucose metabolism, lipid metabolism, protein metabolism and nucleic acid metabolism, such as essential amino acids, non-essential amino acids, vitamins, carbohydrates and inorganic salts, etc. Only by satisfying these basic conditions can cells survive and grow normally *in vitro*. At present, synthetic media such as DMEM and RPMI-1640 have been sold.

According to the types of cultured cells, animal cell culture technology can generally be divided into primary cell culture and passage cell culture. According to different media, animal cell culture methods also can bedivided into liquid culture and solid culture. According to the different culture containers and modes, animal cell culture consists of static culture, rotating culture, stirring culture, microcarrier culture and hollow fiber culture, fixed bed or fluidized bed culture, etc. In terms of production practice, large-scale culture of animal cells can be divided into suspension culture, adherent culture and adherent-suspension culture.

In this section, we mainly introduce the basic knowledge and operation of primary culture, subculture, cryopreservation of animal cells, and general experimental methods of bioactivity determination of biopharmaceuticals using animal cells.

扫码"学一学"

实验三　大鼠主动脉内皮细胞的培养与鉴定

【实验目的】

1. **掌握**　大鼠主动脉内皮细胞的原代培养及传代培养方法。
2. **了解**　细胞培养的一般方法和步骤。
3. 学习内皮细胞鉴定方法。

【实验原理】

细胞培养或组织培养是指将细胞或组织从机体取出，给予必要的生长条件，模拟体内生长环境，使其在体外继续生长和增殖。细胞培养和组织培养并无严格区别。

细胞培养技术具有如下优点：①培养条件便于控制，可用于细胞生命活动规律的研究；②细胞处于体外培养环境中，便于用各种技术方法研究和观察细胞结构和功能的变化；③细胞在体外培养环境中可以长期地存活和传代，便于研究和观察细胞遗传性的改变；④可以比较经济大量地提供生物性状相同的细胞作为研究对象。但是细胞离体后失去与其周围环境的密切联系，其细胞生物学性质会发生改变。

原代培养也称初代培养，是从供体取得组织细胞后在体外进行的首次培养。原代培养细胞刚刚离体，生物学特性未发生很大变化，仍具有二倍体核型，最接近体内生长特性，适合用作药物测试、细胞分化等实验研究。本实验采用大鼠主动脉制备原代内皮细胞，血管内皮细胞是覆盖在血管内膜表面纵向排列的单层扁平细胞。原代培养方法很多，最常用

的有两种，即组织块培养法和消化法。这里介绍组织块培养法。

组织块培养法是将剪碎的组织块直接接种于培养皿或培养瓶中，加入培养基进行培养的方法。组织块培养法适用于内皮细胞产量低、管径较小血管的内皮细胞培养，其方法较消化法简便、经济。缺点是易混有成纤维细胞和平滑肌细胞。

本实验采用 Giemsa 染色进行内皮细胞的鉴定。内皮细胞在光镜下呈梭形、三角形或多角形，融合成片后的细胞可呈鹅卵石状镶嵌排列。

【实验材料】

1. 器材

（1）细胞培养瓶	100ml	1只
（2）培养板		1块
（3）剪刀		2支
（4）镊子		2支
（5）弯头镊		2支
（6）手术刀		1把
（7）平皿		2个
（8）离心管		2支
（9）吸管		数支
（10）二氧化碳培养箱		1台
（11）倒置显微镜		1台
（12）离心机		1台

2. 试剂

（1）胎牛血清（Fetal Bovine Serum，FBS）　经56℃灭活半小时后，小量分装，-20℃保存备用。

（2）无血清 RPMI 1640 培养基　取 RPMI 1640 培养基粉末一袋（规格为1L），加水溶解并稀释至1000ml，加青霉素钠 10^5U 和硫酸链霉素 10^5U，再加碳酸氢钠 2.0g，溶解混合均匀后调节 pH 7.0±0.2，以 0.22μm 微孔滤膜过滤除菌，分装，置于4℃保存。

（3）PBS　称取氯化钠 8.0g，氯化钾 0.20g，磷酸氢二钠 3.28g，磷酸二氢钾 0.24g，加双蒸水溶解并稀释至1000ml，经121℃ 20分钟高压灭菌备用。

（4）0.25%胰蛋白酶溶液　取乙二胺四乙酸二钠 0.02g，胰蛋白酶 0.25g，加 PBS 溶解并定容至100ml，过滤除菌，冷藏保存备用。

（5）Hank's 液　称取 $CaCl_2$ 0.14g，NaCl 8.0g，KCl 0.4g，KH_2PO_4 0.06g，$MgCl_2 \cdot 6H_2O$ 0.10g，$MgSO_4 \cdot 7H_2O$ 0.10g，$NaHCO_3$ 0.35g，$Na_2HPO_4 \cdot 12H_2O$ 0.13g，D-葡萄糖 1.0g，酚红 0.02g。配制方法：先将 $CaCl_2$ 溶于100ml水中；将其他成分依次溶解于约750ml水中，待前一种试剂完全溶解后，再溶解下一种成分；将1缓缓倒入2中，并搅动防止出现沉淀，补足水分，正压过滤灭菌，冷藏保存备用。

（6）Giemsa 储备液　称取 Giemsa 粉 0.5g，甘油 22ml，在研钵内先用少量甘油为 Giemsa 粉充分混合，研至无颗粒，再与剩余甘油混在一起，56℃保温2小时后加入33ml纯甲醇热过滤，棕色瓶保存。

（7）Giemsa 工作液按 1∶9 比例取 Giemsa 储备液和 pH 7.0 磷酸盐缓冲液混合，使用前配制。

【实验方法】

1. 颈椎脱臼处死大鼠，将大鼠浸泡于75%乙醇中消毒约1分钟。在无菌条件下，逐层剪开胸腔，取出大鼠胸主动脉，置于含高抗生素 Hank's 液的培养皿中冲洗2~3次除去残血。

2. 用剪刀和镊子将外膜的脂肪和结缔组织分离干净，并用磷酸缓冲液反复冲洗，直至无血液残留。纵向剪开管腔，平铺于培养皿中，用手术刀轻轻将血管切成边长约0.1cm的正方形，间隔放置于培养皿中。

3. 加入含有10%胎牛血清的RPMI 1640培养液，不能将组织块悬浮，置37℃、5% CO_2 条件下静置培养。

4. 待组织块周围生长的细胞融合成片，即可去除组织块，用胰蛋白酶消化法消化细胞。弃去原培养液，用PBS轻洗细胞表面两次，加入等体积的0.25%胰蛋白酶消化液，使其覆盖细胞表面。室温消化1~2分钟，显微镜下可见细胞收缩，细胞间隙清晰时，立即倾斜培养皿，使细胞与消化液避免接触，弃去消化液，直接加入含有10%血清的1640培养液，用吸管将细胞从瓶壁全部吹打下来，

5. 按内皮细胞以 1×10^5 个/ml 的密度接种于带有玻片的六孔板，培养48小时后染色。

6. 染色步骤　甲醇固定10分钟（或醋酸：甲醇固定30分钟）；固定后的细胞干燥，用Giemsa工作液完全覆盖样本染色10~15分钟；用自来水将玻片上的染色液冲洗，空气干燥，二甲苯透明；显微镜观察。

7. 结果分析　通过显微镜观察画出显微镜观察的形态。

【思考题】

扫码“练一练”

1. 内皮细胞培养应注意哪些问题？
2. 内皮细胞的鉴定方法包括哪些？

EXPERIMENT 3　Culture and identification of rat aortic endothelial cells

【Purpose】

1. To master the primary culture and subculture methods of rat aortic endothelial cells.
2. To understand the general methods and procedures for cell culture.
3. To learn how to identify endothelial cells.

【Principle】

Cell culture or tissue culture refers to taking tissue or cells out of the body, giving necessary growth condition, simulating the growth environment *in vivo*, and continuing to grow and proliferate *in vitro*. There is no strict distinction between cell culture and tissue culture.

Cell culture technology has the following advantages: ①The culture conditions are easy to con-

trol and can be used for the study of the regularity of cell life activities; ②The cells are in an *in vitro* culture environment, which is convenient for studying and observing changes in cell structure and function by various techniques; ③Cells can survive and passage for a long time *in vitro*, so that it is easy to study and observe the cytogenetic changes; ④It is possible to provide cells with the same biological traits as a research object in large quantities. However, the biological properties of the cells will change when they lose close contact with their surroundings.

Primary culture is the first culture *in vitro* after obtaining tissue or cells from donors. The biological characteristics of primary cultured cells have not changed greatly. They still have diploid karyotypes, which can mostly reflect the growth characteristics *in vivo*. They are suitable for experimental research such as drug testing and cell differentiation. In this study, rat aorta was used to prepare primary endothelial cells (ECs), which were monolayer flat cells covering the surface of intima longitudinally. There are many primary culture methods, including the method of enzymatic digestion and the method of tissue block. The method of tissue block was used in this experiment.

Tissue block method is to inoculate shredded tissue directly into Petri dish or flask and add culture medium for culture. Tissue block culture method is suitable for endothelial cell culture with low endothelial cell yield and small vessel diameter. It is simpler and more economical than digestion method. The disadvantage is that fibroblasts and smooth muscle cells are easily mixed.

In this experiment, Giemsa staining will be used to identify endothelial cells. Endothelial cells are fusiform, triangular or polygonal under microscopy, and the cells after fusion into a piece can be arranged in a cobblestone-like mosaic setting.

【Materials】

1. Apparatus

(1) Cell culture flask, 100ml

(2) Culture plate

(3) Scissors

(4) Tweezers

(5) Elbow tweezers

(6) Scalpel

(7) Petri dish

(8) Centrifuge tube

(9) Pipette

(10) Carbon dioxide incubator

(11) Electrophysiological inverted microscope

(12) Centrifuge

2. Reagents

(1) Inactivate fetal bovine serum (FBS) at 56℃ for half an hour, thenaliquoted, and store at -20℃.

(2) Serum-free RPMI-1640 medium、reconstitute RPMI 1640 medium with 1000ml distilled water, add $NaHCO_3$ 2.0g, penicillin 10^5 U, streptomycin 10^5 U, stir at room temperature, completely

dissolve the medium, adjust the pH to 7.0 ± 0.2, then use 0.22μm mixed cellulose microporous membrane to filter the bacteria by positive pressure filtration, then store at 4℃.

(3) PBS. Weigh NaCl 8.0g, KCl 0.2g, KH_2PO_4 0.24g, $Na_2HPO_4 \cdot 12H_2O$ 3.28g, add distilled water to 1000ml, autoclave at 121℃ for 20 minutes.

(4) 0.25% trypsin. 0.02g of disodium edetate, 0.25g of trypsin, dissolved in PBS and made up to 100ml, and sterilized by filtration. Store at 4℃.

(5) Hank's solution. Weigh $CaCl_2$ 0.14g, NaCl 8.0g, KCl 0.4g, KH_2PO_4 0.06g, $MgCl_2 \cdot 6H_2O$ 0.10g, $MgSO_4 \cdot 7H_2O$ 0.10g, $NaHO_3$ 0.35g, $Na_2HPO_4 \cdot 12H_2O$ 0.13g, D - glucose 1.0g, phenol red 0.02g. Preparation method: First dissolve $CaCl_2$ in 100ml water. Dissolve other components in about 750ml water and dissolve the next component after the former reagent is completely dissolved. Slowly pour 1 into 2, stir to prevent precipitation, make up the water, filter by positive pressure filtration, and store at 4℃.

(6) Weigh Giemsa powder 0.5g, glycerin 22ml, mix with a small amount of glycerin for Giemsa powder in the mortar, grind to no granules, mix with residual glycerin, keep at 56℃ for 2 hours, then add 33ml methanol for hot filtration, use brown bottles to store.

(7) The Giemsa application solution is prepared by mixing the Giemsa stock solution and the pH 7.0 phosphate buffer solution at a ratio of 1 : 9.

【Procedures】

1. Kill the rat by cervical dislocation and immerse it in 75% ethanol for about 1 minute. Under sterile conditions, cut the thoracic cavity layer by layer, remove the thoracic aorta, and wash it in a culture dish containing high antibiotic Hank's solution for 2 to 3 times to remove the residual blood.

2. Use scissors and tweezers to separate the adipose tissue and connective tissue of adventitia and rinse them repeatedly with phosphate buffer until no blood remains. Cut the lumen lengthwise and lay it flat in the Petri dish. Cut the blood vessel into a square about 0.1cm in length with a scalpel and place it in the Petri dish at intervals.

3. When the RPMI - 1640 medium containing 10% fetal bovine serum was added, the tissue mass could not be suspended. Culture the dish at 37℃ under 5% CO_2.

4. When the cells growing around the tissue block fuse into pieces, the tissue block can be removed, and the cells can be digested by trypsin digestion method. Discard the culture medium, lightly wash the cell surface twice with PBS. Add an equal volume of 0.25% trypsin to cover the cell surface. Digestion at room temperature for 1 to 2 minutes. When the cell gap is clear, immediately tilt the Petri dish, avoid contact with trypsin, discard the digestive solution, directly add 10% 1640 medium, and use the pipette to remove the cells from the dish.

5. Seed the endothelial cells at a density of 1×10^5 cells/ml into a six - well plate with a slide for 48 hours and then stain.

6. Dyeing step: fix with methanol for 10 minutes (or fix with acetic acid: methanol for 30 minutes). Cover the samples with Giemsa solution for 10 - 15 minutes. Rinse the staining solution with tap water, air dry, and transparent with xylene; Observe with microscope.

7. Analyze the results: Draw what is observed by the microscope.

【Questions】

1. What are the tips for endothelial cell culture?
2. How to identify endothelial cells?

实验四　干扰素生物学活性的测定

扫码“学一学”

【实验目的】

1. 掌握　传代细胞培养及细胞冻存的技术。
2. 学习干扰素生物学活性测定的原理和方法。

【实验原理】

干扰素（Interferon，IFN）具有抗病毒、抗肿瘤及调节机体免疫等活性，目前已广泛的应用于临床治疗多种病毒性疾病和恶性肿瘤。根据其抗原特异性不同，可以分为 IFN－α、IFN－β 及 IFN－γ 三大类型，每一类型干扰素又有若干不同的亚型。干扰素活性的测定对干扰素的生产、研究和临床应用都十分重要。

最常用的测定干扰素活性的方法是细胞病变抑制法，该方法所依据的原理为：干扰素能刺激指示细胞（人羊膜上皮细胞 WISH 株）产生抗病毒蛋白，从而保护人羊膜细胞（WISH）免受水泡性口炎病毒（VSV，Vesicular Stomatitis Virus）的攻击，不同稀释度的干扰素样品保护细胞能力的不同，利用结晶紫对存活的细胞染色，酶标仪测定出相对应的光密度值（OD），干扰素保护细胞效果越好，其 OD 值越高。进而可以通过计算得到干扰素生物学活性单位。这种方法以细胞病变抑制效应为基础，具有操作简便、结果可靠的特点。

结晶紫染色测定的原理是结晶紫染色液可以选择性地被细胞核吸收，活细胞被固定染色，而死亡的细胞几乎不吸收结晶紫染料。用水小心洗去未被吸收的多余染色液，再以有机溶媒提取被细胞核所吸收的结晶紫，提取液的光密度值大小可反映出活细胞的数量。

体外培养的细胞在接种后经过一段潜伏期就进入对数生长期，当细胞密度达到铺满整个瓶底的有效培养基时，细胞生长停止或生长速度大大放缓。此时需进行分瓶培养，也就是传代。作为组织培养的常规操作方法，传代培养可获得大量细胞供实验所需，同时也是一种将细胞种保存下去的方法。传代要在严格的无菌条件下进行，每一步都需要认真仔细的无菌操作。悬浮型细胞的传代直接分瓶即可，而贴壁细胞需经消化后才能分瓶。

随着传代次数的增加和体外环境的变化，细胞会逐渐衰老、生物特性将发生变化，因此，及时进行细胞冻存十分必要。实验证明，细胞储存在液氮（－196℃）中可以最大限度地保存细胞活力。

【实验材料】

1. 器材

（1）光学倒置显微镜　　1 台
（2）生物洁净型工作台　　1 台

(3) 二氧化碳培养箱　　1台
(4) 酶标仪　　1台
(5) 96孔细胞培养板　　2块
(6) 细胞培养瓶(100ml)　　4只
(7) 吸管及移液管　　数支
(8) 微量移液器(100μl)　　1支
(9) 程序降温盒　　1个

2. 试剂

(1) 无血清RPMI 1640培养基　取RPMI 1640培养基粉末一袋(规格为1L),加水溶解并稀释至1000ml,加青霉素钠10^5U和硫酸链霉素10^5U,再加碳酸氢钠2.0g,溶解混合均匀后调节pH 7.0±0.2,以0.22μm微孔滤膜过滤除菌,分装。置于4℃保存。

(2) 新生牛血清　经56℃灭活半小时后,小量分装,-20℃保存备用。

(3) 完全培养液　量取新生牛血清25ml,加RPMI 1640培养液225ml,4℃保存。

(4) 测定培养液　量取新生牛血清10ml,加RPMI 1640培养液190ml,4℃保存。

(5) 攻毒培养液　量取新生牛血清4ml,加RPMI 1640培养液196ml,4℃保存。

(6) 冻存液　含10%~20%的新生牛血清液,10% DMSO(分析纯)的完全培养液。

(7) PBS　称取氯化钠8.0g,氯化钾0.20g,磷酸氢二钠3.28g,磷酸二氢钾0.24g,加双蒸水溶解并稀释至1000ml,经121℃ 20分钟高压灭菌备用。

(8) 0.25%胰酶溶液　取乙二胺四乙酸二钠0.02g,胰蛋白酶0.25g,加PBS溶解并定容至100ml,过滤除菌。冷藏保存备用。

(9) 染色液　取结晶紫50mg,加无水乙醇20ml溶解后,加蒸馏水稀释至100ml。

(10) 脱色液　取无水乙醇50ml,乙酸0.1ml,加水稀释至100ml。

(11) 干扰素活性测定国家标准品　按说明书复溶,用测定培养液稀释成1000U/ml。分装后保存-20℃备用。

【实验方法】

1. WISH细胞的传代和冻存

(1) 进入无菌室之前用肥皂洗手,用75%乙醇擦拭消毒双手。超净台台面应整洁,用0.1%新洁尔灭溶液擦净。培养液置37℃下预热。

(2) 显微镜下观察WISH细胞形态,细胞应饱满并呈多角形。确定细胞是否需要传代。

(3) 倒掉培养细胞的旧培养基。酌情用2~3ml PBS润洗后,用消化液消化细胞,37℃温育1~2分钟,盖好瓶盖后在倒置显微镜下观察,待显微镜下观察到细胞变圆,胞质回缩,细胞不再连成片时,立即翻转培养瓶,使细胞脱离胰酶,然后将消化液倒掉(注意勿使细胞提早脱落入消化液中)。

(4) 加入少量的完全培养液以终止残余的消化液的作用,反复吹打消化好的细胞使其脱壁并分散,成为单细胞悬液。再根据分传瓶数补加一定量的完全培养液(7~10ml/瓶)制成细胞悬液,然后分装到新培养瓶中。盖上瓶盖,适度拧紧后再稍回转,以利于CO_2气体的进入,将培养瓶放回CO_2恒温培养箱中,于37℃ 5% CO_2条件下培养。

(5) 细胞冻存时,将获得的单细胞悬液1000r/min离心3分钟,弃上清,再以冻存液重悬浮细胞,调整细胞密度为1×10^6/ml。将细胞悬液分装至标记好的冻存管中(1~1.5ml/管),封口。使用程序降温盒对冻存细胞进行降温,4小时后可将细胞转移至-80℃

冰箱或液氮罐。

（6）细胞冻存在液氮中可以长期保存，但为妥善起见，冻存后一段时间，最好取出一只复苏培养，观察细胞生长情况和对冻存的适应性，然后再继续冻存。

2. 干扰素的生物学活性测定

（1）取培养的 WISH 细胞弃去培养液，PBS 溶液洗一次后，用 0.25% 胰酶消化细胞，弃去消化液，再加入完全培养液，吹打细胞成为单细胞状态，收集细胞，用完全培养液配制成每 1ml 含 $2.5\times10^5\sim3.5\times10^5$ 个细胞的悬液，接种于 96 孔细胞培养板中，每孔 100μl。于 37℃ 5% CO_2 条件下培养 4～6 小时。

（2）将分装好的干扰素标准品在 96 孔细胞培养板中，做 4 倍系列稀释，共 8 个稀释度，每个稀释度做 2 孔。无菌条件下操作。

（3）待测样品溶液经过预稀释后在 96 孔细胞培养板中，做 4 倍系列稀释，共 8 个稀释度，每个稀释度做 2 孔。无菌条件下操作。

（4）将配制完成的标准品溶液和待测样品溶液移入接种 WISH 细胞的培养板中，每孔 100μl，每个稀释浓度有 2 个复孔，同时分别预留四孔作为细胞对照和病毒对照，不加干扰素。于 37℃ 5% CO_2 条件下培养 18～24 小时。

（5）弃去细胞培养板中的上清液，将保存的 VSV 用攻毒培养液稀释至 100 $TCID_{50}$（组织半数感染剂量）。每孔加病毒液 100μl。细胞对照孔只加攻毒培养液，不加病毒。于 37℃ 5% CO_2 条件下培养 18～22 小时。

（6）在显微镜下观察细胞病变情况，若病毒对照组各孔细胞出现 75%～100% 的明显病变（变圆、颗粒化、脱落、细胞质空泡化），而细胞对照组中的细胞仍生长良好，则表明对照系统合格，结果成立。弃上清，每孔加入结晶紫染液 50μl，室温染色 30 分钟后，用流水小心冲去染色液，并控干残留水分。每孔加入脱色液 100μl，用酶标仪以 630nm 为参比波长，在波长 570nm 处测定吸光度，在下表中记录测定结果。

稀释度	样品组 OD_{570}			标准品 OD_{570}		
	1	2	平均值	1	2	平均值
1/4						
1/16						
1/64						
1/256						
1/1024						
1/4096						
1/16384						
1/65536						

（7）干扰素保护终点是在样品的系列梯度稀释度中，能保护半数细胞免受病毒攻击损害的最高稀释度，其倒数即为干扰素的活性单位。按下式计算试验结果：

$$[OD]_{终点}=\frac{[OD]_{细胞对照}+[OD]_{病毒对照}}{2}$$

$$\log_x(半效稀释倍数)=\frac{高于[OD]_{终点}的OD值-[OD]_{终点}}{高于[OD]_{终点}的OD值-低于[OD]_{终点}的OD值}+\log_x(稀释倍数)$$

因为方法中采用 4 倍系列稀释，所以式中 $\chi=4$，稀释倍数为高于［OD］$_{终点}$所对应的干扰素稀释倍数：

$$\text{IFN 样品生物学活性(U/ml)} = Pr \times \frac{Ds \times Es}{Dr \times Er}$$

式中，*Pr* 为标准品生物学活性，U/ml；*Ds* 为样品预稀释倍数；*Es* 为样品相当于标准品半效量的稀释倍数；*Dr* 为标准品预稀释倍数；*Er* 为标准品半效量的稀释倍数。

扫码“练一练”

【思考题】

1. 试述传代培养的步骤和注意事项，并指出哪些是关键步骤？
2. 干扰素生物学活性测定的原理是什么？
3. 测定细胞存活率的方法有哪些？

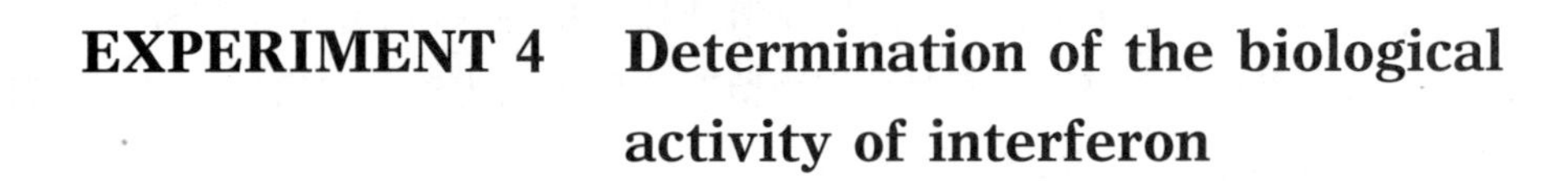

EXPERIMENT 4　Determination of the biological activity of interferon

【Purpose】

1. To master the techniques of cell culture and cell cryopreservation.
2. To learn the principles and methods of measuring the biological activity of interferons.

【Principle】

Interferon(IFN) has antiviral, anti - tumor and immune regulationeffect. It has been widely used in clinical to treat various viral diseases and malignant tumors. According to its antigen specificity, it can be divided into three major types: IFN - α, IFN - β and IFN - γ. Each type of interferon has several different subtypes. The determination of biological activity is important for the production, research and clinical application of interferon.

The most commonly used method for measuring interferon activity is cytopathic inhibition, which is based on the principle that interferon stimulates human amniotic epithelial cells(WISH) to produce antiviral proteins, thereby protecting WISH against vesicular stomatitis virus(VSV). Different dilutions of interferon have different protective abilities. Stain viable cells with crystal violet and determine the corresponding optical density values by microplate reader. The higher OD value indicates the better ability of interferon to protect WISH. The biological activity of interferon can be calculated accordingly. This method is simple and reliable.

The principle of crystal violet staining is that the crystal violet staining solution can be selectively absorbed by the living cells. The dead cells hardly absorb the crystal violet dye. Carefully wash away the excess staining solution that has not been absorbed, and then extract the crystal violet absorbed by the nucleus with an organic solvent. The optical density of the extract can reflect the number of living cells.

The cells cultured *in vitro* enter the logarithmic growth phase after an incubation period. When the cells cover the entire bottom of the flask, the cells stop growing or grow very slowly. At this point, a separate bottle culture is required, that is, passage. As a routine method of tissue culture, subculture can supply a large number of cells for the experiment and also preserve the cell species. Passage should be carried out under strict aseptic conditions, and each step requires careful and sterile operation. For the suspension cells, cells can be directly distributed into different flasks; For the adherent cells, digestion is needed before passage.

As the number ofcell passages increases and the environment *in vitro* changes, the cells will gradually age and the biological characteristics will change. Therefore, it is necessary to perform cell cryopreservation in time. Experiments have shown that cells stored in liquid nitrogen(−196℃) can preserve cell viability to the maximum.

【Materials】

1. Apparatus

(1) Optical inverted microscope

(2) Laminar flow hood

(3) Carbon dioxide incubator

(4) Microplate reader

(5) 96 − well cell culture plate

(6) Cell culture flask(100ml)

(7) Pipettes

(8) Micropipette(100μl)

(9) Programmed cell freezing box

2. Reagents

(1) Serum − free RPMI 1640 medium: reconstitute RPMI 1640 medium with 1000ml distilled water, add $NaHCO_3$ 2.0g, penicillin 10^5 U, streptomycin 10^5 U, stir at room temperature, completely dissolve the medium, adjust the pH to pH 7.0 ± 0.2, then use 0.22μm mixed cellulose microporous membrane to filter the bacteria by positive pressure filtration, then store at 4℃.

(2) Newborn bovine serum: inactivate at 56℃ for half an hour, then aliquoted, and store at −20℃.

(3) Complete culturemedium: add 25ml newborn calf serum and 225ml RPMI 1640 culture medium, store at 4℃.

(4) Assay medium: add 10ml newborn calf serum and 190ml RPMI 1640 culture medium, store at 4℃.

(5) Culture medium for virus: add 4ml newborn calf serum and 196ml RPMI 1640 culture medium, store at 4℃.

(6) Cryopreservationmedium: complete culture medium containing 10% − 20% newborn calf serum and 10% DMSO(analytical grade).

(7) PBS: Weigh − NaCl 8.0g, KCl 0.2g, KH_2PO_4 0.24g, $Na_2HPO_4 \cdot 12H_2O$ 3.28g, add distilled water to 1000ml, autoclave at 121℃ for 20 minutes.

(8) 0.25% trypsin: 0.02g of disodium edetate, 0.25g of trypsin, dissolved in PBS and made up to 100ml, and sterilized by filtration. Store at 4℃.

(9) The crystal violet stain solution: dissolve 50mg crystal violet in 20ml absolute ethanol, and dilute with distilled water to 100ml.

(10) Decolorizing solution: add 50ml absolute ethanol, 0.1ml acetic acid, and dilute with water to 100ml.

(11) Interferon activity National standard: reconstitute according to the instructions and dilute to 1000U/ml with the assay medium. Store at −20℃ after aliquoted.

【Procedures】

1. Passage and cryopreservation of WISH cells

(1) Wash hands with soap before entering the sterile room and sterilize the hands with 75% ethanol. Keep the bench clean and wipe with 0.1% chlorhexidine solution. Preheat the culture medium at 37℃.

(2) Observe WISH cells using the microscope. The cells should be full and polygonal. Determine whether the cells need to be passaged.

(3) Pour off the old medium of the cultured cells and wash the cells with 2 ~ 3ml PBS. Digest the cells with 0.25% trypsin, incubate for 1 − 2 minutes at 37℃, observe using microscope. When the cells are round, the cytoplasm is retracted, and the cells are not connected, immediately turn the culture flask off, remove the trypsin from the cells, and then pour off the digestive solution (prevent the cells from falling off into the digestive solution).

(4) Add a small amount of complete culture medium to terminate the digestion. Blow off the digested cells and disperse them to single cell suspension. Add a certain amount of complete medium (7 ~ 10ml/flask) to make the cell suspension, and then dispense the cells into a new flask. Cover the cap, moderately tighten and then rotate slightly to facilitate the entry of CO_2. Return the flask to the CO_2 incubator and incubate at 37℃, 5% CO_2.

(5) For cryopreservation, centrifuge the obtained single cell suspension at 1000r/min for 3 minutes, discard the supernatant, resuspend the cells with cryopreservation medium to adjust the cell density to 1×10^6/ml. Add the cell suspension into a labeled cryotube (1 to 1.5ml/tube). Cool the cells with programmed cell freezing box. After 4 hours, the cells could be transferred to − 80℃ refrigerator or liquid nitrogen tank.

(6) The cell frozen in liquid nitrogen can be stored for a long time. To insure the viability of frozen cells, after a period of cryopreservation, it is better to resuscitate a tube of cells, observe the cell growth and adaptability to cryopreservation, and then continue to cryopreservation.

2. Determination of the biological activity of interferon

(1) Discard the culture medium of WISH cells, wash the cells with PBS solution, digest the cells with 0.25% trypsin, discard the digestive solution, add the complete culture medium, blow the cells into a single cell suspension, collect the cells. Adjust the density of cells suspension into 2.5×10^5 ~ 3.5×10^5 cells/ml. Seed the cells in a 96 − well plate with 100μl per well. Incubate the plate at 37℃, 5% CO_2 for 4 ~ 6 hours.

(2) Dilute the interferon standard 4 times in a series of 8 dilutions in the 96 − well plate, make duplicates for each dilution. Operate under sterile conditions.

(3) Dilute the pre - diluted sample 4 times in a series of 8 dilutions in the 96 - well plate, make duplicates for each dilution. Operate under sterile conditions.

(4) Add the prepared standard solution and the sample solution into the WISH cells seeded plate, 100μl per well, make duplicate for each dilution. Reserve 4 wells as cell control and virus control without interferon, respectively. Incubate the cells at 37℃, 5% CO_2 for 18 ~ 24 hours.

(5) Discard the supernatant of the cell culture plate. Dilute the preserved VSV to 100 $TCID_{50}$ (50% tissue culture infective doses) with culture medium for virus. Add 100μl virus solution to each well. Add only culture medium without virus as control. Incubate at 37℃, 5% CO_2 for 18 ~ 22 hours.

(6) Observe the cytopathic condition using the microscope. If the cells in the virus control group show 75% ~ 100% obvious lesions (rounding, granulation, shedding, cytoplasmic vacuolization), while the cells in the cell control group grow well, it indicates that the system is qualified and the result is convincing. Discard the supernatant, add 50μl crystal violet dye solution to each well, stain for 30min at room temperature, carefully rinse off the staining solution with running water, and minimize the residual water. Add 100μl the decolorizing solution to each well and measure the absorbance at a wavelength of 570nm with a reference wavelength of 630nm. Record the results in the following table.

Dilution	Sample OD_{570}			Standard OD_{570}		
	1	2	Average	1	2	Average
1/4						
1/16						
1/64						
1/256						
1/1024						
1/4096						
1/16384						
1/65536						

(7) The interferon protection endpoint is the highest dilution that protects half of the cells from viral attack damage in a series of gradient dilutions of the sample, the reciprocal of which is the unit of activity of the interferon. Calculate the results as follows

$$[OD]_{\text{end point}} = \frac{[OD]_{\text{Cell control}} + [OD]_{\text{Virus control}}}{2}$$

$$\log_x(\text{median dilution factor}) = \frac{[OD]\text{value Higher than}[OD]_{\text{end point}} - [OD]_{\text{end point}}}{[OD]\text{value higher than}[OD]_{\text{end point}} - [OD]\text{value lower than}[OD]_{\text{end point}}} + \log_x(\text{dilution factor})$$

4 times serial dilution is used, therefore $\chi = 4$, and the dilution factor is the dilution factor of interferon corresponding to the OD value higher than $[OD]_{\text{end point}}$.

$$\text{biological activity of IFN(U/ml)} = Pr \times \frac{Ds \times Es}{Dr \times Er}$$

Pr: the biological activity of standard, U/ml;

Ds: pre－dilution factor of the sample;

Es: the dilution factor of the sample equivalent to median effective dose the standard;

Dr: pre－dilution factor of the standard;

Er: median dilution factor of the standard.

【Questions】

1. Describe the protocols and precautions of subculture. Which are the key steps?
2. What is the principle of measuring biological activity of interferon?
3. How to estimate the viability of cells?

第三节　固定化酶技术基础实验

固定化酶技术就是利用物理或化学手段将游离的酶或微生物细胞定位于限定的空间领域，并使其保持活性且能反复利用的一项技术。固定化酶技术在发酵工业、食品工业、医药工业、有机合成工业、环境净化等各个领域中都有广泛的应用。

制备固定化酶要根据不同情况（不同酶、不同应用目的和应用环境）来选择不同的方法，但是无论如何选择，确定什么样的方法，都要遵循基本原则：①必须注意维持酶的催化活性及专一性。酶蛋白的活性中心是酶的催化功能所必需的，酶蛋白的空间构象与酶活力密切相关。因此，在酶的固定化过程中，必须注意酶活性中心的氨基酸残基不发生变化，也就是酶与载体的结合部位不应当是酶的活性部位，而且要尽量避免那些可能导致酶蛋白高级结构破坏的条件。由于酶蛋白的高级结构是凭借氢键、疏水键和离子键等弱键维持，所以固定化时要采取尽量温和的条件，尽可能保护好酶蛋白的活性基团。②固定化应该有利于生产自动化、连续化。为此，用于固定化的载体必须有一定的机械强度，不能因机械搅拌而破碎或脱落。③固定化酶应有最小的空间位阻，尽可能不妨碍酶与底物的接近，以提高产品的产量。④酶与载体必须结合牢固，从而使固定化酶能回收贮藏，利于反复使用。⑤固定化酶应有最大的稳定性，所选载体不与废物、产物或反应液发生化学反应。⑥固定化酶成本要低，以利于工业使用。

制备固定化酶化方法有多种，传统的酶固定化方法大致可分为 4 类：吸附法、交联法、包埋法、共价结合法。

吸附法是最早出现的酶固定化方法，包括物理吸附和离子结合法。这类方法条件温和，酶的构象变化较小或基本不变，因此对酶的催化活性影响小，但酶和载体之间结合力弱，在不适 pH、高盐浓度、高温等条件下，酶易从载体脱落并污染催化反应产物等。

交联法是利用双功能或多功能交联试剂，在酶分子和交联试剂之间形成共价键，采用不同的交联条件和在交联体系中添加不同的材料，可以产生物理性质各异的固定化酶。

包埋法的基本原理是载体与酶溶液混合后，借助引发剂进行聚合反应，通过物理作用将酶限定在载体的网格中，从而实现酶固定化的方法。该法不涉及酶的构象及酶分子的化学变化，反应条件温和，因而酶活力回收率较高。但包埋法固定化酶易漏失，常存在扩散限制等问题，催化反应受传质阻力的影响，不宜催化大分子底物的反应。

共价结合法是指酶分子的非必须基团与载体表面的活性功能基团通过形成化学共价键实现不可逆结合的酶固定方法，又称共价耦联法。共价耦联法所得的固定化酶与载体连接

牢固，有良好的稳定性及重复使用性，成为目前研究最为活跃的一类酶固定化方法。但该法较其他固定方法反应剧烈，固定化酶活性损失更加严重。

近年来，一些通过辐射、光、等离子体、电子等制备高活性固定化酶的新方法不断出现。Mohy 等以 ^{137}Cs 为辐射源，通过 γ－射线引发将甲基丙烯酸甲酯接枝共聚于尼龙膜表面，经进一步活化，用于青霉素酰化酶的固定。光耦联法是以光敏性单聚体聚合物包埋固定化酶或带光敏性基团的载体共价固定化酶，由于条件温和，可获得酶活力较高的固定化酶。这些新方法实现了在较为温和的条件下进行酶的固定化，尽量减少或避免酶活力的损失。

要想制得理想的固定化酶，既要选用合理有效的固定化方法，同时又要有良好的载体。理想的载体要具备良好的机械强度、高结合能力、物理和化学稳定性及抗微生物降解等特性。常用载体归纳起来主要可分为二类：一类是天然高分子凝胶载体，如琼脂、角叉菜胶(卡拉胶)、海藻酸钠等；另一类是有机合成高分子凝胶载体，如聚丙烯酰胺凝胶、聚乙酰醇凝胶、光硬化树脂、聚丙烯酸凝胶等。

在本节中，主要总结了我校长期开设的生物制药工艺学理论和实验课以及科研工作中的部分实践经验，着重介绍了几种传统的固定化方法，希望通过这几个实验使大家对该项技术有较全面的了解和掌握。

Section 3　Basic experiments of immobilized enzyme technology

Immobilized enzyme technology is a technology that make free enzyme or microorganism cells located in limited space by physical or chemical means, and keep them active and be used repeatedly. Immobilized enzyme technology has been widely used in fermentation industry, food industry, medicine industry, organic synthesis industry, environmental purification and other fields.

The preparation of immobilized enzyme depends on different conditions (different enzymes, different application purposes and application environment) to choose different methods, but no matter how to choose and determine what kind of method, must follow the basic principles:

①Care must be taken to maintain the catalytic activity and specificity of the enzyme. The activity center of enzyme protein is necessary for the catalytic function of enzyme. The spatial conformation of enzyme protein is closely related to enzyme activity. Therefore, during the immobilization of the enzyme, it must be noted that the amino acid residues in the enzyme activity center do not change, that is, the binding site of the enzyme and the carrier should not be the active site of the enzyme, and conditions that may lead to the destruction of the high－level structure of the enzyme protein should be avoided as far as possible. Since the advanced structure of the enzyme protein is maintained by weak bonds such as hydrogen bond, hydrophobic bond and ionic bond, the immobilization should be conducted under mild conditions to protect the active groups of the enzyme protein as much as possible. ②Immobilization should be conducive to production automation and continuity. For this reason, the carrier used for immobilization must have a certain mechanical strength. The carrier can not be broken or fall off because of mechanical agitation. ③Immobilized enzyme should be the minimum steric hindrance, as far as possible not to hinder the enzyme and substrate close, in order to improve the production of products. ④Enzyme and carrier must be combined firmly, so that the immobilized enzyme can be recycled storage, conducive to repeated use. ⑤Immobilized enzyme should havc thc maximum stability. The selected carrier does not react chemically with waste, product or reaction solution. ⑥Immobilized enzyme cost is low to facilitate industrial use.

There are many methods to prepare immobilized enzyme, and the traditional immobilized enzyme can be divided into four categories: adsorption, crosslinking, embedding and covalent binding.

Adsorption is the earliest enzyme immobilization method, including physical adsorption and ion binding. Under mild conditions, the conformational change of the enzyme is small or basically unchanged, so it has little influence on the catalytic activity of the enzyme. However, the binding force between the enzyme and the carrier is weak, and the enzyme is easy to fall off from the carrier and contaminate the catalytic reaction product under the conditions of unfavorable pH, high salt concentration and high temperature.

Crosslinking method is to form covalent bonds between enzyme molecules and crosslinking reagents by using bifunctional or multifunctional crosslinking reagents. Immobilized enzymes with different physical properties can be produced by using different crosslinking conditions and adding different materials in the crosslinking system.

The basic principle of embedding method is that after the carrier and the enzyme solution aremixed, the polymerizing reaction is carried out with the help of initiator, and the enzyme is limited in the grid of the carrier through physical action, so as to realize the immobilization of the enzyme. The method does not involve the conformation of the enzyme or the chemical changes of the enzyme molecules. The reaction conditions were mild, so the recovery of enzyme activity was high. However, immobilized enzyme by embedding method is easy to lose, and there are often problems such as diffusion limitation. The catalytic reaction is affected by mass transfer resistance, so it is not suitable to catalyze the reaction of macromolecular substrates.

Covalent coupling method, also known as covalent coupling method, refers to an enzyme fixation method in which the non – essential groups of enzyme molecules and the active functional groups on the surface of the carrier can realize irreversible binding through the formation of chemical covalent bond. The immobilized enzyme obtained by covalent coupling is firmly connected to the carrier, has good stability and reusability, and has become one of the most active immobilized enzyme methods. However, compared with other fixation methods, the reaction of immobilized enzyme was more severe. The loss of immobilized enzyme activity is more serious.

In recent years, some new methods for the preparation of highly active immobilized enzymes by radiation, light, plasma, electron, etc. With ^{137}Cs as the radiation source, Mohy et al. grafted methyl – methacrylate onto the surface of nylon membrane through gamma – ray initiation and further activated it for fixation of penicillin acylase. The photocoupling method is to imbed immobilized enzyme or carrier with photosensitive group by photosensitive monomer polymer. These new methods immobilized the enzyme under mild conditions and reduced or avoided the loss of enzyme activity.

In order to obtain the ideal immobilized enzyme, reasonable and effective immobilized method and good carrier should be selected. The ideal carrier should have good mechanical strength, high binding ability, physical and chemical stability and anti – microbial degradation. Commonly used

carriers can be classified into two types, one is natural polymer gel carrier, such as AGAR, carrageenan, sodium alginate, etc. The other is the organic synthesis of polymer gel carrier, such as polyacrylamide gel, polyacetyl alcohol gel, light hardening resin, polyacrylic gel and so on.

In this section, it mainly summarizes the theory and experimental course of biopharmaceutical technology and some practical experience in scientific research, and mainly introduces several traditional immobilization methods, hoping that through these experiments we can have a more comprehensive understanding and grasp of this technology.

实验五　共价耦联法制备固定化胰蛋白酶

扫码“学一学”

【实验目的】

1. 学习用耦联法制备固定化酶的基本方法。
2. 学习测定亲和吸附剂耦联量的方法。
3. 学习测定胰蛋白酶活力的方法。

【实验原理】

共价耦联法是制备固定化酶的重要手段之一，该法是借助共价键将酶的非活性侧链基团和载体的功能基团进行耦连，即先将载体活化，再将酶通过共价键结合于亲水性的不溶性载体。其优点是酶和载体的结合牢固，不易脱落，得到的固定化酶利于连续使用。缺点是载体活化的操作复杂，反应条件激烈，酶活力的回收率较低，需要严格控制条件才可以获得较高活力的固定化酶。

共价耦联法的主要因素有：载体亲水性好，并且有一定的机械强度和稳定性，同时具备在温和条件下与酶结合的功能基团；耦联反应的反应条件必须在温和 pH、中等离子强度和低温的缓冲溶液中；所选择的耦联反应要尽量考虑到对酶的其他功能基团副反应尽可能少；要考虑到酶固定化后的构型，尽量减少载体的空间位阻对酶活力的影响。

本实验中使用琼脂糖凝胶（Sepharose 4B）为固相载体，在碱性条件下用溴化氰（CNBr）活化，引入活泼的“亚氨基碳酸盐”，在弱碱的条件下直接耦联胰蛋白酶的游离氨基，形成氨基碳酸盐和异尿衍生物，获得固定化酶。其反应机理如下：

-OH, -OH —cyanogen hromide→ [-O-C≡N, -OH] → -OCONH$_2$, -OH　氨基甲酸酯（活性）
→ 环状 -O-C(=NH)-O-　亚胺碳酸（活性）

-O-C(=NH)-NH-蛋白质, -OH　异脲衍生物

coupling steps: -O-C(=NH)-O- (imine earbonate) —H$_2$N-pratein→ -O-C(=N-蛋白质)-O-　N-取代的亚胺碳酸

-O-C(=O)-N-蛋白质, -OH　N-取代的氨基甲酸

溴化氰极易挥发，有剧毒。其分解产物也有毒。活化反应必须在通风橱内进行。因活化基团极易被水解，活化中间体的稳定性差，所以操作必须紧凑、迅速，反应完成后的洗涤液应快速。

蛋白耦联量一般有两种测定方法：一种是直接测定法，即通过定氮或氨基酸组成分析来直接获得蛋白耦联量的数据；另一种方法是间接测定法，即将耦联时所用蛋白量减去耦联后所残存的蛋白量。本实验为了简便起见，采用间接测定法。亲和吸附剂的活性可以通过胰蛋白酶活性来表示。

胰蛋白酶能催化蛋白质的水解，对于由碱性氨基酸（如精氨酸、赖氨酸）的羧基所组成的肽键具有高度的专一性。胰蛋白酶不仅能水解肽键而且也能水解酰胺键和酯键，因此可用人工合成的酰胺及酯类化合物为底物来测定胰蛋白酶的活力。本实验采用人工合成的苯甲酰 - L - 精氨酸 - β - 萘酰胺（benzoyl - L - arginine naphthylamide，BANA）为底物测定胰蛋白酶活力。

【实验材料】

1. 器材

（1）磁力搅拌器		1 台
（2）玻璃漏斗	4 ~ 5cm	1 只
（3）砂芯漏斗		1 只
（4）三颈烧瓶	250ml	1 只
（5）乳钵	10cm	1 只
（6）751 紫外分光光度计		1 台
（7）层析柱	2cm × 15cm	1 只
（8）温度计		1 只
（9）量筒	100ml	1 只
（10）抽吸瓶	1000 ~ 2000ml	1 只
（11）吸量管	0. 5ml	11 支
	1ml	5 支
	2ml	2 支
（12）烧杯	2000ml	1 只
	500ml	2 只
	250ml	1 只
	125ml	1 只

2. 试剂

（1）溴化氰

（2）琼脂糖凝胶 4B

（3）胰蛋白酶结晶

（4）2mol/L NaOH 溶液

（5）2mol/L HCl 溶液

（6）0. 025mol/L、pH 10. 2 硼酸缓冲液（含 0. 2mol/L $CaCl_2$） 四硼酸钠 2. 38g、氯化钙 2. 22g 和 2mol/L NaOH 3. 0ml 溶解于水，定容至 1000ml，pH 计校正。

（7）0.1mol/L 乙醇胺　取 6.1g（6.0ml）乙醇胺溶于 1000ml 水。

（8）0.1mol/L pH 8 硼酸缓冲液（含 0.02mol/L $CaCl_2$）　四硼酸钠 2.862g、氯化钙 2.78g 和硼酸 4.328g，溶于水，定容 1000ml，pH 计校正。

（9）0.1mol/L pH 4 醋酸缓冲液（含 1mol/L NaCl）　醋酸钠 2.45g、氯化钠 58.4g 和 36% 醋酸溶液 13.66ml，溶于水，定容 1000ml，pH 计校正。

（10）0.01mol/L pH 4 醋酸缓冲液（含 0.1mol/L NaCl 和 0.025mol/L $CaCl_2$）：醋酸钠 0.245g 和 36% 醋酸溶液 1.37ml、氯化钠 5.84g 和氧化钙 2.78g 溶于水，定容至 1000ml，pH 计校正。

（11）0.05mol/L pH 7.8 磷酸缓冲液：$Na_2HPO_4 \cdot 12H_2O$（Mw 358.16）16.476g 和 0.625g $NaH_2PO_4 \cdot 2H_2O$（M_W 156.03），溶于水，定容至 1000ml，pH 计校正。

（12）底物 0.06% 苯甲酰-L-精氨酸-β-萘胺（BANA）溶液：取 BANA 60mg，加 95% 乙醇 20ml，使溶解，用 pH 7.8、0.2mol/L 磷酸缓冲液定容至 100ml。

（13）0.05% 萘基乙二胺盐酸盐（NEDA）乙醇溶液

（14）0.5% 氨基磺酸铵溶液

（15）0.1% 亚硝酸钠溶液

（16）标准酶液　称取胰蛋白酶结晶 1mg，用 pH 7.8 磷酸缓冲液定容至 100ml。

【实验方法】

1. 载体活化　将市售 Sepharose 4B 用蒸馏水多次漂洗，以布氏漏斗抽干。称取约 7g（或自沉积 10ml）置于三颈瓶中。加水 10ml 并用 2mol/L NaOH 调 pH 11。将此烧瓶移入通风橱，开动电磁搅拌（低速）搅匀，并加热或冷却至 18℃。开动抽风机，迅速称取 0.5g 溴化氰，置乳钵中加入少量水研磨，将溶液由中口通过小漏斗倾入三颈瓶，固体部分继续加水研磨，直至全部转移。共用水 10ml。将溴化氰加入后反应立即开始。此时控制温度在 20～22℃，并不断从小分液漏斗滴加 2mol/L NaOH 调整 pH 维持 11～12，每半分钟检查并记录瓶内 pH 值和反应温度。等 pH 不再降低时，继续反应 5 分钟（整个反应控制在 15 分钟以内）停止搅拌，取出烧瓶，投以小冰块迅速冷却。反应液以三号砂芯漏斗抽滤，并以 5～10 分钟以内 4℃ 以下 1000～2000ml 0.1mol/L $NaHCO_3$ 溶液连续抽洗，随即再用 150ml 0.025mol/L pH 10.2 硼酸缓冲液分三次抽洗。然后移入 125ml 烧杯加入 15～20ml 上述缓冲液，使呈 40% 悬液。

2. 耦联胰蛋白酶　迅速在上述载体悬液中加入 20mg 用 5ml 硼酸缓冲液溶解的胰蛋白酶结晶，在低温下置磁力搅拌器搅拌反应 4～6 小时。将反应物装入层析柱，先用 7 倍体积 pH 10.2 硼酸缓冲液洗柱，流速 2ml/min，收集洗脱液，测 A_{280} 并计算体积，再用 5 倍体积 pH 10.0 0.1mol/L 乙醇胺溶液洗柱，流速 2ml/min。然后分别用 0.1mol/L pH 8 硼酸缓冲液及 0.1mol/L pH 4 醋酸缓冲液交替洗柱各 2 次，每次用 50ml，流速 2ml/min，直至 $A_{280} < 0.02$。洗完后用 0.01mol/L pH 4 醋酸缓冲液 50ml 缓慢过柱予以平衡，流速 1ml/min，直至流干。出柱，所得固相化胰蛋白酶称重，取样后以含 0.01% NaN_3 的 0.01mol/L pH 4 醋酸缓冲液悬浮，置冰箱保存。

3. 固定化胰蛋白酶活力测定　用试管分别称取抽干的固定化胰蛋白酶 10mg 和 20mg，按下表操作：

	0	1	2	3	4
标准酶液（ml）		0.5	0.5		
固定化酶（mg）				0.5	0.5
pH 7.8 磷酸缓冲液（ml）	0.5			0.5	0.5
2mol/L HCl（ml）	0.5				
底物（BANA）（ml）	0.5	0.5	0.5	0.5	0.5
37℃准确反应 15 分钟					
2mol/L HCl（ml）	0	0.5	0.5	0.5	0.5
0.1% $NaNO_2$（ml）	1	1	1	1	1
摇匀，放置 3 分钟以上					
0.5% 氨基磺酸铵（ml）	1	1	1	1	1
摇匀，放置 2 分钟以上					
NEDA（ml）	2	2	2	2	2
摇匀，放置 0.5 小时，离心 2500r/min × 5 分钟					
A_{580}					
比活力（U/ml）					

胰蛋白酶活力单位定义是在一定条件下，每分钟能水解底物 BANA 导致吸光度变化 0.01 的酶活力称为一个酶活力单位。

效价计算：

$$P = \frac{M}{W \times 0.01 \times 15}$$

式中，P 为每 mg 供试品中的胰蛋白酶单位；M 为样品液的吸光度；W 为反应液中供试品的浓度（mg/ml）。

由此求出：（1）每克固定化酶的活力

（2）固定化酶总活力

（3）固定化酶活力回收率

（4）耦联效率

扫码“练一练”

【思考题】

1. 本实验的关键环节是什么？应采取什么措施，为什么？
2. 在本实验中测定固相化胰蛋白酶活性时应该注意哪些问题？

EXPERIMENT 5 Preparation of immobilized trypsin by covalent coupling

【Purpose】

1. To learn the basic method of preparing immobilized enzymes by coupling method.
2. To learn the method to determine the coupling amount of affinity adsorbents.
3. To learn how to measure trypsin activity.

【Principle】

Covalent coupling method is one of the important means to prepare immobilized enzymes. The method is to couple the inactive side chain group of the enzyme with the functional group of the carrier by means of a covalent bond, that is, to activate the carrier first, and then the enzyme binds to the hydrophilic insoluble carrier by a covalent bond. The advantage is that the combination of the enzyme and the carrier is firm and not easy to fall off, and the obtained immobilized enzyme can be continuously used. The disadvantage is that the operation of the carrier activation is complicated, the reaction conditions are fierce, and the recovery rate of the enzyme activity is low, and it is necessary to strictly control the conditions to obtain a highly viable immobilized enzyme.

The main factors of the covalent coupling method are: the carrier is hydrophilic, has certain mechanical strength and stability, and has a functional group that binds to the enzyme under mild conditions. The reaction conditions of the coupling reaction must be in a buffer solution of mild pH, medium ionic strength and low temperature. The coupling reaction chosen should be as far as possible to consider as little as possible side reactions to other functional groups of the enzyme. The configuration of the enzyme after immobilization should be considered to minimize the influence of the steric hindrance of the carrier on the enzyme activity.

In this experiment, agarose gel (Sepharose 4B) was used as a solid phase carrier and activated with cyanogen bromide (CNBr) under alkaline conditions. Introduce an active "iminocarbonate", which is directly coupled to the free amino group of trypsin under weak base conditions. And the aminocarbonate and the isoflavone derivative are formed to obtain an immobilized enzyme.

Cyanogen bromide is extremely volatile and highly toxic. Its decomposition products are also toxic. The activation reaction must be carried out in a fume hood. Since the activating group is easily hydrolyzed and the stability of the activated intermediate is poor, the operation must be compact and rapid, and the washing after completion of the reaction should also be fast.

There are generally two methods for determining the amount of protein coupled. One is the direct measurement method, which directly obtains the data of the protein coupling amount by nitrogen or amino acid composition analysis; the other method is the indirect measurement method, that is, the amount of protein used in the coupling minus the protein remaining after the coupling the amount. For the sake of simplicity, this experiment uses an indirect assay. The activity of the affinity adsorbent can be expressed by trypsin activity.

Trypsin catalyzes the hydrolysis of proteins with a high degree of specificity for peptide bonds composed of carboxyl groups of basic amino acids such as arginine and lysine. Trypsin not only hydrolyzes peptide bonds but also hydrolyzes amide and ester bonds. Therefore, the activity of trypsin can be determined by using synthetic amide and ester compounds as substrates. In this experiment, trypsin activity is to be determined by using synthetic benzoyl – L – arginine naphthylamide (BANA) as substrate.

【Materials】

1. Apparatus

(1) Magnetic stirrer

(2) Glass funnel

(3) Sand core funnel

(4) Three - necked flask

(5) Milk Thistle

(6)751 UV spectrophotometer

(7) Column

(8) Thermometer

(9) Measuring cylinder

(10) Suction bottle

(11) Pipette

(12) Beaker

2. Reagent

(1) Cyanogen bromide

(2) Agarose gel 4B

(3) Trypsin crystallization

(4)2mol/L NaOH solution

(5)2mol/L HCl solution

(6)0. 025mol/L, pH 10. 2 borate buffer(containing 0. 2mol/L $CaCl_2$): 2. 38g sodium tetraborate, 2. 22g calcium chloride and 3. 0ml of 2mol/L NaOH dissolved in water, to a volume of 1000ml, pH meter calibration.

(7)0. 1mol/L ethanolamine: 6. 1g(6. 0ml) of ethanolamine dissolved in 1000ml of water.

(8)0. 1mol/L pH 8 boric acid buffer(containing 0. 02mol/L $CaCl_2$): 2. 862g of sodium tetraborate, 2. 78g of calcium chloride and 4. 328g of boric acid, dissolved in water, fixed to a volume of 1000ml, corrected by pH meter.

(9)0. 1mol/L pH 4 acetate buffer(containing 1mol/L NaCl): sodium acetate 2. 45g, sodium chloride 58. 4g and 36% acetic acid solution 13. 66ml, dissolved in water, constant volume 1000ml, pH meter correction.

(10)0. 01mol/L pH 4 acetate buffer(containing 0. 1mol/L NaCl and 0. 025mol/L $CaCl_2$): sodium acetate 0. 245g and 36% acetic acid solution 1. 37ml, sodium chloride 5. 84g and calcium oxide 2. 78g, dissolved in water, make up to 1000ml, pH meter calibration.

(11) 0. 05mol/L pH 7. 8 phosphate buffer: $Na_2HPO_4 \cdot 12H_2O$ (M_w 358. 16) 16. 476g and 0. 625g $NaH_2PO_4 \cdot 2H_2O$ (M_W 156. 03), dissolved in water, to a volume of 1000ml, pH meter calibration.

(12) substrate 0. 06% benzoyl - L - arginine - β - naphthylamine(BANA) solution: take BANA 60mg, add 95% ethanol 20ml, dissolve, use pH 7. 8, 0. 2mol/L phosphate buffermake up to 100ml.

(13)0. 05% naphthylethylenediamine hydrochloride(NEDA) ethanol solution

(14)0. 5% ammonium sulfamate solution

(15)0. 1% sodium nitrite solution

(16) Standard enzyme solution: 1mg of trypsin crystal is weigh and made up to 100ml with pH 7. 8 phosphate buffer.

【Procedures】

1. Carrier activation

Rinse commercially available Sepharose 4B several times with distilled water and drain with a Buchner funnel. Approximately 7g (or 10ml from the deposition) that is weighed into a three – necked flask. Add 10ml of water and adjust to pH 11 with 2mol/L NaOH. Transfer the flask was to a fume hood, stired by electromagnetic stirring(low speed), and heat or cool to 18℃. Start the exhaust fan, quickly weigh 0. 5g of cyanogen bromide, add a small amount of water to the mortar, grind the solution, pour the solution from the middle mouth through the small funnel into the three – necked flask, and continue to add water to the solid part until all the transfer. A total of 10ml of water. The reaction begins immediately after the addition of cyanogen bromide. At this time, control the temperature at 20 to 22℃, and continuously add 2mol/L NaOH added from a small separatory funnel to adjust the pH to maintain 11 – 12. Check and record the pH value and reaction temperature of the bottle every half minute. When the pH is no longer lowered, continue the reaction for 5 minutes(the whole reaction is controlled within 15 minutes), stop the stirring, take out the flask, and quickly cool the small ice. Suction filter the reaction solution with a No. 3 sand core funnel, and continuously pump with a solution of 1000 to 2000ml of 0. 1mol/L $NaHCO_3$ below 4℃ for 5 to 10 minutes, and then washed three times with 150ml of 0. 025mol/L pH 10. 2 borate buffer. Then, add 15 to 20ml of the above buffer to a 125ml beaker to make a 40% suspension.

2. Coupling trypsin

To the above carrier suspension, quickly add 20mg of trypsin crystals dissolved in 5ml of boric acid buffer, and stir the mixture at a low temperature for 4 to 6 hours. Charge the reaction into a chromatography column, and wash the column with 7 volumes of a pH 10. 2 borate buffer at a flow rate of 2ml/min, and collect the eluate, and measure the A_{280} and calculate the volume. Then wash the column with 5 volumes of a pH 10. 0 0. 1mol/L ethanolamine solution at a flow rate of 2ml/min. Then, the column is alternately washed twice with 0. 1mol/L pH 8 borate buffer and 0. 1mol/L pH 4 acetate buffer, 50ml each time, and the flow rate was 2ml/min until $A_{280} < 0.02$. After washing, the column slowly equilibrate with 50ml of 0. 01mol/L pH 4 acetate buffer, and at the flow rate of 1ml/min until it is drained. After the column is taken out, weigh the obtained solid phase trypsin, sample, suspended in 0. 01mol/L pH 4 acetate buffer containing 0. 01% NaN_3, and store in a refrigerator.

3. Immobilized trypsin activity assay

The dried immobilized trypsin is weighed 10mg and 20mg separately in a test tube, and the following operation is performed:

	0	1	2	3	4
Standard enzyme solution(ml)		0. 5	0. 5		
Immobilized Enzyme(mg)				0. 5	0. 5
pH 7. 8phosphate buffer(ml)	0. 5			0. 5	0. 5

continued

	0	1	2	3	4
2mol/L HCl(ml)	0.5				
Substrate(BANA)(ml)	0.5	0.5	0.5	0.5	0.5
Accurately react at 37℃ for 15 minutes					
2mol/L HCl(ml)	0	0.5	0.5	0.5	0.5
0.1% $NaNO_2$(ml)	1	1	1	1	1
Shake well and place for more than 3 minutes					
0.5% Ammonium sulfamate(ml)	1	1	1	1	1
Shake well and place for more than 2 minutes					
NEDA(ml)	2	2	2	2	2
Shake well, place for half an hour, centrifuge 2500r/min ×5 minutes					
A_{580}					
Specific activity(U/ml)					

The unit of trypsin activity is defined as the enzyme activity that, under certain conditions, can hydrolyze the substrate BANA per minute, resulting in a change in absorbance of 0.01, called an enzyme activity unit.

Potency calculation:

$$P = \frac{M}{W \times 0.01 \times 15}$$

In the formula:

P: Trypsin unit per mg of test sample.

M: The absorbance of the sample solution.

W: Concentration of the test sample in the reaction solution(mg/ml).

From this, we find:

(1) Activity of immobilized enzyme per gram.

(2) Total activity of immobilized enzyme.

(3) Immobilized enzyme activity recovery rate.

(4) Coupling efficiency.

【Questions】

1. What is the key part of this experiment? What measures should be taken, and why?

2. What should be noted when determining the activity of solid phase trypsin in this experiment?

扫码“学一学”

实验六　包埋法制备固定化延胡索酸酶产生菌

【实验目的】

1. 掌握包埋法制备固定化细胞的原理和方法。

2. 学习延胡索酸酶的活力测定方法。

【实验原理】

扫码“看一看”

微生物固定化方法中，以包埋法最为常用。包埋法的原理是将微生物细胞截留在水不溶性的凝胶聚合物孔隙的网络空间中。通过聚合作用或通过离子网络形成，或通过沉淀作用，或改变溶剂、温度、pH 值使细胞截留，凝胶聚合物的网络可以阻止细胞的泄露，同时能让底物物质渗入进去和产物扩散出来。

包埋材料可分为天然高分子多糖类和合成高分子化合物两大类，如卡拉胶、海藻酸钙、壳聚糖、聚丙烯酰胺等。天然高分子多糖类的海藻酸钠和卡拉胶应用最多，它们具有固化、成形方便、对微生物毒性小及固定化密度高等优点。但它们抗微生物分解性能较差，机械强度较低。合成高分子化合物中常用的载体物质的突出优点是抗微生物分解性能好，机械强度高，化学性能稳定。但是聚合物网络的形成条件比较剧烈，对微生物细胞的损害较大，而且成形的多样性和可控性不好。

延胡索酸酶又称反丁烯二酸酶或延胡索酸水化酶（fumarate hydratase），酶学分类号：E. C. 4. 2. 1. 2。具有立体异构性，只催化反丁烯二酸（又称延胡索酸、富马酸）而对顺丁烯二酸（又称马来酸）没有作用。延胡索酸酶的催化机制是通过将 H_2O 立体特异添加到延胡索酸双键上，催化延胡索酸水化生成 L－苹果酸，反应是可逆的。

本实验首先培养含有高延胡索酸酶活力的微生物，以卡拉胶为包埋载体，采用凝胶包埋法制备固定化延胡索酸酶产生菌，并以延胡索酸氨为底物在适当条件下转化生成 L－苹果酸。采用这种生物转化法制备 L－苹果酸是目前国内外生产 L－苹果酸的主要方法之一。L－苹果酸是生物体三羧酸循环的中间体，易被人体吸收，因此作为性能优异的食品添加剂和功能性食品被广泛应用于食品、化妆品、医疗和保健品等领域。

【实验材料】

1. 器材

（1）水浴振荡培养箱		1 台
（2）721 分光光度计		1 台
（3）高压蒸汽灭菌锅		1 台
（4）离心机		1 台
（5）恒温水浴锅		1 台
（6）磁力搅拌器		1 台
（7）真空泵		1 台
（8）布氏漏斗	5cm	1 个
（9）抽滤瓶	100ml	1 个
（10）摇瓶	250ml	5 个
（11）锋利小刀		1 把
（12）烧杯	500ml	2 个
	200ml	2 个
	100ml	2 个
（13）容量瓶	50ml	1 个

2. 试剂

（1）菌种

（2）延胡索酸

（3）β-萘酚溶液

（4）0.3mol/L 氯化钾

（5）1mol/L 延胡索酸钠钾（pH 7.0） 含 0.6% 胆酸的 1mol/L 延胡索酸钠钾（pH 7.0）。

（6）卡拉胶

（7）L-苹果酸标准品

（8）浓硫酸

（9）4mol/L 盐酸

（10）微生物培养基 2.0% 丙二酸、2.0% 玉米浆、0.5% 柠檬酸二钠、0.2% 硫酸二氢钾、0.05% 七水合硫酸镁，调节 pH 7.0，115℃湿热灭菌 30 分钟。

【实验方法】

1. 菌体的制备

接种 1ml 对数期菌种至 250ml 摇瓶中，内含 50ml 灭菌培养基，与 30℃振荡培养 24 小时，培养液 3000r/min 离心 20 分钟，倾出上清液，将菌体用生理盐水洗涤 1 次，离心收集菌体。

2. 固定化细胞的制备 将 4g 湿细胞悬浮于 4ml 生理盐水中，置 45℃恒温水浴保温，再将 0.8g 卡拉胶溶于 17ml 生理盐水中，加热使之溶解，将两者混合均匀后置 4℃冰箱放置 40 分钟。再将凝胶在 100ml 0.3mol/L 氯化钾溶液中浸泡 4 小时。然后将凝胶切成 $3mm^3$ 颗粒，放入内含 0.2% 胆酸的 1mol/L 延胡索酸钠钾溶液中，在 37℃恒温处理 24 小时备用。

3. 固定化细胞内延胡索酸酶活力的测定

（1）标准曲线的制备

苹果酸在浓硫酸中和 β-萘酚反应，产生淡黄色化合物，在 370nm 处有特征吸峰，在一定范围内与苹果酸含量成线性关系，且延胡索酸、琥珀酸、柠檬酸等有机酸对此反应均无影响。

配制 60μg/L 浓度的 L-苹果酸标准液，按下表依次加入试剂测定 L-苹果酸的标准曲线：

	1	2	3	4	5	6	7
L-苹果酸标准液/ml	0	0.2	0.4	0.5	0.6	0.8	1.0
蒸馏水/ml	1.0	0.8	0.6	0.5	0.4	0.2	0
β-萘酚试剂/ml	0.1	0.1	0.1	0.1	0.1	0.1	0.1
浓硫酸/ml	6	6	6	6	6	6	6

混匀后，沸水中加热 20 分钟，取出置冷水中冷却至室温，于 370nm 处测定吸收值，并做出标准曲线。

（2）酶活力的测定

延胡索酸酶的酶活定义：将每小时相当于 1g 游离湿细胞的固定化细胞转化生成 1μmol

苹果酸定义为一个酶活力单位，酶活力单位为 μmol/h · g · cells。

取6g 固定化细胞，加 40ml 1.0mol/L 延胡索酸钠钾（pH 7.0）于 100ml 小烧杯中，于 37℃恒温下搅拌反应 30 分钟，抽滤除去固定化细胞，反应液加 20ml 4mol/L 的盐酸酸化，使未反应完的延胡索酸沉淀析出，过滤除去。取样稀释至标准曲线范围内，测定 L-苹果酸的含量，计算延胡索酸酶活力。

【思考题】

1. 固定化细胞的方法主要有哪些？
2. 延胡索酸酶催化反应过程中加入胆酸的作用是什么？

扫码“练一练”

EXPERIMENT 6　Preparation of immobilized fumarate-producing bacteria by embedding

【Purpose】

1. To learn the principle and method of preparing immobilized cells by embedding.
2. To learn how to determine the activity of fumarase.

【Principle】

Embedding is the most commonly used method among the microorganism immobilization methods. The principle of embedding is to trap microbial cells in the pores of the water insoluble gel polymer. By polymerizing or forming an ionic network, or by precipitation, or changing the solvent, temperature, pH to trap cells, gel polymers can prevent cell leakage, while allowing substrates and products penetrate to diffuse.

The embedding material can be divided into two types, natural high molecular polysaccharides and synthetic high molecular compounds, such as carrageenan, calcium alginate, chitosan, and polyacrylamide. Natural high molecular weight polysaccharides such as sodium alginate and carrageenan are most widely used, and they have the advantages of curing, convenient formation, low toxicity to microorganisms, and high immobilization density. However, their antimicrobial decomposition performance is poor and the mechanical strength is low. The outstanding advantages of the vector materials commonly used in synthetic polymer compounds are good antimicrobial decomposition performance, high mechanical strength and chemical stability. However, the formation conditions of the polymer network are relatively severe, and the damage to the microbial cells is grievous, and the diversity and controllability of the formation are unmanageable.

Fumarate is also known as fumarate hydratase, and the enzymology classification number is E. C. 4. 2. 1. 2. It has catalytic stereoisomerism and only catalyzes fumaric acid, has no effect on maleic acid. The catalytic mechanism of fumarase is to form L-malic acid by stereospecificity adding H_2O to the fumaric acid double bond to catalyze the hydration of fumaric acid, and the reaction is reversible.

In this experiment, the microorganism containing high fumarase activity was firstly cultured, and carrageenan was used as the embedding vector. The immobilized fumarate-producing bacteria were

prepared by gel embedding method, and the fumarate ammonia was used as substrate to convert L-malic acid. The preparation of L-malic acid by biotransformation is one of the main methods. L-malic acid is an intermediate of the bio-tricarboxylic acid cycle and is easily absorbed by the human body. Therefore, it is widely used in food, cosmetics, medical field and health care products with excellent performance.

【Materials】

1. Apparatus

(1) Water bath shaker incubator

(2) 721 spectrophotometer

(3) High pressure steam sterilization pot

(4) Centrifuge

(5) Constant temperature water bath

(6) Magnetic stirrer

(7) Vacuum pump

(8) Brinell funnel

(9) Filter bottle

(10) Shake bottle

(11) Sharp knife

(12) Beakers

(13) Volumetric flask

2. Reagents

(1) Strain

(2) Fumaric acid

(3) β-naphthol solution

(4) 0.3mol/L potassium chloride

(5) 1mol/L potassium fumarate (pH 7.0): 1mol/L potassium fumarate (pH 7.0) containing 0.6% cholic acid.

(6) Carrageenan

(7) L-malic acid standard sample

(8) Concentrated sulfuric acid

(9) 4mol/L hydrochloric acid

(10) Microbial culture medium: 2.0% malonic acid, 2.0% corn syrup, 0.5% disodium citrate, 0.2% potassium dihydrogen sulfate, 0.05% magnesium sulfate heptahydrate (adjusted to pH 7.0, moist heat sterilization at 115℃ for 30 minutes).

【Procedures】

1. Preparation of bacterial cells Inoculate 1ml of logarithmic strain into 250ml shake flask, which is containing 50ml of sterilized medium. Incubate the flask at 30℃ for 24 hours, centrifuge at 3000r/min for 20 minutes, and pour the supernatant to remove the bacteria. The precipitation was

washed once with physiological saline, and the cells were collected by centrifugation.

2. Preparation of immobilized cells

4g of wet cells suspended in 4ml of physiological saline were placed in a constant temperature water bath at 45℃. Then 0. 8g of carrageenan was dissolved in 17ml of physiological saline, heated to dissolve. These two materials were uniformly mixed and placed at 4℃ for 40 minutes. The gel was immersed in 100ml of 0. 3mol/L potassium chloride solution for 4h. The gel was then cut into $3mm^3$ pellets and placed in a 1mol/L sodium potassium fumarate solution which is containing 0. 2% bile acid at 37℃ for 24 hours.

3. Determination of fumarase activity in immobilized cells

(1) Preparation of standard curve

Malic acid reacts with β – naphthol in concentrated sulfuric acid to produce a pale – yellow compound with a characteristic peak at 370nm. It has a linear relationship with malic acid content in a certain range, and organic acids such as fumaric acid, succinic acid and citric acid have no effect in this reaction.

Prepare a standard solution of L – malic acid at a concentration of 60μg/L, and add a reagent to determine the standard curve of L – malic acid according to the following table:

	1	2	3	4	5	6	7
L – malic acid standard solution/ml	0	0. 2	0. 4	0. 5	0. 6	0. 8	1. 0
Distilled water/ml	1. 0	0. 8	0. 6	0. 5	0. 4	0. 2	0
β – naphthol/ml	0. 1	0. 1	0. 1	0. 1	0. 1	0. 1	0. 1
Concentrated sulfuric acid/ml	6	6	6	6	6	6	6

After mixing, it was heated with boiling water for 20 minutes, and cooled to room temperature in cold water. The absorption value was measured at 370nm, and a standard curve was prepared.

(2) Determination of enzyme activity

Enzymatic activity definition of fumarase: The immobilized cells containing 1g of free wet cells inverting to 1μmol of malic acid in an hour is defined as an enzyme activity unit, and the unit of enzyme activity is μmol/h · g · cells.

Add 40ml of 1. 0mol/L sodium potassium fumarate (pH 7. 0) in 6g of immobilized cells, stir at 37℃ for 30 minutes, and remove the immobilized cells by suction filtration. Unreacted fumaric acid was precipitated after added 20ml of 4mol/L hydrochloric acid and removed by filtration. The sample was diluted to the concentrations within the range of the standard curve, and the content of L – malic acid could be measured to calculate the fumarase activity.

【Questions】

1. What are the main methods of immobilizing cells?
2. Why is the cholic acid added during the catalytic reaction of fumarate?

第四节　基因工程技术基础实验

基因工程（genetic engineering）是在现代生物学、化学和化学工程学以及其他数理科学的基础上产生和发展起来的。基因工程技术是指利用现代分子生物学技术，特别是酶学技术，在分子水平上按照人们的设计方案将遗传物质 DNA 片段（目的基因）插入载体（vector）DNA 分子（如质粒、病毒等），从而实现 DNA 分子体外重组操作与改造，产生新的自然界从未有过的重组 DNA 分子，然后再将之引入特定的宿主（host）细胞进行扩增和表达，使宿主细胞获得新的遗传性状的技术。基因工程的核心是构建重组体 DNA 的技术，所以基因工程有时也称为基因操作（gene manipulation）或重组 DNA 技术（recombinant DNA technology）。

基因工程的出现标志着人类已经能够按照自己意愿进行各种基因操作上规模生产基因产物，并自主设计和创建新的基因、新的蛋白质和新的生物物种。自 1979 年美国基因技术公司用人工合成的人胰岛素基因重组转入大肠埃希菌中合成人胰岛素（insulin），至今，我国已有人干扰素、人白介素 2、人集落刺激因子、重组人乙型肝炎病毒疫苗、基因工程幼畜腹泻疫苗等多种基因工程药物和疫苗进入生产或临床试用，世界上亦有几百种基因工程药物及其他基因工程产品在研制中，这是当今农业和医药业发展的重要方向。

基因工程的基本过程是将一个含目的基因的 DNA 片段经体外操作与载体连接，并转入宿主细胞，使之扩增、表达的过程。可概括为切、接、转、增、检。主要步骤如下。

1. 目的基因（靶基因）的制备 方法主要有四种：由纯化 mRNA，反转录成 cDNA 的反转录法，内切酶切割直接分离法和人工合成法（60 ~ 100bp 长度为宜），基因通过 DNA 聚合酶链式反应（Polymerase Chain Reaction，PCR）在体外进行扩增法。

2. 载体的选择与制备 载体必须具备三个条件：①具有能使外源 DNA 片段组入的克隆位点。②能携带外源 DNA 进入受体细胞，或游离在细胞质中进行自我复制，或整合到染色体 DNA 上随染色体 DNA 的复制而复制。③必须具有选择标记，承载外源 DNA 的载体进入受体细胞后，以便筛选克隆子。

3. 酶切 用限制性内切酶分别将目的基因和载体分子切开。

4. 目的基因与载体的连接，形成重组 DNA 分子 连接方法主要有黏性末端连接法和钝性末端连接法。

5. 传染、扩增 将重组 DNA 分子转入受体细胞，并在其中进行复制、扩增，使重组 DNA 分子在受体细胞内的拷贝数大量增加：常用的受体细胞有大肠埃希菌、枯草杆菌、土壤农杆菌、酵母菌和动植物细胞等。用人工的方法使体外重组的 DNA 分子转移到受体细胞，主要是借鉴细菌或病毒侵染细胞的途径。

6. 重组子的筛选与鉴定 方法有很多种，如抗生素抗性基因筛选，进行限制性酶切位点切割或通过 PCR 扩增特定基因片段，进行琼脂糖凝胶电泳基因片段分析，从而确定哪些质粒具有目的 DNA 插入片断或通过诱导蛋白质表达结合 SDS - PAGE 电泳检测目的蛋白，最后通过 DNA 序列测定对克隆片段进行鉴定等。

本部分实验主要参照本学院科研成果——构建大肠埃希菌外分泌表达重组水蛭素Ⅲ的整个过程，实验内容包括碱裂解法制备质粒 DNA、酶切和鉴定、PCR 扩增目的基因、氯化钙法制备感受态细胞、目的基因与载体的连接、转化以及通过抗生素标记筛选转化子或限制性酶谱分析基因、PCR 扩增特定基因片段或电泳检测表达蛋白筛选转化子等基因工程的基本操作技术。

Section 4 Basic experiments of genetic engineering technology

Gene engineering is based on modern biology, chemistry, chemical engineering and other mathematical and physical sciences. Gene engineering technology refers to the use of modern molecular biology technology, especially enzymology technology, to insert DNA fragments (target genes) into vectors (such as plasmids, viruses, etc.) according to people's designation at the molecular level. In this way, the recombination operation and modification of DNA molecule in vitro can be realized, and new recombinant DNA molecule never existed in nature can be produced. Then, the recombinant DNA molecule can be introduced into specific host cells for amplification and expression, so that the host cells can acquire new genetic characteristics. The core of genetic engineering is the technology of constructing recombinant DNA, so genetic engineering is also called gene manipulation or recombinant DNA technology.

The emergence of genetic engineering indicates that human beings have been able to produce gene products on a large scale according to their own wishes, and independently design and create new genes, new proteins and new biological species. In 1979, American Gene Technology Company transferred synthetic human insulin gene recombination into Escherichia coli to synthesize human insulin. Up to now, there have been many genetic engineering drugs in China, such as human interferon, human interleukin - 2, human colony stimulating factor, recombinant human hepatitis B virus vaccine, genetic engineering juvenile animal diarrhea vaccine and so on. Vaccines and genetic engineering drugs already enter the production or clinical trials. There are also hundreds of genetically engineered drugs and other genetically engineered products in the world, which is an important direction for the development of agriculture and medicine.

The basic process of genetic engineering is to connect a DNA fragment containing the target gene with the vector in vitro and transfer it to the host cell for amplification and expression.

The main steps are as follows:

1. Preparation of target gene: There arefour main methods: reverse transcription, direct separation by endonuclease cleavage, chemical synthesis and PCR.

2. Selection and preparation of vector. The vector must have three conditions: ①It should have cloning site that can make foreign DNA fragments integrate. ②It can carry exogenous DNA into recipient cells, or self - replicate in cytoplasm, or integrate into chromosomal DNA and replicate with the replication of chromosomal DNA. ③It should have selective markers in order to screen recombinant clones.

3. The target gene and vector molecule were cut by restriction endonuclease.

4. The methods of link the target genes and vectors to form recombinant DNA molecule: sticky end - to - end ligation and passive end - to - end ligation.

5. Transforming recombinant DNA molecule to replicate and amplify in receptor cells. The commonly used receptor cells include *E. coli*, *Bacillus subtilis*, *Agrobacterium soils*, yeasts, animal and plant cells, etc. Artificial transfer of recombinant DNA molecules to recipient cells in vitro mainly refers to the way of bacteria or viruses infecting cells.

6. Screening and identification of recombinants: There are many methods, such as screening antibiotic resistance genes, restriction cleavage, amplification of specific gene fragments by PCR, *etal.* Combined with agarose gel electrophoresis, gene fragments were analyzed to determine which plasmids had a target DNA insertion fragment. Or the target protein could be detected by inducing protein expression and SDS – PAGE electrophoresis. Finally, the cloned fragments were identified by DNA sequencing.

This part of the experiment mainly refers to the whole process of constructing recombinant hirudin Ⅲ secreted by Escherichia coli. The whole experiment include preparation of plasmid DNA by alkaline lysis, digestion and identification, amplification of target genes by PCR, preparation of competent cells by calcium chloride, connection between target genes and vectors, transformation and screening of transformants or restriction zymogram analysis genes or PCR by antibiotic markers. Basic operation techniques of gene engineering, such as amplification of specific gene fragments and screening transformants by electrophoresis detection of expressed proteins.

扫码“学一学”

实验七　碱裂解法制备质粒 DNA、限制性内切酶酶切和琼脂糖凝胶电泳鉴定

【实验目的】

1. 掌握　以大肠埃希菌 *E. coil* 为材料，用碱裂解法小量制备质粒 DNA、限制性酶酶切消化质粒 DNA 和琼脂糖凝胶电泳进行鉴定的分子生物学中微量操作技术方法

2. 了解　碱裂解法小量制备质粒 DNA、限制性酶酶切消化质粒 D 和琼脂糖凝胶电泳进行鉴定的工作原理和技术。

【实验原理】

质粒 DNA 是基因工程研究中常用的克隆与表达外源基因的载体，也是分子水平上研究 DNA 结构、功能和复制的较理想的实验模型。它的提取和分离是基因工程与分子生物学研究中最常用和基本的实验操作技术。

质粒 DNA 是微生物细胞中相对分子质量比染色体 DNA 小得多的双链、共价、闭合环状 DS 分子，是一种存在于染色体外但能够自主复制的遗传因子。质粒通常携带有染色体上所不存在的基因，与细胞的主要代谢活动无关但表现出一些有用的性状，如抗生素抗性、抗细菌等。根据质粒所携带的基因以及这些基因所赋予的宿主细胞的性状进行质粒分类，包括抗性质粒（R）、致育因子（F）、col 以及毒性质粒。R 质粒编码抗生素抗性，可将其抗性转移到缺乏该质粒的适宜的受体细胞，使后者也获得同样的抗生素抗性能力以指示载体或重组 DNA 分子是否进入宿主细胞。质粒拥有自己的复制原点，可以不依赖于染色体面进行独立复制。

质粒的提取方法有很多种，主要有碱裂解法、煮沸裂解法、氯化铯一溴乙锭密度梯度法和柱层析法等。而碱裂解法是制各质粒 DNA 最为常用的一种简便快速的方法之一。在本实验中，从少量细胞中分离质粒 DNA，采用这种方法分离的 DNA 常被称为小量制备。这种方法提取的质粒 DNA 纯度很高，无须进一步纯化，就可以用于限制性酶酶切消化、PCR 扩

增、序列测定等。

本方法是依据共价闭合环状质粒DMA与染色体DNA在变性和复性特性之间存在差异进行的当细胞悬浮于NaOH和十二烷基硫酸钠（SDS）溶液中时，在pH 12.0～12.6的碱性条件下细胞发生裂解，蛋白质和染色体DNA氢键断裂发生变性，而双链共价闭环质粒DNA分子变性程度小，改变条件能很快恢复自然状态，加入中和溶液（酸性乙酸钾），调节pH至中性并保持高盐状态，离心后，大分子染色体DNA不能复性缠成网状结构，染色体DNA、蛋白质在去垢剂SDS作用下就会与细胞碎片一起沉淀下来，而双链共价闭环质粒DNA可溶，则留在上清液中。其实，当加入碱溶液时，质粒DNA也发生变性，但其两条链仍然靠得很近，就像一条链上的两个环链，当加入酸性溶液进行中和时，质粒DNA的两条链则分别与其互补链重新退火，进而形成原始状态的质粒。用酚－氯仿抽提上清中质粒，进一步纯化，最后用乙醇沉淀。

限制性内切酶Ⅱ可以识别并结合到双链DNA的特异核酸序列上，同时对其进行水解，如EcoRⅠ的识别序列为GAATTC，一旦遇到相应序列，即可水解DNA链中的磷酸二酯键。在进行限制性酶消化时，将双链DNA分子与适量的限制性酶以及供应商推荐的相应的缓冲液中进行混合保温，并在该酶的最适温度下进行反应。一个酶单位通常定义为（依据供应商）1小时内在合适的温度下，尤其是37℃下，用于完全消化1μg双链DNA的酶的数量。使用中一般以1μg DNA对1～3U酶保温1～4小时为宜。

琼脂糖电泳是生物化学和分子生物学中应用最为广泛的技术之一，常用于质粒DNA或DNA片断的分离、纯化与鉴定。在中性pH的电泳缓冲体系中，DNA分子带负电，只能向阳极泳动。琼脂糖是一种海藻多糖，直链，由D－半乳糖和3，6－脱水L－半乳糖的残基交替排列组成。在水溶液中，琼脂糖分子间由于氢键作用而形成凝胶。琼脂糖凝胶浓度越大，凝胶就越硬。DNA分子在电场中通过凝胶泳动的速率除与其所带电荷、分子大小有关外，还与其凝胶浓度、电泳电压、构象密切相关。相同分子量的线状双链DNA分子以不同速度通过不同浓度的琼脂糖凝胶，通过凝胶的速率与其相对分子质量对数成反比。固利用不同浓度的凝胶可分辨大小不同的DNA片段。在低电压时，线状DNA片段的迁移率与所用电压成正比；但电压增高时，大分子量DNA片段迁移率的增大与电压不成正比关系，分辩率下降。为了获得良好的分离效果，凝胶电泳时电压不应超过5V/cm。双链的质粒DNA分子具有三种不同的构象：共价闭合环状DNA（cccDNA）、开环DNA（ocDNA）和线性DNA（cDNA），具有不同的电泳迁移率，可在琼脂糖凝胶电泳中被分开。一般情况下，迁移率cccDNA > cDNA > ocDNA。线性质粒的出现意味着制备过程中核酸酶的污染或操作方法不当。

琼脂糖胶浓度一般在0.5%～2%之间，通过观察示踪染料的迁移距离可以判断DNA的迁移距离。溴酚蓝和二甲苯青染料在琼脂糖凝胶中的迁移速率大致分别与300bp和4000bp大小的双链DNA片段相同，迁移足够距离后，就可以通过溴化乙锭染色来观察DNA片段。溴化乙锭是一种荧光染料，它可以在制胶时混入其中，在电泳时进行染色，也可以待电泳完成后将凝胶浸泡在稀释的溴化乙锭溶液中进行染色45分钟，它嵌插在DNA和RNA碱基之间，在紫外灯光下会显出橙黄色光。其灵敏度很高，用低浓度的荧光染料（0.5μg/ml溴化乙锭）染色，在紫外灯下可直接观察检测少至1ng的DNA。

为了判断目的DNA片段的大小，常在同一凝胶的目的DNA旁加一分子量参照物，同时电泳并染色后，就能在紫外灯下很快知道目的片段的大小。最常用的相对分子质量参照物是HindⅢ消化物，各片段的大小以bp表示，分别为：23130、9416、6557、4361、2322、

2027、564、125bp。

【实验材料】

1. 器材

(1) 培养皿 (直径 9cm) 若干
(2) 电热恒温培养箱 1台
(3) 标准净化工作台 1台
(4) 高压灭菌锅 1台
(5) 恒温水浴锅 1台
(6) 冷冻离心机 1台
(7) 微量移液器 (20μl 或 100μl) 若干
(8) 旋涡混合器 1台
(9) 摇床 1台
(10) 水平琼脂糖凝胶电泳槽 1台
(11) 稳压电泳仪 1台
(12) 核酸蛋白紫外检测仪 1台
(13) 微波炉 1台
(14) 照相器材 1台
(15) 灭菌的 Eppendorf 管 (1.5 和 0.5ml 微量离心管) 和 Eppendorf 离心管架 若干
(16) 盛冰的容器
(17) 灭菌的移液器枪头 若干
(18) 真空干燥装置 1台
(19) 菌株

2. 试剂

(1) 碱裂解法小量制备质粒 DNA 试剂

1) 培养基

①LB (Luria - Bertani) 液体培养基　1% 胰蛋白胨 (Tryptone), 0.5% 酵母提取物 (Yeast Extract), 1% NaCl, 加蒸馏水配制, 分装, 高压蒸汽 (1.03×10^5Pa) 灭菌 20 分钟。

②含抗生素的 LB 固体培养基　在 LB 液体培养基中加入 1.8% (W/V) 琼脂, 1.03×10^5Pa 高压蒸汽灭菌 20 分钟, 降至 65℃左右, 在无菌条件下, 加入抗生素, 氨苄青霉素的终浓度为 100μg/ml, 卡拉霉素的终浓度为 50μg/ml。趁热分装在灭菌的培养皿内 (30 ~ 35ml 培养基), 凝固后倒置, 4℃保存备用, 使用前 2 小时取出。

2) 氨苄青霉素 (Amp)　无菌水配置或 0.22μm 滤器过滤除菌 (100mg/ml), 1ml/分装储存于 -20℃。

3) STE　0.1mol/L NaCl, 10mmol/L Tris - HCl (pH 8.0), 1mmol/L EDTA (pH 8.0)。

4) 溶液Ⅰ　50mmol/L 葡萄糖, 25mol/L Tris - HCl (pH 8.0), 10mmol/L EDTA (pH 8.0); 在 6.895×10^4Pa 高压下蒸汽灭菌 15 分钟, 贮存于 4℃。临用时加热至室温。

5) 溶液Ⅱ　0.2mol/L NaOH (临用前用 10mol/L 贮存液现用现稀释) 加 1% SDS。

6) 溶液Ⅲ　5mol/L 乙酸钾 60ml, 冰醋酸 11.5ml, 水 28.5ml; 配好的溶液中含 3mol/L 钾盐, 含 5mol/L 乙酸根 (pH 4.8)。

7）苯酚/氯仿溶液　重蒸苯酚以1mol/L Tris-HCl，pH 8.0平衡后，以1∶1比例与氯仿混合。该氯仿为氯仿和异戊醇（24∶1 V/V）的混合物。

8）70%乙醇和无水乙醇。

9）TE缓冲液　10mmol/L Tris-HCl，pH 8.0，1mmol/L EDTA，含有20μg/ml无DNA酶的RnaseA。

（2）限制性内切酶双酶切质粒DNA试剂

1）共价闭合环状质粒DNA。

2）限制性内切酶Nhe Ⅰ和Hind Ⅲ。

3）10×限制性内切酶缓冲液（含牛血清白蛋白）。

（3）琼脂糖凝胶电泳鉴定质粒DNA和酶切DNA片段的试剂

1）琼脂糖。

2）50×TAE电泳缓冲液　242g Tris；57.1ml冰醋酸；46.5g EDTA，加入600ml去离子水溶解，调pH 8.2，加水定溶至1000ml。用时稀释。

3）溴化乙锭（EB）储备液（10mg/ml）　在100ml水中加入1g溴化乙锭，用磁搅拌器搅拌数小时，转移到黑色瓶中，4℃保存。用时稀释至终浓度0.5μg/ml。（溴化乙锭是强诱变剂，在称取时务必戴上手套、面具。一旦接触到，要立即用大量的水冲洗）。

4）6×凝胶加样缓冲液　0.25%溴酚蓝，30%甘油水溶液，4℃保存长期使用。

5）标准DNA。

【实验方法】

1. 破裂解法小量制备质粒DMA与鉴定

（1）从LB平板上挑取单菌落，接种于50ml含相应抗生素的LB培养液中，37℃振荡（约225转/分钟）培养过夜

（2）取1.2ml培养液至Eppendorf管中，4℃，12000×g离心30秒。弃上清。

（3）取1.2ml STE溶液中漂洗菌体，12000×g离心30秒。弃去上清，尽可能去净。

（4）加入100ml溶液Ⅰ（冰浴中预冷的），在旋涡混合器上剧烈振荡重新悬浮混匀菌体。

（5）加入200ml溶液Ⅱ，盖紧管口，快速温和倒离心管数次，以混匀内容物。不要振荡，冰浴3分钟。

（6）加入150ml冰冷的溶液Ⅲ，盖紧管口，温和颠倒离心管数次，混合内容物，冰浴3~5分钟。有白色絮状沉淀物形成。

（7）4℃，12000×g离心5分钟，将上清液移至新的离心管中，注意避免带入白色沉淀物

（8）加入等体积苯酚：氯仿（1∶1）混合物。旋涡混匀，4℃，12000×g离心2分钟，将上清移入新的离心管中。

（9）加入等体积氯仿溶液，旋涡混匀，4℃，12000×g离心2分钟，将上清移入新的离心管中。

（10）加入2倍体积的冰冷的无水乙醇，混匀后于室温下放置2分钟，4℃，12000×g离心2分钟，弃上清。

（11）加1.0ml冰冷的70%乙醇振荡漂洗沉淀。

（12）4℃，12000×g 离心2分钟小心吸去上清液。将离心管倒置于一张纸巾上，以使所有液体流出，再用消毒的滤纸小条将附于管壁的液滴除尽。于室温蒸发痕量的乙醇或对沉淀进行真空干燥。

（13）将沉淀重新悬浮在40μl TE 缓冲液（含无 DNA 酶的 Rnase A，20μg/ml）中。37℃保温5分钟，溶解 DNA。

（14）将小量制备的质粒试管标上日期和内容，-20℃保存备用或立即用于酶切实验。

（15）取5~10μl 产物准备琼脂糖电泳。

2. 限制性内切酶双酶切质粒 DNA（包括酶切片段的回收）与鉴定

（1）20℃冰箱中取出实验的样品，让其升至室温。

（2）限制性酶消化是在特定的缓冲液中进行的。缓冲液以10×浓缩贮存液供应，使用时必须稀释成1×缓冲液。

（3）反应可以在20μl 总体积中进行。在0.5ml Eppendorf 管中依次加入：dd H_2O 6μl，10×缓冲液2μl，质粒 DNA 10μl，Nhe Ⅰ1μl，HindⅢ1μl。

（4）加入所有组分后，盖紧管口，轻弹混匀并在离心机上离心2~3秒钟，将试剂甩至管底。

（5）37℃保温4小时。

（6）取5~10μl 产物准备琼脂糖电泳。与此同时，铺琼脂糖胶。

（7）电泳跑完后请助理教师给凝胶拍照。将限制性酶切割的质粒泳道与未切割质粒的泳道进行比较。它们应有所不同。否则，消化反应很可能就有问题。假如不能断定，可以向实验老师请教。

（8）如果凝胶未出结果，或许还要进行重新消化/重新电泳。

3. 琼脂糖凝胶电泳鉴定凝胶的制备

（1）搭好凝胶电泳槽。梳齿应该保持笔直，在梳齿底部与胶槽之间应保持数毫米的间隙。

（2）称取0.7g 琼脂糖（或根据需要确定琼脂糖用量），加入1×TAE 电泳缓冲液100ml，加热溶解成清澈、透明的溶液。

（3）将琼脂糖溶液缓缓倒入插有梳齿的胶模（梳齿与胶模之间保持1~2mm 的距离），避免在齿梳两侧产生气泡。凝胶厚度一般为0.3~0.5cm。

（4）在琼脂糖凝固过程中，准备样品。一般取5μl 样品与6×凝胶加样缓冲液1μl 混匀，准备点样。

（5）室温下放置30~60分钟后，琼脂糖凝胶凝固，去除梳齿，将胶模板移至电泳冰槽上，加样孔在靠近阴极的一端（黑色端）。向槽中加入适量的1×TAE 缓冲液，通常应没过胶面1cm，小心移走梳齿。

（6）按顺序用移液枪缓慢将 DNA 样品垂直加入加样孔直至其开口下方，并在笔记本上记录下点样顺序。一般 DNA 量最好在0.5~1μg。

（7）加完所有样品后，将电泳槽与电源正确连接（黑色对阴极，红色对阳极）。打开电源之前要调好电压和时间。以1~5V/cm 的电压进行电泳。

（8）当指示剂溴酚蓝迁移至胶三分之二时，停止电泳，切断电源，取出胶模。

（9）凝胶中已经加入 EB，则可以用紫外灯观察分析。凝胶中没有 EB，取出凝胶后，用0.5μg/ml 的 EB 溶液染色30分钟，取出，再用紫外灯（观察分析凝胶摄影）。

（10）拍照记录结果。

4. 结果处理

（1）绘图或照片。

（2）凝胶（图象）解释。

【思考题】

1. 质粒载体的特点（基本条件）、遗传标记基因、类型和常用质粒载体。
2. 质粒的提取方法及其基本原理。
3. 简述限制性内切酶的定义、命名、特点和用途。
4. 简述琼脂糖电泳检测质粒 DNA 的过程。

EXPERIMENT 7　Isolation of plasmid DNA by the alkaline – detergent method, and identification by restriction endonuclease analysis and agarose gel electrophoresis

【Purpose】

1. Learn the basic principle and technique for plasmid DNA extraction.

2. Master the method of DNA identification by restriction endonuclease analysis and agarose gel electrophoresis.

【Principle】

Plasmid DNA is a commonly used vector to clone and express foreign genes in gene engineering research, and it is also an ideal experimental model to study DNA structure, function and replication at the molecular level. Its extraction and separation are the most common and basic experimental techniques in the research of gene engineering and molecular biology.

Plasmid DNA is a double stranded, covalent, closed – loop DS molecule with much smaller molecular weight than chromosome DNA in microbial cells. It is a genetic factor that exists outside the chromosome but can replicate independently. Plasmids usually carry genes that do not exist on chromosomes, which are not related to the main metabolic activities of cells, but show some useful properties, such as antibiotic resistance, bacterial resistance, etc. Plasmids are classified according to the genes carried by plasmids and the characteristics of host cells given by these genes, including resistant plasmids(R), fertility factors(f), Col and toxic plasmids. R plasmid encodes antibiotic resistance. R plasmid can transfer its resistance to the appropriate receptor cells lacking the plasmid, so that the latter can also obtain the same antibiotic resistance ability to indicate whether the vector or recombinant DNA molecules enter the host cells. Plasmids have their own replication origin, which can replicate independently without relying on the chromosomal surface.

There are manymethods to extract plasmids, such as alkali cracking, boiling cracking, density gradient of cesium bromide ethidium chloride and column chromatography. Alkaline lysis is one of

the most convenient and fast methods for the preparation of plasmid DNA. In this experiment, plasmid DNA was isolated from a small number of cells, and DNA isolated by this method is often called small amount preparation. The plasmid DNA extracted by this method has high purity and can be used for restriction enzyme digestion, PCR amplification and sequencing without further purification.

This method is based on the difference of denaturation and renaturation between covalently closed loop plasmid DMA and chromosome DNA. When the cell is suspended in NaOH and SDS solution, it will split under the basic condition of pH 12.0 - 12.6, and the hydrogen bond break of protein and chromosome DNA will denature, while the denaturation degree of double chain covalent closed - loop plasmid DNA molecule is small, and the natural state can be restored quickly by changing the conditions. The neutralization solution (acid potassium acetate) is added to adjust the pH to neutral and maintain high salt state. After centrifugation, the macromolecular chromosome DNA cannot be rewound into a network structure, and the chromosome DNA and protein will precipitate together with the cell fragments under the action of the detergent SDS, while the double strand covalent closed - loop plasmid DNA is soluble and remains in the supernatant. when alkali solution is added, the plasmid DNA will also denature, but the two chains are still close to each other. When acid solution is added to neutralize, the two chains of the plasmid DNA will be re annealed with their complementary chains, and then the original plasmid will be formed. The plasmids in the supernatant were extracted with phenol chloroform, further purified, and finally precipitated with ethanol.

The typeII restriction endonucleases are DNA - cutting enzymes that recognize and cut DNA only at a particular sequence of nucleotides. For example, EcoR Ⅰ cuts DNA wherever it encounters the sequence GAATTC. Once it encounters its particular specific recognition sequence, it will bond to the DNA molecule and makes one cut in each of the two sugar phosphate backbones of the double helix. When in experiments, DNA should be mixed with proper restriction endonuclease and buffer which supplied by seller, then put the mixture to the optimum temperature. One unit of the enzyme is required for complete digestion of I ug of double - stranded DNA in an hour under proper temperature, especially 37℃, Generally 1μg DNA with 2 - 3U enzyme should be keep proper temperature for 1 - 4 hours.

Electrophoresis is a technique used to separate and sometimes purify macromolecules, especially proteins and nucleic acids, which differ in size, charge or conformation. As such, it is one of the most widely used techniques in biochemistry and molecular biology. In neutral pH electrophoresis buffer system, nucleic acids have a consistent negative charge imparted by their phosphate backbone, and migrate toward the anode. Agarose is a polysaccharide, straight chain, composed of D - galactose and 3,6 - dehydrated L - galactose residues alternately arranged. In aqueous solution, the agarose molecules form a gel by hydrogen bonding. The greater the concentration of agarose gel, the harder the gel. The rate of DNA molecule swimming through the gel in the electric field is closely related to its gel concentration, electrophoresis voltage and conformation besides its charge and molecular size. A linear double stranded DNA molecule of a given size is passed through agarose gel with different concentrations at different speeds. It is inversely proportional to the relative molecular mass of the gel through the rate of gel and the gel of different concentrations can be used to distinguish the different DNA fragments. At low voltage, the mobility of linear DNA fragments is directly propor-

tional to the applied voltage, but when the voltage is increased, the mobility of large molecular weight DNA fragments is not directly proportional to the voltage, and the resolution is decreased. In order to achieve good separation effect, the voltage of gel electrophoresis should not exceed 5V/cm. The double stranded plasmid DNA molecule has three different conformations: covalently closed circular DNA (cccDNA), open loop DNA ocDNA, linear DNA (cDNA), and has different electrophoretic mobility, which can be separated from agarose gel electrophoresis. In general, mobility cccDNA > cDNA > ocDNA. The appearance of linear plasmids means the contamination or improper operation of nuclease in the preparation process.

The concentration of agarose is generally between 0.5% and 2%. The migration distance of DNA can be determined by observing the migration distance of tracer dyes. The migration rates of-bromphenol blue and xylene blue dyes in agarose gel are approximately the same as those of double stranded DNA fragments of 300bp and 4000bp size. After migration enough, DNA fragments can be observed by cthidium bromide staining. Ethidium bromide is a fluorescent dye. It can be mixed into gel when it is used for gel. It can be dyed in electrophoresis. After electrophoresis, it can be dipped in diluted ethidium bromide solution to dye 45min. It is embedded between DNA and RNA bases. It will show orange yellow light under ultraviolet light. Its sensitivity is very high, and it can be dyed with low concentration fluorescent dye (0.5μg/ml ethidium bromide). Under UV light, DNA of as little as 1ng was observed and detected directly.

In order to determine the size of the DNA fragment, a molecular weight reference was added to the target of the same gel, and electrophoresis and dyeing were performed. Then the size of the target fragment could be known quickly under the ultraviolet lamp. DNA The most commonly used reference of relative molecular weight is HindⅢ digest, and the size of each fragment is expressed in bp, which is 23130, 9416, 6557, 4361, 2322, 2027, 564, 125。

【Materials】

1. Apparatus

(1) 9cm dishes

(2) Incubator

(3) Clean bench

(4) Autoclave

(5) Water bath

(6) Microcentrifuge

(7) Pipette

(8) Vortex mixer

(9) Shaker

(10) Electrophoresis tank

(11) Electrophoresis apparatus

(12) Nucleic acid/protein UV analyzer

(13) Microwave oven

(14) Gel Imaging System

(15) Sterile microcentrifuge tubes

(16) Ice tank

(17) Sterile pipette tips

(18) Vacuum drying apparatus

(19) Strains

2. Reagents

To prepare plasmid DNA reagent in small quantity by alkali lysis method:

(1) culture medium

①LB Liquid medium

1% Tryptone, 0. 5% yeast extract, 1% NaCl, add distilled water to prepare, sub pack, and sterilize with high - pressure steam (1.03×10^5 Pa) for 20 minutes.

②LB solid medium containing antibiotics

Add 1. 8% (*W/V*) agar to the LB liquid medium, then sterilized with 1.03×10^5 Pa high - pressure steam for 20 minutes. Add the antibiotics at 65℃. The final concentration of ampicillin and kalamycin was 100μg/ml and 50μg/ml respectively. Sub packed in a sterilized culture dish (30 - 35ml culture medium), then inverted after coagulation, stored at 4℃ for standby, and taken out 2 hours before use.

(2) Ampicillin (AMP)

Sterile water configuration or 0. 22μm filter (100mg/ml), 1ml/sub package storage at -20℃.

(3) STE solution

0. 1mol/L NaCl, 10mmol/L Tris - HCl (pH 8. 0), 1mmol/L EDTA (pH 8. 0);

(4) Solution Ⅰ

25M Tris - HCl (pH 8. 0), 50mM glucose, 10mM EDTA. Sterilized with 6.895×10^4 Pa high - pressure steam for 15 minutes. Store at 4℃ but warm to room temperature when ready to use.

(5) Solution Ⅱ

0. 2M NaOH, 1. 0% SDS, made fresh just prior to use.

(6) Solution Ⅲ

5M potassium acetate 60ml, glacial acetic acid 11. 5ml, water 28. 5ml; the prepared solution contains 3M potassium salt and 5M acetate (pH 4. 8)

(7) Phenol : Chloroform

After the equilibrium of 1mol/L Tris HCl and pH 8. 0, the autoclaved phenol was mixed with chloroform in the proportion of 1 : 1. The chloroform is a mixture of chloroform and isopentyl alcohol (24 : 1 *V/V*).

(8) 70% ethanol and absolute ethanol.

(9) TE buffer

10mmol/L Tris - HCl, pH 8. 0, 1mmol/L EDTA, containing 20μg/ml RNaseA.

Restriction endonuclease double enzyme digestion plasmid DNA reagent:

(1) plasmid DNA.

(2) Nhe Ⅰ and Hind Ⅲ.

(3) 10 × enzyme buffer (contain BSA).

Agarose gel electrophoresis reagent:

(1) Agarose.

(2) 50 × TAE buffer

Tris base 242g, glacial acetic acid 57.1ml, 46.5g EDTA, add 600ml dH_2O, adjust the pH to 8.2, bring to 1000ml with dH_2O.

(3) Ethidium bromide stork (10mg/ml)

Add 1g of ethidium bromide to 100ml of H_2O, stir with a magnetic stirrer for several hours and transfer to a dark bottle and store at 4℃ (ethidium bromide is a powerful mutagen. Wear gloves when weighting it out. In case of contact, immediately flush with copious amounts of water).

(4) 6 × loading buffer

0.25% bromophenol blue, 30% glycerin solution, 4℃ for long - term use.

(5) Standard DNA.

【Procedures】

1. Extraction of plasmid DNA

1) Inoculate a single colony from an LB plate in 50ml of LB medium containing the appropriate antibiotic Incubate overnight at 37℃ with shaking (approximately 225r/min).

2) Place 1.2ml of the overnight culture into a microcentrifuge tube and centrifuge at 12000r/min, 4℃ for 30 seconds. Remove the supernatant.

3) Wash the pellet with 1.2ml STE solution, centrifuge at 12000r/min, 4℃ for 30 seconds. Remove the supernatant.

4) Resuspend the pellet in 100μl of solution Ⅰ containing 4mg/ml lysozyme (pre - cooling). Vortex cells into suspension, making sure there are no clumps of cells.

5) Add 200μl of solution Ⅱ at room temperature and mix by inversion. Return tube to the ice bucked for another three minutes.

6) Add 150μl of ice - cold solution Ⅲ, mix and return to the ice bucket for 3 - 5 minutes. A fluffy white precipitate will form.

7) Centrifuge at 12000r/min for 5 minutes in a microcentrifuge, 4℃. Transfer the supernatant to a fresh tube, avoiding the white precipitate.

8) Add an equal volume of phenol : chloroform (1 : 1), mixed by vortex. Then centrifuge the mixture at 12000r/min for 2 minutes at 4℃. Transfer the supernatant to a fresh tube.

9) Add an equal volume of phenol, mixed by vortex again. Centrifuge the mixture at 12000r/m for 2 minutes at 4℃. Transfer the supernatant to a fresh tube.

10) Add 2 volumes of ice - cold 100% ethanol. Mix and place the mixture at room temperature for 2 minutes. Discard the supernatant.

11) Rinse the pellet with ice - cold 70% ethanol.

12) Centrifuge the pellet at 12000r/min for 2 minutes at 4℃. Pour off the supernatant carefully. Remove as much fluid as possible. Vacuum drying of precipitates.

13) Resuspend the precipitates in 40ul of TE buffer containing 20μg/ml of Rnase A. Incubate the solution at 37℃ for 5 minutes to dissolve the RNA.

14) Label the tube with date and contents, and store at -20℃. Or the product can be used to for restriction endonuclease digestion.

15) Take out 5 ~ 10μl sample for agarose gel electrophoresis.

2. DNA identification by restriction endonuclease.

1) Take out the sample from refrigerator, heat it at room temperature.

2) The digest of restriction endonucleases should be kept in specifically buffer. The storing buffer is supplied as 10 times concentrate. It should be diluted to 1 × buffer.

3) The reaction is occurred in 20μl volume. Ending up these ingredients in the same tube: ddH_2O 6μl, 10 × buffer 2μl, plasmid DNA 10μl, Nhe Ⅰ 1μl, Hind Ⅲ 1μl.

4) After adding all the ingredients, close the tube, Centrifuge for 2 - 3 seconds.

5) Incubate at 37℃ for 4 hours.

6) Loading DNA samples and gel running.

7) After the electrophoresis runs, take photos of the gel. The restriction enzyme cleavage sample lane should be compared with the uncut sample lane. The results should be different. Otherwise, digestive reactions are likely to be problematic. If you are not sure, you can ask the experiment teacher for advice.

8) If no results are obtained, the digestion and electrophoresis steps may need to be redone.

3. Agarose Gel Electrophoresis

1) Set up the gel apparatus as demonstrated. The comb should be straight, and there should be a few millimeters of clearance between the bottom of the comb and the bottom of the gel tray.

2) Weigh 0.7g agarose and dissolve in 1 × TAE buffer 100ml by heating.

3) Pour the agarose slowly onto a gel bed with the comb insert. The clearance between the comb and the gel should be 1 - 2mm, and the height of gel should be 0.3 - 0.5cm.

4) Prepare the DNA sample while the gel is solidifying. 5ul DNA sample can be mixed with 1ul 6x loading buffer containing the dye such as bromophenol blue.

5) Remove the comb after the agarose gel has solidified and place the gel bed onto the electrophoresis tank with the wells near the cathode. Fill the tank with proper amount of 1xTAE buffer. Usually the buffer is about 1cm above the gel.

6) Holding the pipette perpendicular to the well and add DNA sample slowly with the pipette tip just beneath the opening of the well. Record all sample loading sequence. The amount of DNA-sample should between 0.5 - 1μg.

7) Connect the gel tank with the power supply properly (The negative pole is black, while the positive pole is red). Adjust the voltage and time before turn on the power. The recommend voltage should between 1 - 5V/cm.

8) Cut down the power while the strips move to 2/3 of the gel, then take out the gel.

9) If the gel already contained EB, it can be observed by UV lamp directly. If not, the gel should be dyed with 0.5μg/ml EB for 30 minutes. Then it can be observed by UV lamp.

10) Take pictures and record the results.

4. analysis result.

1) Picture the map.

2) Analyzing the gel map.

【Questions】

1. Expound the characteristic of the plasmid, the inherited marking genes and the plasmids that in common use.

2. The principle of the plasmid DNA extraction and basic method.

3. The definition, the rules of nomenclature, the characteristics and the application of restriction endonuclease.

4. Expound the method of DNA identification by restriction endonuclease and Agarose Gel E-lectrophoresis.

实验八　PCR 扩增目的基因、酶切与鉴定

扫码“学一学”

扫码“看一看”

【实验目的】

1. 掌握　PCR 引物设计方法；PCR 反应、限制性酶酶切消化 PCR 产物和琼脂糖凝胶电泳进行鉴定的操作技术方法。

2. 了解　PCR 反应的原理及 PCR 反应条件优化需要考虑的因素。

【实验原理】

聚合酶链反应（Polymerase Chain Reaction）也称多聚酶链式反应，简称 PCR 技术，是 20 世纪分子生物学领域最重要的发明之一。PCR 是体外酶促合成特异 DNA 片段的一种方法，由高温变性、低温退火（复性）及适温延伸等反应组成一个周期，循环进行，使目的 DNA 得以迅速扩增，具有特异性强、灵敏度高、操作简便、省时等特点。

PCR 反应体系由模板 DNA、引物、4 种 dNTP、DNA 聚合酶以及反应缓冲液组成。典型的扩增反应包括：①模板变性（94℃以上，1 ~ 2 分钟），使 DNA 双链解离；②在相对较低的温度下引物退火（50 ~ 55℃，1 ~ 2 分钟）；③适中温度下的引物延伸（72℃，1 ~ 2 分钟）。此三步为一个循环，每个循环的产物可以作为下一循环的模板，n 个循环后，DNA 产量可达到 2^n 拷贝。

合适的引物设计是 PCR 反应成功的关键之一。以下是高特异性扩增的引物需要考虑的主要因素：①引物长度，一般认为引物的最佳长度在 18 ~ 22bp 左右，这个长度可以同时保证特异性识别和结合模板的高效性。②引物的 Tm 值，即 DNA 的熔解温度，指把 DNA 的双螺旋结构降解一半时的温度。Tm 值在 52 ~ 58℃ 的引物往往具有最佳扩增效果，Tm 值大于 65℃的引物容易产生二聚体。③GC 含量，即 G 和 C 碱基在引物中占全部碱基的含量，应该在 40% ~ 60% 之间。

在进行 PCR 扩增时，需要事先给引物两端设计好相应的酶切位点，一般说来，限制酶的选择非常重要，尽量选择粘端酶切和那些酶切效率高的限制酶，如 BamHⅠ、HindⅢ，提前看好各种双切酶所用共用的缓冲液以及各酶在共用缓冲液里的效率。选好酶切位点后，需要在各个酶的两边加上保护碱基。需要强调的是，很多人建议酶切过夜，其实完全没有必要，一般酶切 3 个小时，其实 1 个小时已经足够。尽量选择应用大体系，如 100μl。

【实验材料】

1. 器材

（1）PCR扩增仪　　1台

（2）其他同实验七。

2. 试剂　PCR扩增rHV3基因试剂。

（1）模板DNA（0.1～1ng质粒DNA）。

（2）灭菌水。

（3）10×扩增缓冲液，含15mmol/L $MgCl_2$。

（4）2.5mmol/L dNTPs。

（5）10μmol/L上游引物1：5′AATGCTAGCTATCACCTACACTGACTGCACC 3′。

（6）10μmol/L下游引物2：5′CCAAGCTTCCTGCAGCTTACTATTCATCGTACGCGTCTTCCG 3′。

（7）5 U/μl Taq DNA聚合酶。

（8）限制性内切酶NheⅠ和HindⅢ。

（9）琼脂糖凝胶电泳鉴定试剂同实验七。

【实验方法】

1. PCR扩增rHV3基因

PCR反应体系如下：

10×PCR Buffer	5μl
dNTP Mixture（各2.5mmol/L）	4μl
TaKaRarTaq（5U/μl）	0.5μl
$MgCl_2$	4μl
Forward Primer（20μmol/L）	2μl
Reverse Primer（20μmol/L）	2μl
DNA template（pET22b－AnsB）	1μl
无菌水	总体积至50μl

DNA的模版为质粒pET22b－Rhv3（预先准备）。

PCR扩增程序：95℃，1.5分钟；95℃，30秒，50℃，30秒，72℃，1.5分钟，30循环；72℃，10分钟。

2. 电泳鉴定

（1）电泳检测条件　1%琼脂糖凝胶电泳。

（2）配制方法　50ml TAE缓冲液+0.5g琼脂糖，微波炉加热沸腾至完全澄清，加入5μl GOLDWAVE染色液体，插入小梳齿。

（3）点样　2.5μl样品+1μl 6×或者10×Loading Buffer点样。

（4）电泳时间　120V恒压30分钟。

（5）分子量Marker

1）DL2000　每次取5μl电泳时，750bp的DNA片段量约为150ng，其余条带的DNA量约为50ng。

2）λ DNA digest DNA　在电泳前进行热处理（60℃，5 分钟），能使 Marker 的电泳图像变得更为清晰。

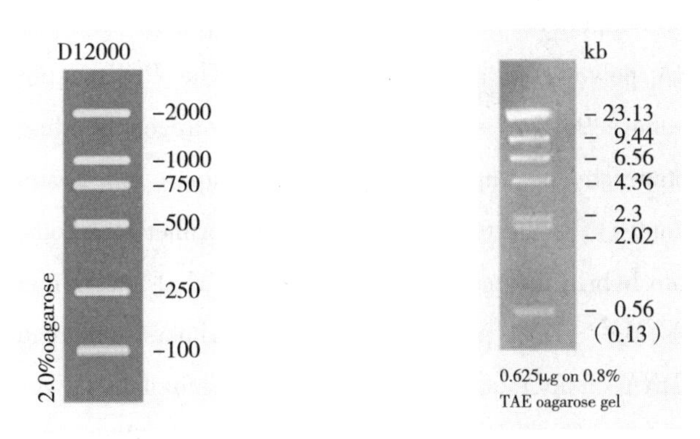

3. 限制性内切酶双酶切 PCR 产物（包括酶切片段的回收）

10 × Buffer	5μl
NcoI	1μl
Hind Ⅲ	1μl
Plasmid or PCR product	20μl
Nuclease - Free Water to final volume	50μl

酶切条件：37℃温育 3 小时。

保留 5μl 载体产品作为转化实验的阳性参照，收集其余样品。

4. 结果处理

拍照记录并解释凝胶（图象）。

【思考题】

1. PCR 条件优化需要考虑的因素。
2. 如何设计 PCR 引物？

扫码“练一练”

EXPERIMENT 8　Amplification of DNA by the polymerase chain reaction and identification by restriction endonuclease

【Purpose】

1. To master the basic techniques of plasmid extraction, PCR technology and digested method.
2. To learn the principle of constructing a recombinant plasmid

【Principle】

Polymerase Chain Reaction (PCR) is widely used as one of the most important inventions of the 20th century in molecular biology. PCR enzymatic synthesis of specific DNA fragments *in vitro*, using a DNA denaturation and refolding characteristics, denatured by the high temperature, low tempera-

ture annealing(renaturation) and the optimum temperature extension reaction steps to form a cycle, rapidly amplify of the target DNA with specificity, high sensitivity, easy to operate and saving time.

The PCR reactions components are constituted by template DNA, primers, four kinds of deoxyribonucleotides, DNA polymerase and reaction buffer. The PCRreactions include three basic steps: ①Denaturation (above 94℃, 1 - 2 minutes), the hydrogen bonding between the template double chain DNA ruptured by heating, and double chain apart into a single chain. ②Annealing (50℃ - 55℃, 1 - 2 minutes), as the temperature falls, the primer and complementary area of template DNA combined into hybrid molecules. ③Extension(72℃, 1 - 2 minutes), in the presence of DNA polymerase, dNTPs, Mg^{2+}, DNA polymerase catalytic primers outspread according to the 5′ to 3′ direction, synthesis DNA chain that complementary with template DNA chain. The above three steps for a cycle, the product of each loop can be us a template for the next cycle. After n cycles, the amount of DNA plus to 2^n copy.

Good primer design is essential for successful reactions. The important design considerations described below are a key to specific amplification with high yield. ①Primer length. It is generally accepted that the optimal length of PCR primers is 18 - 22bp. This length is long enough for adequate specificity and short enough for primers to bind easily to the template at the annealing temperature. ②Primer Melting Temperature. Primer Melting Temperature(Tm) by definition is the temperature at which one half of the DNA duplex will dissociate to become single stranded and indicates the duplex stability. Primers with melting temperatures in the range of 52 - 58℃ generally produce the best results. Primers with melting temperatures above 65℃ have a tendency for secondary annealing. ③GC Content. The GC content(the number of G and C in the primer as a percentage of the total bases) of primer should be 40% - 60%.

Performing PCR amplification time, designed restriction sites to both ends of the primer, the choice of restriction enzyme is very important. Generally, try to choose the sticky ends digested and high efficiency restriction enzyme, such as BamH Ⅰ, Hind Ⅲ. You should know the shared buffer in advance, as well as the efficiency of each enzyme in the shared buffer. Plus, a protective base on both sides of the respective enzyme after cleavage sites has been selected. Many people recommend digested overnight. It's not necessary. Generally digested 3 hours, in fact, one hour is sufficient if application of large systems, such as 100μl.

【Materials】

1. Apparatus

(1) PCR Instrument

(2) See experiment 7 for more information

2. Reagents

(1) the pET22b - AnsB plasmid(0.1 - 10ng)

(2) ddH_2O

(3) 10 × PCR buffer

(4) 2.5mmol/L dNTPs

(5) 10μmol/L Forward Primer 1: 5′AATGCTAGCTATCACCTACACTGACTGCACC3′

(6) 10μmol/L Reverse Primer 2: 5′CCAAGCTTCCTGCAGCTTACTATTCATCGTACG－CGTCTTCCG3′

(7) 5 U/μl Taq DNA polymerase

(8) Enzyme: Nhe Ⅰ and Hind Ⅲ

(9) See experiment 7 for more information

【Procedures】

1. PCR

PCR reaction system is as follows:

10 × PCR Buffer	5μl
dNTP Mixture (2.5mmol/L)	4μl
TaKaRarTaq (5U/μl)	0.5μl
$MgCl_2$	4μl
Forward Primer (20μmol/L)	2μl
Reverse Primer (20μmol/L)	2μl
DNA template (pET22b－AnsB)	1μl
Sterile distilled water	up to 50μl

PCR amplification program: 95℃, 1.5 minutes; 95℃, 30 seconds, 50℃, 30 seconds, 72℃, 1.5 minutes, 30 cycles; 72℃ 10 minutes.

2. Electrophoresis

(1) Electrophoresis conditions: 1% agarose gel electrophoresis.

(2) Preparation method: 50ml TAE buffer + 0.5g agarose, microwave heating to boiling to fully clarified, adding 5μl GOLDWAVE staining liquid, insert a small comb.

(3) Spotting: 2.5μl sample + 1μl 6 × or 10 × Loading Buffer.

(4) Electrophoresis time: 120V constant voltage 30 minutes.

(5) Marker:

①DL2000: each 5μl of electrophoresis, a 750 bp DNA fragment is about 150ng, display bright band, the remaining bands of DNA about 50ng.

②λDNA digest DNA: Heat before electrophoresis (60℃, 5 minutes), enabling the Marker electrophoretic image becomes more clear.

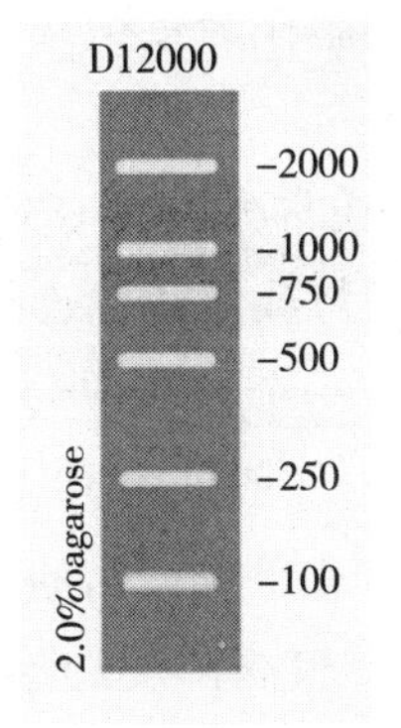

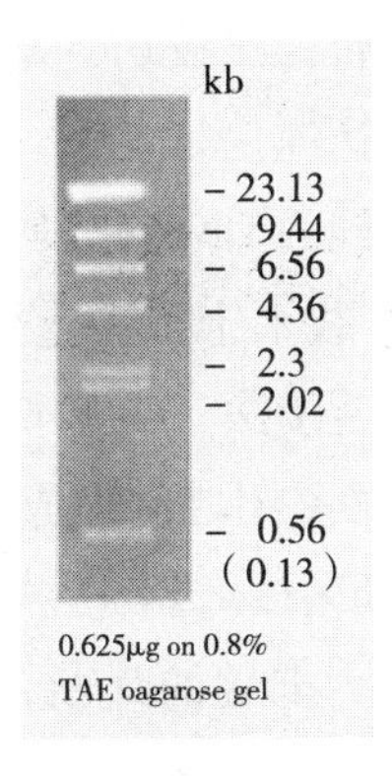

3. Digestion

Digest plasmid and PCR product in accordance with the following systems in a PCR tube.

10 × Buffer	5μl
NcoI	1μl
Hind Ⅲ	1μl
Plasmid or PCR product	20μl
Nuclease – Free Water to final volume	50μl

Digestion conditions：37℃ incubation 3 hours.

Take 5μl vector as positive control used for transformation experiment and collect the rest.

4. analysis result

(1) Picture the map.

(2) Analyzing the gel map.

【Questions】

1. What factors should be considered in order to optimize of PCR conditions?

2. How to design the PCR primers?

扫码“学一学”

扫码“看一看”

实验九　感受态细胞制备、DNA 片段连接、转化及重组克隆的 SDS – PAGE 电泳检测

【实验目的】

1. 掌握　氯化钙制备大肠杆菌感受态细胞；DNA 片段的连接及感受态细胞转化的操作技术方法。

2. 了解　大肠埃希菌感受态细胞的制备；DNA 片段的连接及感受态细胞转化的原理。

【实验原理】

感受态细胞（competent cell）是指通过理化方法诱导，使其处于最适摄取和容纳外来 DNA 的生理状态的细胞。在分子生物学实验中，细胞通过特殊处理后可变成感受态细胞（competent cells），即可以接受外源 DNA。将构建好的载体转入感受态细胞进行表达，不仅可以检验重组载体是否构建成功，最主要的是感受态细胞作为重组载体的宿主可以进行后续实验，如蛋白质表达纯化等工作。

连接是将两个不同的 DNA 片段“黏合”在一起，通常为目的 DNA 片段与载体片段的连接，这在重组 DNA 技术中至关重要。目的 DNA 片段与载体片段连接方法有：①同聚尾连接法；②人工接头连接法。更确切地说，连接包括在一个核苷酸的 3’端羟基与另外一个核苷酸的 5’端磷酸基团之间形成磷酸二酯键的过程。一般采用 T4 DNA 连接酶连接 DNA 片断，它来源于 T4 噬菌体。T4 DNA 连接酶可以连接具有单链突出的黏性末端片断，但同时也可以连接具有平齐末端的片断，只是此时通常需要较高浓度的连接酶。除了水之外，连接反应通常还需要三种组分：两个或两个以上的具有匹配的黏性末端或平齐末端的 DNA

片断、含有 ATP 的缓冲液及 T4 DNA 连接酶。

转化常用 $CaCl_2$ 法：将快速生长的大肠埃希菌置于经低温（0℃）预处理的低渗氯化钙溶液中，便会造成细胞膨胀，同时 Ca^{2+} 会使细胞膜磷脂双分子层形成液晶结构，促使细胞外膜与内膜间隙中的部分核酸酶解离开来，离开所在区域，诱导细胞成为感受态细胞。细胞膜通透性发生变化，极易与外源 DNA 相黏附并在细胞表面形成抗脱氧核糖核酸酶的羟基－磷酸钙复合物。此时，将该体系转移到 42℃下做短暂的热刺激（90 秒），细胞膜的液晶结构会发生剧烈扰动，并随机出现许多间隙，外源 DNA 就可能被细胞吸收。进入细胞的外源 DNA 分子通过复制、表达，实现遗传信息的转移，使受体细胞出现新的遗传性状。将转化后的细胞在选择性培养基上培养，筛选出带有外源 DNA 分子的阳性克隆。

SDS－PAGE 电泳分析原理：蛋白质在聚丙烯酰胺凝胶中电泳时，它的迁移率取决于它所带净电荷以及分子的大小和形状等因素。如果在 PAGE 胶系统中加入阴离子去污剂十二烷基硫酸钠（SDS），在蛋白质溶液中加入 SDS 和巯基乙醇后，巯基乙醇能使蛋白质分子中的二硫键还原各种蛋白质－SDS 的复合物都近似于长椭圆，带相同密度的负电荷，掩盖了不同种类蛋白质间原有的电荷差别。这样的蛋白质－SDS 复合物，在凝胶电泳中的迁移率不再受蛋白质原有电荷和形状的影响，而主要取决于其相对分子质量的大小。因而 SDS－PAGE 可以依据蛋白质的相对分子质量的不同来对蛋白质进行分析。

【实验材料】

1. 器材

（1）微量移液器	若干
（2）标准净化工作台	1 台
（3）恒温水浴锅	1 台
（4）无菌 Eppendorf 离心管	若干
（5）灭菌的移液器枪头	若干
（6）摇床	1 台
（7）分光光度计	1 台
（8）制冰机	1 台
（9）恒温培养箱	1 台
（10）电磁搅拌器	1 台
（11）高速台式离心机	1 台
（12）稳压稳流定时电泳仪	1 台
（13）夹心式垂直板型电泳槽	1 台
（14）旋涡混合器	1 台

2. 试剂

（1）酶切并回收后的 PCR 产物

（2）T4 DNA 连接酶

（3）10×连接缓冲液

（4）大肠杆菌 BL21

（5）LB 液体培养基

（6）LB 固体培养基

(7) 氨苄青霉素(Amp)

(8) 100mmol/L 的 $CaCl_2$溶液

(9) 0.5mol/L 乳糖

(10) 2×SDS 电泳上样缓冲液

(11) N, N, N′, N′-四甲基乙二胺(简称 TEMED) 密封于4℃保存。

(12) 样品处理液 50mmol/L Tris-HCl(pH 7.4)。

(13) 10%过硫酸铵,新鲜配制。

(14) 10%SDS 称10g SDS,加重蒸水至100ml,微热使其溶解,置试剂瓶中贮存。

(15) 2×SDS-PAGE 加样缓冲液(30ml) 2.4ml 浓缩胶 pH 6.8 缓冲液,二硫苏糖醇 DTT 0.9g(或β-巯基乙醇 1.8ml),10% SDS(电泳级)6ml,溴酚蓝 0.1mg,甘油 3.75ml,ddH_2O 稀释至30ml。分为1ml/支储存于-20℃。

(16) 10×Tris-Gly 电泳缓冲液 30.2g Tris,144.0g 甘氨酸,10.0g SDS。用水定容到1000ml。无须调节 pH。

(17) 考马斯亮蓝 R-250 染色液(1000ml) 考马斯亮蓝 0.5g,甲醇 450ml,水 450ml,冰醋酸 100ml。

(18) 凝胶脱色液(1000ml) 甲醇450ml,水450ml,冰醋酸100ml。

(19) 低分子量标准蛋白(14.4~94kDa)。

注意:丙烯酰胺是一种强力可积累的神经毒素。

【实验方法】

1. 连接体系 在 PCR 管中按照下列体系连接。

T4 DNA 连接酶(400U/μl,NEB)	0.5μl
10×连接酶缓冲液	2μl
质粒	~100ng
PCR 产物	100-300ng
加去核酶的水至	20μl

冰浴3小时或16℃恒温过夜。

2. 转化

(1) 大肠杆菌感受态的制备($CaCl_2$法) 提前将大肠杆菌 BL21(DE3)划线分离单菌落。从甘油管中取300μl,接种到 LB 液体培养基中,培养过夜。第二天按照1∶50的比例转接到100ml LB 培养基中,当 OD_{600}达到0.5~1.0之间时,按照以下流程制备感受态细胞:

1) 将1.5ml 菌液移入离心管中,冰上放置10分钟,然后于4℃下,5000r/min 离心5分钟。

2) 弃去上清(轻轻倾倒掉上清液,并用移液器移除残余的过多水分)。

3) 加入1ml 预冷的0.05mol/L 的 $CaCl_2$溶液,轻轻拨动管底使细胞完全悬浮,冰上放置15分钟,4℃下5000r/min 离心5分钟。

4) 弃去上清,加入0.2ml 预冷的0.05mol/L 的 $CaCl_2$溶液,轻轻悬浮细胞,即制成感受态细胞。将感受态细胞悬液分装成2份,每份100μl(注意悬液密度要均匀)。冰上放置备用。暂时不用的感受态细胞可置于-80℃保存。

注:$CaCl_2$处理后的细胞比较脆弱,尽量轻柔操作。

（2）连接产物的转化

1）用无菌吸头分别向3管100μl感受态细胞中加入10μl连接产物、10μl去离子水（阴性对照）和1μl空载体（阳性对照），轻轻旋转或轻弹管壁以混匀内容物，在冰上放置30min。

2）将管放到42℃水浴中热激90秒，不要摇动离心管。

3）快速将离心管转移到冰浴中，冷却1－2分钟。

4）每管中加入800μl LB液体培养基后，37℃振荡培养1小时。

5）吸取100μl已转化的感受态细胞加到含选择抗性LB平板上，用L型玻璃棒轻轻地把细胞均匀地涂到琼脂平板表面。

6）倒置培养皿于37℃培养12－16小时，观察单菌落并拍照。

3. 重组克隆的SDS－PAGE电泳分析

（1）重组蛋白诱导表达

1）挑取阳性菌落于10ml含氨苄的LB培养基37℃培养4小时（150r/min），每个克隆一管。

2）加入乳糖至终浓度10mmol/L，继续培养4小时。

3）6000r/min离心15分钟收集菌体细胞。

（2）电泳样品准备　把菌体和2×SDS上样缓冲液在1.5ml离心管中混合，煮沸10分钟，短暂离心取上清。

（3）制备电泳胶　用Easy－Casting SDS－PAGE gel rapid preparation Kit制备电泳胶。以1mm厚度的胶为例，其他厚度的按照比例调整用量。

1）分别取3ml 2×SDS－PAGE resolving gel和2×resolving gel buffer在烧杯中混合。

2）加入60μl新鲜配制的10%过硫酸铵，混合均匀。

3）灌制分离胶，并在胶上加入1.5cm高度的水层。

4）等胶和水分层后，倾去水层。

5）向烧杯中加入1.5ml Stacking gel premix，添加15μl新鲜配制的10%过硫酸铵和1.5μl TEMED，混合均匀，灌制浓缩胶，插入梳齿。室温静置10分钟，让胶充分聚合。

（4）SDS－PAGE电泳分析　上样，电泳2～3小时后，取出凝胶，用0.15%考马斯亮蓝－R250进行染色，过夜后倾出染色液，不断脱色，拍照观察电泳区带。

【思考题】

1. 转化实验中设立的阴性对照和阳性对照分别用于验证什么问题？
2. 为什么SDS－PAGE电泳的浓缩胶和分离胶的pH不同？

扫码“练一练”

EXPERIMENT 9　Preparation of competent cell, DNA fragment linking and preliminary identification of the recombinant plasmid

【Purpose】

1. To master the operation methods of preparing competent cells of Escherichia coli by calcium

chloride, DNA fragment connection and competent cell transformation.

2. To learn the preparation of competent cells of E. coli, the connection of DNA fragments and the principle of competent cell transformation.

【Principle】

Competent cell

Induce cells using physical and chemical methods to make the cells in the optimum physiological state of intaking and accommodating foreign DNA. In the molecular biology experiments, cells can turn into competent cells which can accept foreign DNA by special treatment. Not only we can test whether the recombinant vector was successfully constructed, the most important is competent cells can host a follow - up experiment as a recombinant vector when the constructed vectors were transfected into competent cells for expression, such as protein expression and purification and so on. Two common methods are used:

Connection is the "gluing" of two different DNA fragments, usually for the purpose of connecting DNA fragments to vector fragments. It is a very important technology in molecular biology lab. Objective DNA fragments and vector fragments were linked by ① homopolytail ligation and ② artificial ligation. The connection involves the formation of a phosphodiester bond between the 3′ terminal hydroxyl group of one nucleotide and the 5′ terminal phosphoric group of another nucleotide. T4 DNA ligase is usually used to connect DNA fragments, which are derived from T4 bacteriophage. T4 DNA ligase can connect viscous end segments with single strand prominence, but it can also connect fragments with flat end, only at this time, a higher concentration of ligase is usually required. In addition to water, the ligation reaction usually requires three components: two or more DNA fragments with matched viscous or even ends; buffer containing ATP; and T4 DNA ligase.

The method of $CaCl_2$

Putting the rapid growth of E. coli at a low temperature (0℃) pretreatment of hypotonic solution of calcium chloride, will cause cell expansion. At the same time, Ca^{2+} causes the membrane phospholipid bilayer to form a liquid crystal structure, which prompt the part of the nucleic acid enzymatic hydrolysis of the cells in the adventitia and intima gap left to leave area and induced the cells to become competent cells. The permeability of cell membrane is changed, making it easily adhered with foreign DNA and formed anti - deoxyribonuclease hydroxy - calcium phosphate complexes on the cell surface. Thenthe system was transferred at the temperature of 42℃ being a short thermal stimulation (90 seconds). The liquid crystal structure of the cell membrane violent disturbances, and then many gaps appear. Foreign DNA can be absorbed by cells. The foreign DNA in the cells is in the progress of replication and expression, achieving the transfer of genetic information. Then the receptor cells emerge the new genetic trait. The transformed cells were cultured on selective medium to screen the positive clones having foreign DNA.

SDS - PAGE electrophoresis analysis

SDS - PAGE electrophoresis analysis is usually used to determine the relative protein molecular mass. The migration rate of proteins in polyacrylamide gel depends on its net charge, molecular size, shape and other factors. When anionic detergent sodium dodecyl sodium sulfate(SDS) is added into PAGE gel system meanwhile SDS and mercaptoethanol are added into protein solution, the disulfide bond of various proteins are reduced and SDS – protein complexes are formed with similar long oval shapes and same density of negative charges, concealing the original charge differences between different proteins. The migration rates of such SDS – Protein complexes depend no longer on their original charges and shapes, but mainly on their relative molecular mass. Thus, the relative protein molecular mass could be determined by SDS – PAGE electrophoresis analysis based on the principle mentioned above.

【Materials】

1. Apparatus

(1) Micropipettor

(2) Standard Clean Bench

(3) Constant temperature water bath

(4) Sterilized Eppendorf tube

(5) Sterile pipette tips

(6) Shake

(7) Spectrophotometer

(8) Ice maker

(9) Incubator

(10) Blender

(11) High – speed desk centrifuge

(12) Electrophoresis apparatus with function of voltage stabilizing and steady flow

(13) Vertical Electrophoresis

(14) A vortex mixer

2. Reagents

(1) PCR and plasmid product, digested by enzyme, recovered by gel.

(2) T4 DNA ligase.

(3) 10 × buffer.

(4) *E. coli* BL21.

(5) LB liquid medium.

(6) LB solid medium.

(7) Ampicillin(Amp).

(8) $CaCl_2$ solution.

(9) 0.5mol/L lactose.

(10) Easy – Casting SDS – PAGE gel rapid preparation Kit containing 2 × SDS – PAGE resolving gel and 2 × resolving gel buffer.

(11) N,N,N',N' – tetramethyl – ethylenediamine(TEMED)　saved at 4℃.

(12) 2 × SDS sample loading buffer　50mmol/L Tris – HCl(pH 7.4).

(13) 10% ammonium persulfate: fresh preparation.

(14) 10% SDS　Dissolve 10g SDS with distilled water up to 100ml, slightly heated, stored in reagent bottle.

(15) 2 × SDS - PAGE sample loading buffer (30ml)　2.4ml Stacking gel buffer (pH 6.8), 0.9g dithiothreitol DTT (or 1.8ml β - mercapto ethanol), 6ml 10% SDS (electrophoresis Grade), 0.1mg bromophenol blue, 3.75ml Glycerol, diluted with ddH_2O to 30ml. Divided into 1ml per tube, stored at -20℃.

(16) 10 × Tris - Gly electrophoresis buffer　30.2g Tris, 144.0g glycine, 10.0g SDS. Dissolve with water to 1000ml. Do without pH adjustment.

(17) Coomassie Brilliant Blue R - 250 staining solution (1000ml)　0.5g Coomassie brilliant blue, 450ml methanol, 450ml water and 100ml glacial acetic acid.

(18) Destaining solution (1000ml)　450ml methanol, 450ml water and 100ml glacial acetic.

(19) Premixed protein marker (Low).

Note: The acrylamide is a potent accumulated neurotoxin.

【Procedures】

1. Ligation system

Connecting in the PCR tube as the following systems:

T4 DNA Ligase	0.5μl
10 × Ligase buffer	2μl
Plasmid	50 - 100ng
PCR product	50 - 100ng
Nuclease - Free water to final volume	20μl

Place on ice for 3 hours or 16℃ overnight. When the content of the plasmids and PCR product considerable, both the ends of the molar concentration ratio of about 1 : 5.

2. Transformation

2.1　The preparation of *E. coli* competent cell (the method of $CaCl_2$)

(1) Separate the single bacteria of *E. coli* BL21 (DE3) through streaking in advance. Take 300μl from the glycerol tube. Then inoculate into LB liquid medium and cultured overnight. The next day, transfer to 100ml LB medium as the ratio of 1 : 50. When the OD_{600} reaches 0.5 - 1.0, prepare the competent cells as the following processes.

(2) Transfer 1.5ml broth to a centrifuge tube. And place on ice for 10 minutes. Then centrifuge at 5000r/min for 5 minutes at 4℃.

(3) Discard the supernatant (dumping off the supernatant gently and remove excessive moisture with a pipette).

(4) Adding 1ml precooling 0.05mol/L $CaCl_2$ solution, gently struck the bottom of the tube to make the cells completely suspended. Then placed on ice for 15 minutes, centrifuged at 5000r/min for 5 minutes at 4℃.

(5) Discard supernatant, adding 0.2ml 0.05mol/L pre - cooling $CaCl_2$ solution, then gently

suspended cells. That is competent cells. Distribute the state of cell suspension into two parts, each 100μl(pay attention to the suspension density should be uniform). Placed on ice to spare.

(6) Stored at -80℃ for the unused competent cells.

Attention: The cells which have been treated with $CaCl_2$ are relatively weak, so operate as tenderly as we can.

2.2 The transformation of ligation products.

(1) Add 10μl ligation product, 10μl deionized water(negative control) and 1μl pET22b empty vector(positive control) to three 100μl competent cells, respectively, using sterile suction. And gently rotate or flick wall to mix. The contents were placed on ice for 30 minutes.

(2) Place the tube at 42℃, heat shock for 90 seconds, and without shaking the centrifuge tube.

(3) Transfer the centrifuge tube in an ice bath quickly and cooled to 1 - 2 minutes.

(4) Add 800μl LB liquid medium to each tube, shaking culture 37℃ for 1 hour.

(5) Lessons 100μl competent cells that have been transformed and add to the LB plates containing a selective. Then daub gently cells evenly to the surface of the agar plate using an L - type glass rod.

(6) Culture at 37℃ for 12 ~ 16 hours, observe with the petri dishes and take a picture.

3. identification of the recombinant plasmid

3.1 Induction of recombinant protein expression

(1) Take the positive colons from the LB plates by tips, put them into 10ml LB liquid medium (including Amp 100μg/ml), one colon for one tube, 37℃ for 4 hours(150r/min).

(2) Add the lactose to final concentration of 10mmol/L(10ml add 0.2ml 0.5mol/L lactose), culture for another 4 hours.

(3) Centrifuged 6000r/min 15 minutes for collection cells.

3.2 Prepare the sample

(1) mixed the cells collections and 2 × SDS sample loading buffer in Eppendorf tube.

(2) Heat samples for 10 minutes at 100℃ to fully denature the protein

(3) Briefly centrifuge sample tubes before loading.

3.3 Preparation of electrophoresis gel

Prepare electrophoresis gel with Easy - Casting SDS - PAGE gel rapid preparation Kit. Take 1mm thick gel for example:

(1) Mix 3ml 2 × SDS - PAGE resolving gel and 2 × resolving gel buffer together in a beaker.

(2) Add 60μl freshly prepared 10% ammonium persulfate, and mix well.

(3) Prepare the resolving gel and cover the gel with 1.5cm water layer.

(4) Remove the water layer after gel and water layer are separated.

(5) Add 1.5ml stacking gel premix, 15μl fresh prepared 10% ammonium persulfate into a breaker, mix well. Prepare the stacking gel, and insert comb. Let gel polymerized well at room temperature for 10min.

3.4 SDS - PAGE electrophoresis analysis

Analyze the prepared fermentation cells by SDS - PAGE electrophoresis.

After electrophoresis for 2 - 3 hours, remove the gel, stain it with 0.15% Coomassie brilliant blue - R250 overnight, continuously decolorized, and then photograph it for observation.

【Questions】

1. What is the problem established using the negative control and positive control in the conversion experiments?

2. Please explain the reason why SDS – PAGE stacking gel and separating gel have different pH value.

第五节　生物大分子分离纯化技术基础实验

分离纯化是生化物质制备的核心操作。由于生化物质种类成千上万，因此分离纯化的实验方案也千变万化，没有一种分离纯化方法可适用于所有物质的分离纯化。一种物质也不可能仅有一种分离纯化方法。所以选择合理的分离纯化方法是非常必要的，而它应根据目的物的理化性质与生物学性质依具体实验条件而定。

生物大分子分离纯化的主要原理有：①根据分子形状和大小不同进行分离。如离心法，膜分离（透析、电渗析）与超滤法，凝胶过滤法等。②根据分子电离性质（带电性）的差异进行分离。如离子交换法、电泳法、等电聚焦法等。③根据分子极性大小及溶解度不同进行分离。如溶剂提取法、逆流分配法、分配层析法、盐析法、等电点沉淀法及有机溶剂分级沉淀法等。④根据物质吸附性质的不同进行分离。如选择性吸附与吸附层析法等。⑤根据分子功能专一性进行分离。如亲和层析法。

一般在分离纯化的初期，由于提取液中的成分复杂，目的物浓度较低，与目的物理化性质相似的杂质多，所以不宜选择分辨能力较高的纯化方法，而选用萃取、沉淀、吸附等一些分辨率低的方法较为有利，这些方法负荷能力大，分离量多兼有分离提纯和浓缩作用，为进一步分离纯化创造良好的基础。在分离操作的后期，宜选用高分辨率的纯化方法，如高选择性的离子交换层析及亲和层析等。同时必须注意避免产品的损失，主要损失途径是器皿的吸附、操作过程样品液体的残留、空气的氧化和某些无法预知的因素。为了取得足够量的样品，常常需要加大原料的用量，并在后期纯化工序中注意保持样品溶液有较高的浓度，以防止制备物在稀溶液中的变性，常加入一些电解质以保护生化物质的活性。

本节主要总结了我校长期开设的生物制药工艺学理论和实验课以及科学研究中的部分工作经验，并参考国内外有关教材和资料，以生化品种为依托着重介绍了盐析、凝胶过滤、离子交换、亲和层析及超滤等生物大分子的分离纯化技术的实验。

Section 5　Basic experiments on separation and purification technology of biomacromolecules

Separation and purification are the core operations of biochemical preparation. Due to the thousands of biochemical substances, the experimental schemes for separation and purification are also different. There are none of methods that can be applied to the separation and purification of all substances. It is also impossible for a substance to have only one separation and purification method. Therefore, it is necessary to choose a reasonable separation and purification method, and it should be determined according to the physical and chemical properties and biological properties of

the target according to the specific experimental conditions.

The main principles of separation and purification of biological macromolecules are as follows: ①Separation according to the shape and size of the molecules. Such as centrifugation, membrane separation (dialysis, electrodialysis), ultrafiltration and gel filtration, etc. ②Separation according to the differences in molecular ionization properties (chargeability). Such as ion exchange, electrophoresis and isoelectric focusing, etc. ③Separation according to the differences in the polarity and solubility of the molecules. Such as solvent extraction, countercurrent distribution, distribution chromatography, salting out method, isoelectric precipitation and organic solvent fractionation, etc. ④Separation according to the differences in the adsorption properties of the substance. Such as selective adsorption and adsorption chromatography, etc. ⑤Separation according to the functional specificity of molecules. Such as affinity chromatography.

Generally, in the initial stage of separation and purification, it is not preferable to select a purification method with higher resolution due to the complex components in the extract, the low concentration of the target substances and many impurities with similar physicochemical properties to the target substances. It is advantageous to use some methods with low resolution such as extraction, precipitation and adsorption. These methods have large load capacity, and the separation amount has many separation and purification effects, which creates a good foundation for further separation and purification. In the later stage of the separation operation, high – resolution purification methods such as highly selective ion exchange chromatography and affinity chromatography should be used. By the way, attention must be paid to avoiding product loss. The main loss paths are the adsorption of the vessel, the residual liquid in the process, the oxidation of the air and some unpredictable factors. In order to obtain enough samples, it is necessary to increase the amount of raw materials and maintain the high concentration of the sample solutions in the later purification process for preventing denaturation of the preparations in dilute solutions. In addition, some electrolytes are often added to protect the activities of the biochemicals.

This section mainly summarizes the long – term biopharmaceutical technology theories and experimental courses and some work experience in scientific researches in our university, and refers to relevant textbooks and materials at home and abroad. Based on these contents, it introduces the experiments on separation and purification of biological macromolecules such as salting out, gel filtration, ion exchange, affinity chromatography and ultrafiltration.

实验十 盐析法制备免疫球蛋白

扫码“学一学”

【实验目的】

1. 掌握 血清免疫球蛋白的溶解特性和基本制备方法。

2. 了解 蛋白质等生物大分子的盐析行为以及盐析法在生化制备中的应用。

【实验原理】

盐析（Salting out）是利用各种生物分子在浓盐溶液中溶解度的差异，通过向溶液中引

入一定数量的中性盐，使目的物或杂蛋白以沉淀析出，达到纯化目的的方法。这是一种经典的分离方法，早在19世纪，盐析法就被用于从血液中分离蛋白质。由于它经济、不需特殊设备，操作简便、安全，应用范围广，较少引起变性（有时对生物分子具稳定作用），至今仍广泛用来回收或分离蛋白质（酶）等生物大分子物质。

高浓度的中性盐能够中和溶液中蛋白质分子表面的电荷，同时夺取溶液中的水，降低溶液中自由水的浓度，从而破坏蛋白质分子表面起稳定作用的水化层结构，使蛋白质的溶解度大大降低。在蛋白质盐析中，$(NH_4)_2SO_4$是最常用的一种盐析剂，主要因为它价格低廉，在水中溶解度大，而且溶解度随温度变化小，在低温下仍具较大的溶解度，因此常可得到较高离子强度的溶液，甚至在低温下也能盐析。利用不同浓度的中性盐将各种因分子量及表面电性不同而溶解度有差异的蛋白质分开，这就是盐析作用。

以血浆蛋白为例（蛋白浓度2%）：

硫酸铵饱和度20%　　纤维蛋白原析出

硫酸铵饱和度28%～30%　　α－球蛋白析出

硫酸铵饱和度33%　　γ－球蛋白析出

硫酸铵饱和度40%　　β－球蛋白析出

硫酸铵饱和度>50%　　白蛋白析出

免疫球蛋白又称γ－球蛋白（γ－globular protein），其主要成分为IgG，分子量16×10^4 Da，等电点为6.85～7.3。本实验利用盐析作用从血浆中制备γ－球蛋白，并利用聚丙烯酰胺凝胶电泳进行纯度鉴定。

【实验材料】

1. 器材

（1）量筒：50ml，100ml　　各1只

（2）烧杯：100ml或250ml　　各1只

（3）皮头滴管　　2根

（4）玻璃棒　　数根

（5）冰箱　　1台

（6）离心机　　1台

（7）圆盘电泳仪　　（全套）

（8）层析柱：1.2cm×25cm　　1根

（9）培养皿：9cm　　2只

2. 试剂

（1）血浆（或血清）

（2）饱和硫酸铵溶液　取500ml蒸馏水，加热至70～80℃，将400g固体硫酸铵粉加入该水中，搅拌20分钟。自然冷却后用浓氨水调节至pH 7.1，放置过夜便可使用。

（3）0.05mol/L pH 7.1磷酸缓冲液：取0.2mol/L磷酸氢二钠溶液67ml与0.2mol/L磷酸二氢钠溶液33ml混合，再加蒸馏水300ml，用pH计校正。

（4）0.5%氨基黑染色液　以7%醋酸溶液配制。

（5）0.85%氯化钠溶液

（6）聚丙烯酰胺凝胶全套试剂

（7）10% $BaCl_2$溶液

（8）双缩脲试剂

【实验方法】

1. 制备免疫球蛋白

（1）取血浆 20ml，与 20ml 0. 05mol/L pH 7.1 磷酸缓冲液混匀，逐滴加入 10ml 饱和硫酸铵溶液，边加边搅拌。加完后继续搅拌 3 分钟，随后会有极少量沉淀产生，放置于冰箱过夜。取出上述溶液，在 3500r/min 下离心 15 分钟，倾出上清液。

（2）在不断搅拌下慢慢加 40ml 饱和硫酸铵于上述清液中，继续搅拌 15 分钟，于冰箱放置 30 分钟，3500r/min 离心 30 分钟。

（3）倾出上清液，沉淀用 0. 05mol/L pH 7. 1 磷酸缓冲液 20ml 搅拌溶解，离心除去不溶物（3500r/min，10 分钟）。

（4）测量上清液体积后逐滴加入 10ml 饱和硫酸铵，使终浓度为 33%，加完后继续搅拌 5 分钟，于冰箱放置 45 分钟使沉淀完全。

（5）取出后，3500r/min 离心 30 分钟，沉淀即为粗品 γ－球蛋白，用 2ml 0. 05mol/L pH 7. 1 磷酸缓冲液洗下。

（6）上清液倾入 50ml 量筒测量体积，计算欲达 40% 饱和所需的饱和硫酸铵量，在搅拌下缓慢滴入该上清液中，继续搅拌 3 分钟，冰箱放置 5 分钟，离心（3000r/min，30 分钟），沉淀即粗品 β－球蛋白，用 1ml 0. 05mol/L pH 7. 1 磷酸缓冲液洗下。

2. 脱盐

（1）将溶胀的 Sephadex G－50 装柱，用生理盐水过柱，平衡柱床，床面最好覆盖相同直径的圆形快速滤纸片。

（2）粗品 γ－球蛋白溶液经离心除去不溶物质后上 Sephadex G－50 柱，收集流出液，每管 1～1. 5ml。收到 10 管后，每管取出约 0. 2ml，加入双缩脲试剂，检测蛋白质含量，同时再取 0. 2ml 流出液用 10% $BaC1_2$溶液检查是否存在硫酸铵，画出洗脱曲线。

3. 电泳检查

（1）按常规制备聚丙烯酰胺凝胶，分离胶浓度为 7%。

（2）取蛋白浓度最高的流出液 0. 1ml 加样进行电泳，每管 3mA，并以 1∶5 稀释的血清作对照电泳。

（3）电泳 1. 5～2 小时后，取出凝胶，用 0. 5% 氨基黑染液进行染色。10 分钟后倾出染液，不断以 7% 醋酸脱色，起初每 20 分钟换液 1 次，1 小时后每小时换液 1 次，直到区带明显，画出电泳区带图谱。

4. 结果处理

（1）用箭头图表示盐析法制备免疫球蛋白的操作流程。

（2）由洗脱曲线计算免疫球蛋白的收率。

（3）绘制电泳图谱。

扫码“练一练”

【思考题】

1. 影响免疫球蛋白纯度和收率的因素有哪些？如何加以控制？

2. 试说明电泳图谱，并用电泳图谱判断所制得的免疫球蛋白的纯度。

3. 根据免疫球蛋白与血浆中其他蛋白组分的性质不同，还有什么方法可将它们分离？

EXPERIMENT 10　Preparation of immunoglobulin by salting out

【Purpose】

1. To master the solubility characteristics and basic preparation methods of serum immunoglobulin.

2. To know the salting out behavior of biological macromolecules such as proteins and the application of salting out methods in biochemical preparation.

【Principle】

Salting out is a method for purifying the target protein or impurities by introducing a certain amount of a neutral salt into the solution based on the different solubility of various biomolecules in a concentrated salt solution. This is a classic method of separation. As early as the 19th century, salting out was used to separate proteins from the blood. Because it is economical and does not require special equipment, it is easy to operate, safe, and has a wide range of applications, and it is less likely to cause degeneration(sometimes stable to biomolecules). It is still widely used to recover or separate biological macromolecules such as proteins(enzymes).

The high concentration of neutral salt can neutralize the charge on the surface of the protein molecules in the solution, while capturing the water in the solution, reducing the concentration of free water in the solution, thereby destroying the hydration layer structure that stabilizes the surface of the protein molecule, so that the solubility of the protein Greatly reduced. For salting out, $(NH_4)_2SO_4$ is the most commonly used agent. Because it is cheap with a large solubility in water, also it has a small variety of solubility with temperature, and a large solubility at low temperatures, so it is often a solution with a higher ionic strength can be obtained, which can be salted out even at low temperatures. Separation of proteins with different molecular weights and surface electrical properties and different solubility is carried out by using different concentrations of neutral salts, which is salting out.

Take plasma protein as an example(protein concentration 2%):

Ammonium sulfate saturation 20%	fibrinogen precipitation
Ammonium sulfate saturation 28% ~30%	α - globulin precipitation
Ammonium sulfate saturation 33%	γ - globulin precipitation
Ammonium sulfate saturation 40%	β - globulin precipitation
Ammonium sulfate saturation >50%	albumin precipitation

Immunoglobulin, also known as γ - globular protein, is mainly composed of IgG, molecular weight 16×10^4, and isoelectric point of 6.85 - 7.3. In this experiment, γ - globulin will be prepared from plasma by salting out, and the purity is identified by polyacrylamide gel.

【Materials】

1. Apparatus

(1) Measuring cylinder: 50ml, 100ml each

(2) Beaker: 100ml or 250ml each

(3) leather dropper

(4) Glass rods

(5) Refrigerator

(6) Centrifuge

(7) Disc electrophoresis instrument

(8) Column: 1. 2cm × 25cm

(9) Petri dish: 9cm

2. Reagent。

(1) plasma (or serum)。

(2) Saturated ammonium sulfate solution

Take 500ml of distilled water, heat to 70 to 80℃, and add 400g of solid ammonium sulfate powder to the water and stir for 20 minutes. After natural cooling, adjust to pH 7. 1 with concentrated ammonia water, and use it overnight.

(3) Mix, 0. 05mol/L pH 7. 1 phosphate buffer

67ml of 0. 2mol/L disodium hydrogen phosphate solution and 33ml of 0. 2mol/L sodium dihydrogen phosphate solution, and add 300ml of distilled water, and use the pH meter for calibration.

(4) 0. 5% amino black staining solution

Prepared with 7% acetic acid solution.

(5) 0. 85% sodium chloride solution

(6) A complete set of reagents for polyacrylamide gel

(7) 10% $BaCl_2$ solution

(8) Biuret reagent

【Procedures】

1. Preparation of immunoglobulin

(1) Take 20ml of plasma, mix with 20ml of 0. 05mol/L pH 7. 1 phosphate buffer, add 10ml of saturated ammonium sulfate solution dropwise, and stir while stirring. Continue stirring for 3 minutes after the addition, and then a very small amount of precipitate will be formed and placed in the refrigerator overnight. Take out the above solution, centrifuge at 3500r/min for 15 minutes, and decant the supernatant.

(2) Slowly add 40ml of saturated ammonium sulfate to the above supernatant under constant stirring, continue stirring for 15 minutes, place in the refrigerator for 30 minutes, and centrifuge at 3500r/min for 30 minutes.

(3) Let the supernatant decanted, and then dissolve precipitate by stirring with 20ml of 0. 05mol/L pH 7. 1 phosphate buffer and remove the insoluble matter by centrifugation (3500r/min,

10 minutes).

(4) After measuring the volume of the supernatant, add dropwise 10ml of saturated ammonium sulfate to make a final concentration of 33%. After the addition, continue stirring for 5 minutes, and allow the mixture to stand in the refrigerator for 45 minutes to complete the precipitation.

(5) After taking out, centrifuge at 3500r/min for 30 minutes, and the precipitate crude γ - globulin, then wash with 2ml of 0. 05mol/L pH 7. 1 phosphate buffer.

(6) Pour the supernatant into a 50ml measuring cylinder to measure the volume and calculate the amount of saturated ammonium sulfate required to achieve 40% saturation. Slowly drip into the supernatant under stirring, continue stirring for 3 minutes, and the refrigerator is placed for 5 minutes, and centrifuged (3000r/min, 30 minutes), the precipitate will be crude β - globulin, and wash with 1ml 0. 05mol/L pH 7. 1 phosphate buffer.

2. Desalting

(1) Pack the swollen Sephadex G - 50 into a column, equilibrate the column with physiological saline, and then cover bed surface preferably with a circular quick filter paper of the same diameter.

(2) Centrifuge the crude γ - globulin solution to remove insoluble matter, place on a Sephadex G - 50 column and collect the effluent (each tube should be 1 to 1. 5ml). After receiving 10 tubes, take out about 0. 2ml of each tube, and add the biuret reagent to detect the protein content. At the same time, take 0. 2ml of the effluent and use 10% $BaCl_2$ solution to check whether ammonium sulfate is present, and draw the elution curve.

3. Electrophoresis check

(1) Prepare the Electrophoresis as usual, and the polyacrylamide gel concentration should be 7%.

(2) Apply 0. 1ml of the effluent with the highest protein concentration for electrophoresis, 3mA per tube, and use serum diluted 1 : 5 as a control electrophoresis.

(3) After electrophoresis for 1. 5 to 2 hours, take out the gel and stain with a 0. 5% amino black stain solution. After 10 minutes, pour out the dyeing solution, and decolorize the mixture with 7% acetic acid. At first, change the liquid once every 20 minutes. After 1 hour, change the liquid once every hour until the zone is obvious, and draw the electrophoresis zone map.

4. Result processing

(1) An arrow diagramis used to indicate an operation procedure for preparing an immunoglobulin by salting out.

(2) The yield of immunoglobulinto be calculated from the elution curve.

(3) Draw an electropherogram.

【Questions】

1. What are the factors that affect the purity and yield of immunoglobulins? How to control?

2. The electrophoresis pattern was explained, and the purity of the prepared immunoglobulin was judged by electrophoresis pattern.

3. Depending on the nature of the immunoglobulin and other protein components in the plasma, is there any else way to separate them?

扫码“学一学”

实验十一　溶菌酶结晶的制备及活力测定

【实验目的】

1. 掌握　蛋白质结晶的一般方法，并由此了解蛋白质结晶的基本原理。

2. 了解　酶或蛋白质分离纯化的基本原理。

【实验原理】

溶菌酶（Lysozyme）是糖苷水解酶，分子量14307Da，由129个氨基酸残基构成。由于其中含有的碱性氨基酸残基比酸性氨基酸残基多，所以它的等电点高达11左右。

已经知道，几乎所有的细菌都有坚韧的细胞壁。革兰阳性细菌细胞壁的主要化学成分是肽聚糖。肽聚糖是由N－乙酰葡萄糖胺（NAG）与N－乙酰胞壁酸（NAM）通过β－1，4糖苷键连接形成骨架，并通过NAM部分的乳酰基与寡肽交联而成。

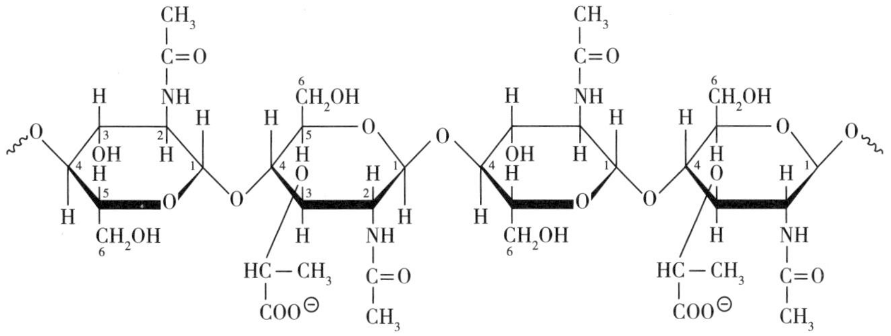

NAG　　NAM　　NAG　　NAM

溶菌酶之所以能破坏革兰阳性细菌的细胞壁而具有溶菌作用，原因就在于它能水解N－乙酰葡萄糖胺与N－乙酰胞壁酸之间的β－1，4糖苷键。

溶菌酶在蛋清中含量较高，在一定条件下能直接从蛋清中结晶出来。蛋壳膜中的溶菌酶含量虽然较少，但仍然可以用于制备，此外，它还广泛存在于哺乳动物的唾液、泪、血浆、乳汁、白细胞及其他组织中。

本实验以蛋清为原料制备溶菌酶结晶，因溶菌酶为碱性蛋白，中性环境中被阳离子交换树脂吸附。利用该性质，可将它和鸡蛋清中的其他蛋白质分离，然后再经盐析、纯化处理所得到的溶菌酶即可结晶。溶菌酶比较容易结晶，其结晶形状随结晶条件而异，有棱形八面体、正方形六面体及棒状结晶等。

溶菌酶的活力测定过去常用细胞溶菌法，即用对溶菌酶敏感的溶性微球菌 *Micrococcuslysodeikticus* 作为底物进行比浊测定。但是此法往往不能得到重复的结果。也有用寡聚和多聚N－乙酰葡萄糖胺的衍生物为底物，通过黏度法或还原糖分析法进行测定，但操作比较复杂。本实验采用的比色测定法，是以活性染料艳红K－2BP所标记的 *M. lysodeikticus* 为底物，由于活性染料的标记部位并不是酶的作用位点，因此当溶菌酶将这种底物水解以后即产生染料标记的水溶性碎片，除去未经酶作用的多余底物，溶液颜色的深浅就能代表酶活力的相对大小，在540nm波长处可直接进行比色测定。本法简便专一，比较准确。

【实验材料】

1. 器材

（1）721型分光光度计　1台
（2）抽滤瓶及布氏漏斗　各1只
（3）研钵　1只
（4）恒温水浴　1台
（5）离心机　1台
（6）透析袋
（7）1cm×35cm色谱柱　1根
（8）吸量管：0.1ml，0.2ml，1ml，5ml　各1只
（9）真空干燥器　1只

2. 试剂

（1）10%硫酸铵，固体硫酸铵
（2）丙酮
（3）pH 6.5 0.15mol/L磷酸缓冲液　先分别配制0.15mol/L NaH_2PO_4及0.15mol/L Na_2HPO_4液，取前者20.5ml加后者97.5ml，混匀，即得（pH计校正）。
（4）pH 6.2 0.1mol/L磷酸缓冲液：称取$NaH_2PO_4 \cdot 2H_2O$ 11.70g，$Na_2HPO_4 \cdot 12H_2O$ 7.86g，EDTA 0.392g，1000ml容量瓶中加水至刻度，摇匀，即得（pH计校正）。
（5）鸡蛋清（鸡蛋）
（6）底物干菌粉
（7）“724”树脂

【实验方法】

1. 蛋清准备　由4~5只新鲜鸡蛋中小心取出蛋清约80~100ml，充分打匀，用纱布滤去杂质，计量体积，并用试纸测量pH，用冰块预冷至0℃备用。

2. 树脂吸附　将处理好的“724”树脂用布氏漏斗抽干，取湿树脂20g（约为蛋清量的1/5~1/4），在不断搅拌下加入预冷的蛋清中，再继续搅拌3小时使充分吸附，静置过夜（0~5℃）。

3. 洗涤　倾出上层蛋清，用蒸馏水清洗树脂二、三次，再用pH 6.5的0.15mol/L磷酸缓冲液40ml搅拌清洗两次，用布氏漏斗抽干。

4. 洗脱　将树脂移入烧杯，取10%硫酸铵溶液30~40ml（树脂量2倍，不可多用！）分三次加入搅拌（15分钟）洗脱，抽干树脂，合并洗脱液（滤液），树脂保存供再生。

5. 盐析　测量洗脱液总体积，按33%量（*W/V*）在搅拌下逐渐加入研细的固体硫酸铵，使终浓度达40%。静置，等沉淀结絮下沉后，小心吸去上清液，将沉淀置离心管离心10分钟（3000r/min），或用布氏漏斗抽滤，收集沉淀。

6. 脱盐　沉淀用1ml蒸馏水溶解．转入透析袋，用蒸馏水透析24小时（0~5℃冰箱），

中途换水 3 ~5 次，或流水（搅拌）透析 24 小时。因该酶分子量小，所以透析时间不可太长，防止酶渗出。

7. 去除碱性杂蛋白　将上述透析液用 1mol/L NaOH（最后用 0.1mol/L NaOH）溶液调至 pH 8.0 ~8.5。如有沉淀，离心除去。

8. 结晶　用药勺在搅拌下慢慢向酶液中加入 5%（W/V）研细的固体 NaCl，注意防止局部过浓。加完后用 NaOH 溶液慢慢调至 pH 9.5 ~10.0，室温下静置 48 小时。

9. 结晶观察与收取　肉眼观察有结晶形成后，用滴管吸取结晶液 1 滴置于载玻片上，在低倍显微镜下观察并画出结晶图形。离心或过滤收集酶晶体，用少量丙酮洗涤晶体两次，以 P_2O_5 真空干燥后称重。

10. 活力测定

（1）酶液配制　准确称取溶菌酶样品 5mg，用 0.1mol/L pH 6.2 磷酸缓冲液配成 1mol/ml 的酶液，再将酶液稀释成 50μg/ml。

（2）底物配制　取干菌粉 5mg；加上述缓冲液少许，在乳钵中（或匀浆器中）研磨 2 分钟，倾出，稀释到 15 ~25ml，此时在光电比色计上的吸光度最好在 0.5 ~0.7 范围内。

（3）活力测定　先将酶和底物分别放入 25℃ 恒温水浴预热 10 分钟，吸取底物悬浮液 4ml 放入比色杯中，在 450nm 波长读出吸光度，此为零时读数。然后吸取样品液 0.2ml（相当于 10μg 酶），每隔 30 秒读一次吸光度，到 90 秒时共计下四个读数。

11. 结果处理

（1）计算　活力单位的定义是：在 25℃，pH 6.2，波长为 450nm 时，每分钟引起吸光度下降 0.001 为 1 个活力单位。

$$\text{每 1mg 酶活力单位数} = \text{吸光度} \times \frac{1000}{\text{样品（μg）}}$$

（2）计算溶菌酶的收率并由其效价计算总活力回收率。

$$\text{收率} = \frac{\text{干燥的酶重量}}{\text{蛋清总重量}} \times 100\%$$

$$\text{总活力回收率} = \frac{\text{酶重量} \times \text{效价}}{\text{蛋清总重量}}$$

附：

1. 底物的制备　菌种 *Micrococcuslysodeikticus* 接种于培养基上，28℃ 培养 48 小时，用蒸馏水将菌体冲洗下来，经纱布过滤，滤液离心（4000r/min，10 分钟），倾去上清液。用蒸馏水洗菌体数次，离心除去混杂其中的培养基，然后将菌体用少置水悬浮，冰冻干燥。如无冻干设备，可将菌体刮在玻璃板上成一薄层，冷风吹干，置于干燥器中。

2. 树脂处理　市售“724”树脂先用清水漂洗，除去细微杂质，加入 1mol/L NaOH 溶液，放置 4 ~8 小时，并间歇搅拌，然后抽去碱液，用蒸馏水洗至 pH 7.5 左右。再用 1mol/L 盐酸如上法浸泡树脂，所用盐酸须过量，搅拌。保证树脂完全转变为氢型，然后抽去酸液。用蒸馏水洗至 pH 5.5，平衡过夜，如 pH 不低 5.0，抽干，用 2mol/L NaOH 把树脂转变为钠型，但须控制 pH 不超过 6.5，抽干。将树脂用 pH 6.5、0.15mol/L 磷酸缓冲液浸泡过夜，如 pH 下降再用 2mol/L NaOH 溶液调回 pH 6.5，冷藏（使不结冰）备用。

扫码"练一练"

【思考题】

1. 利用724树脂分离纯化溶菌酶的原理是什么?
2. 溶菌酶结晶的影响因素有哪些?

EXPERIMENT 11　Preparation and viability determination of lysozyme crystal

【Purpose】

1. To master the general methods of protein crystallization and understand the basic principles of protein crystallization.

2. To know the basic principles of an enzyme or protein separation and purification.

【Principle】

Lysozyme, a glycoside hydrolase, consists of 129 amino acid residues with a molecular weight of 14307Da. Since it contains more basic amino acid residues than acidic amino acid residues, its isoelectric point is as high as about 11.

It is known that almost all bacteria have tough cell walls. The main chemical component of the cell wall of Gram - positive bacteria is peptidoglycan. Lysozyme is able to destroy the cell wall of Gram - positive bacteria has lysis, the reason is that it can hydrolyzing N - acetylglucosamine and N - acetylmuramyl β - 1,4 glycosidic bonds between the acid.

NAG　NAM　NAG　NAM

NAG stands for 2 - N - acetylamino - D - glucose; NAM stands for N - acetylmuramic acid or 2 - N - acetylamino - 3 - O - lactoyl - D - glucose.

High content in egg white lysozyme can be crystallized under certain conditions from egg white. Although the lysozyme content in the eggshell membrane is small, it can still be used for preparation. In addition, it is widely found in mammalian saliva, tears, plasma, milk, white blood cells and other tissues.

In this experiment, lysozyme crystals were prepared from egg white as a raw material, and lysozyme was a basic protein, which was adsorbed by a cation exchange resin in neutral environment. Using this property, it can be separated from other proteins in egg white, Then, the lysozyme obtained by salting out and purification treatment can be crystallized. Lysozyme is relatively easy to crystallize, and its crystal shape varies depending on the crystallization conditions, and has a prismatic octahedron, a square hexahedron, and a rod crystal.

In the past, cell lysis method was commonly used for lysozyme activity assay. using turbidimetric determination of lysozyme – sensitive *Micrococcuslysodeikticus* as a substrate. However, this method often does not lead repeated results. Derivatives of oligomeric and poly – N – acetylglucosamine are also useful as substrates, which are determined by viscosity method or reducing sugar analysis, but the operation is complicated. The colorimetric assay used in this experiment is based on *M. lysodeikticus* labeled with reactive dye brilliant red K – 2BP. Since the labeled site of reactive dye is not the point of action of the enzyme, when lysozyme will use this substrate After hydrolysis, the dye – labeled water – soluble fragments are produced, and the excess substrate without enzyme action is removed. The color depth of the solution can represent the relative size of the enzyme activity, and the colorimetric measurement can be directly performed at a wavelength of λ540nm. This method is simple and specific and relatively accurate.

【Materials】

1. Apparatus

(1) 721 spectrophotometer

(2) suction filter bottle and Buchner funnel

(3) Mortar

(4) thermostat water bath

(5) Centrifuge

(6) Dialysis bag

(7) 1cm × 35cm column

(8) Pipette: 0.1ml, 0.2ml, 1ml, 5ml

(9) Vacuum dryer

2. Reagent

(1) 10% ammonium sulfate, solid ammonium sulfate

(2) Acetone

(3) pH 6.5 0.15mol/L phosphate buffer

First prepare 0.15mol/L NaH_2PO_4 and 0.15mol/L Na_2HPO_4 solution separately, take 20.5ml of the former and 97.5ml of the latter, mix well, then get (pH meter correction).

(4) pH 6.2 0.1mol/L phosphate buffer

Weigh $NaH_2PO_4 \cdot 2H_2O$ 11.70g, $Na_2HPO_4 \cdot 12H_2O$ 7.86g, EDTA 0.392g, add water to the 1000ml volumetric flask to the mark, shake well, then get (pH meter correction).

(5) Egg white (egg)

(6) Substrate dry powder

(7) "724" resin

【Procedures】

1. Egg white preparation

Carefully remove the egg white from about 4 to 5 fresh eggs by about 80 to 100ml, mix well, filter the impurities with gauze, measure the volume, and measure the pH with a test paper, pre – cool with ice cubes to 0℃ for use.

2. Resin adsorption

The handled "724" resin is drained with a Buchner funnel, taking 20g wet resin (about 1/5 to 1/4 of the amount of egg white), and add the pre – cooled egg white under constant stirring, and continue stirring for 3 hours. Adequate adsorption, allow to stand overnight (0 ~ 5℃).

3. washing

Pour out the upper layer of egg white, wash the resin twice with distilled water, and then wash twice with 40ml of pH 6.5 0.15mol/L phosphate buffer, and drain with a Buchner funnel.

4. Elution

Move the resin into the beaker, take 30 ~ 40ml of 10% ammonium sulphate solution (2 times of resin, don't add too much!), add and stir (15 minutes) in three times, drain the resin, combine the eluent (filtrate), and store the resin for regeneration.

5. Salting out

Measure the total volume of the eluate, and the finely gradually add divided solid ammonium sulfate under stirring at a percentage of 33% (W/V) to a final concentration of 40%. After standing, wait for the precipitated floccule to sink, carefully aspirate the supernatant, centrifuge the pellet for 10 minutes (3000r/min), or filter with a Buchner funnel to collect the precipitate.

6. Desalting

Dissolve the precipitate in 1ml of distilled water. Transfer to dialysis bag, dialyze with distilled water for 24 hours (0 ~ 5℃ refrigerator), change water 3 to 5 times in the middle, or dialyze for 24h in running water (stirring). Since the molecular weight of the enzyme is small, the dialysis time should not be too long to prevent the enzyme from oozing out.

7. Removing basic hybrid proteins

Adjust The dialysate to pH 8.0 – 8.5 with 1mol/L NaOH (final 0.1mol/L NaOH) solution. If there is a precipitate, remove it by centrifugation.

8. crystallization

Slowly add 5% (w/v) fine solid NaCl to the enzyme solution with stirring with a spoon, taking care to prevent local over – concentration. After the addition, gradually adjust the pH to 9.5 – 10.0 with a NaOH solution, and allow to stand at room temperature for 48 hours.

9. Crystal observation and collection

After crystal formation is observed by the naked eye, take up a drop of the crystal solution by a dropper and placed on a glass slide, and observed under a low magnification microscope and draw a crystal pattern. Collect the enzyme crystals by centrifugation or filtration and wash the crystals twice

with a small amount of acetone, dry under vacuum in P_2O_5, and weigh.

10. Vitality measurement

(1) Enzyme solution preparation

Accurately weigh 5mg of lysozyme sample, formulate 1mol/ml enzyme solution with 0.1mol/L pH 6.2 phosphate buffer, and dilute the enzyme solution to 50μg/ml.

(2) Substrate preparation

Take 5mg of dry powder, add a little buffer, grind in the mortar (or homogenizer) for 2 minutes, pour out, dilute to 15 ~ 25ml, the absorbance on the photoelectric colorimeter at this time is preferably in the range of 0.5 to 0.7.

(3) Vitality measurement

Separately pre – heat the enzyme and the substrate in a constant temperature water bath at 25℃ for 10 minutes, aspirate 4ml of the substrate suspension into a cuvette, and read the absorbance at a wavelength of λ450nm for time 0. Then draw 0.2ml of sample solution (equivalent to 10μg enzyme) and read the absorbance every 30 seconds. A total of four readings to be taken by 90 seconds.

11. Result processing

(1) Calculation

Activity unit is defined: At 25℃, pH 6.2, the wavelength of 450nm, the absorbance decreased by 0.001 per minute for 1 vitality unit.

$$\text{Each 1mg enzyme activity unit number} = \text{Absorbance} \times \frac{1000}{\text{sample}(\mu g)}$$

(2) Calculate the yield of lysozyme and calculate the total activity recovery rate from its potency.

$$\text{Yield} = \frac{\text{Dry enzyme weight}}{\text{Total egg white weight}} \times 100\%$$

$$\text{Total activity recovery} = \frac{\text{Enzyme weight} \times \text{potency}}{\text{The total weight of egg whites}}$$

Attached:

1. Substrate preparation

Inoculate the strain *Micrococcuslysodeikticus* on the medium, culture at 28℃ for 48 hours, wash thethallus with distilled water, filter through gauze, and the filtrate is centrifuged (4000r/min, 10 minutes), and decant the supernatant. Wash the thallus several times with distilled water, centrifuged to remove the medium mixed therein, and then suspend the thallus in less water and lyophilized.

If there is no lyophilization equipment, the thallus can be scraped on a glass plate to form a thin layer, which is blown dry and placed in a desiccator.

2. Resin treatment

The commercially available "724" resin is first rinse with water to remove fine impurities, add with 1mol/L NaOH solution, place for 4 – 8 hours, and stir intermittently, then remove the alkali solution, and wash with distilled water to a pH of about 7.5. Then, soak the resin with 1mol/L

hydrochloric acid as above, and excessively stir the hydrochloric acid used and stir to ensure that the resin is completely converted into a hydrogen form, and then remove the acid solution. Wash with distilled water to pH 5.5 and equilibrate overnight. If the pH is not lower than 5.0, drain and convert the resin to sodium with 2mol/L NaOH, but control the pH to not exceed 6.5 and drain. Soak the resin was in phosphate buffer solution of pH 6.5 and 0.15mol/L overnight. If the pH is lowered, adjust the pH to 6.5 with 2mol/L NaOH solution, and chill the mixture (not frozen).

【Questions】

1. What is the principle of separating and purifying lysozyme using 724 resin?
2. What are the influencing factors of lysozyme crystallization?

扫码“学一学”

实验十二　凝胶色谱法测定蛋白质的分子量

【实验目的】

1. **掌握**　凝胶层析法测定蛋白质分子量的原理。
2. **了解**　凝胶层析的一般操作方法。

【实验原理】

凝胶层析（Gel chromatography）的分离过程是在装有多孔物质（交联聚苯乙烯、多孔玻璃、多孔硅胶、交联葡聚糖等）填料的柱中进行的。柱的总体积为 V_A，它包括填料的骨架体积 V_{GM}，填料的孔体积 V_i 以及填料颗粒之间的体积 V_0。

$$V_A = V_i + V_0 + V_{GM}$$

孔体积 V_i 中的溶剂为固定相，而在粒间体积 V_0 中的溶剂称为流动相。一个填料的颗粒含有许多不同大小的孔。这些孔对于溶剂分子来说是很大的，它们可以自由地扩散出入。如果对溶质分子大小合适的话，则可以不同程度地往孔中扩散，大个的溶质分子只能占有比较少的孔，而小个的溶质分子则除去能占有大孔外还可以占有另外一些较小的孔。所以随着溶质分子尺寸的减小可以占有的孔体积迅速增加。当具有一定分子量分布的高聚物溶液从柱中通过时，较小的分子在柱中停留时间比大分子停留的时间要长，于是样品各组分即按分子大小顺序而分开，最先淋出的是最大的分子。其定量关系是：

$$V_e = V_0 + V_{i,ace}$$

式中，V_e是淋出体积，V_0是粒间体积，$V_{i,ace}$是对某种大小的溶质分子来说可以渗透进去的那部分孔体积，$V_{i,ace}$是总的孔体积 V_i 的一部分，是溶质分子量的函数，它和 V_i之比等于分配系数 Kd。

从以上公式得到：　$V_e = V_0 + \mathrm{Kd}\ V_i$

用凝胶过滤层析测定生物大分子的分子量，操作简便，仪器简单，消耗样品也少，而且可以回收。测定的依据是不同分子量的物质，只要在凝胶的分离范围内（渗入限与排阻限之间），洗脱体积 V_e 随分子量增加而下降。对于一个特定体系（凝胶柱），待测定物质的洗脱体积 V_e 与分子量 M 的关系符合公式：

$$V_e = -K\log M + C$$

式中，K 与 C 是常数，分别为直线方程的斜率和外推截距。由图可见，物质的洗脱体积与分子量成负相关。

扫码“看一看”

用凝胶层析法测定蛋白质的分子量，方法简单，有时不需要纯物质，用粗制品即可。例如在一粗酶制剂中，为了测定某一酶组分的分子量，只要测定洗脱液中具有该酶最大活性的部分，然后确定其洗脱体积，即可通过标准曲线计算出其分子量。凝胶层析法测定分子量也有一定的局限性，在 pH 6 ~8 的范围内，线性关系比较好，但在极端 pH 时，一般蛋白质有可能因变性而偏离。糖蛋白在含糖量超过 5% 时，测得分子量比真实的要大，铁蛋白则与此相反，测得的分子量比真实的要小。有一些酶底物是糖，如淀粉酶、溶菌酶等会与葡聚糖凝胶形成络合物，这种络合物与酶 - 底物络合物相似，因此在葡聚糖凝胶上层析时表现异常。用凝胶层析法所测分子量的结果，要和其他方法的测定相对照，由此才可得出较可靠的结论。

【实验材料】

1. 器材

（1）层析柱　柱管 1.2cm ×100cm　　1 支

（2）核酸蛋白检测仪（或紫外分光光度计）

（3）部分收集器

（4）下口瓶

（5）滴管　　1 支

（6）试管（1.5cm ×10cm）　　100 支

（7）试管架　　1 个

（8）100ml 烧杯　　1 个

2. 试剂

（1）标准蛋白质　牛血浆清蛋白，鸡卵清蛋白，胰凝乳蛋白酶原 A，牛胰岛素等，均要层析纯。

（2）蓝色葡聚糖 - 2000

（3）0.025mol/L 氯化钾 - 0.2mol/L 乙酸溶液

（4）Sephadex G - 100

（5）待测蛋白质样品

【实验方法】

1. 凝胶的准备与装柱　称取 8g Sephadex G - 100 加蒸馏水 200ml 溶解，沸水浴溶胀 5 小时。用倾斜法除去凝胶水及细小颗粒，反复以蒸馏水洗涤直至无细小颗粒为止。装柱前将处理好的凝胶置真空干燥器中抽真空，以除尽凝胶中的空气。取 1.2cm ×100cm 的洁净层析柱，先于底部填少许玻璃棉，加入 20ml 洗脱液（本实验用 0.025mol/L 氯化钾 - 0.2mol/L 乙酸溶液），关闭柱开口，然后将溶胀后的凝胶搅匀并加入柱中，待柱底凝胶沉积高度为

2cm 时，打开柱的开口，继续装柱至柱床高度为 95cm 左右，关闭出口。装柱过程严禁产生气泡及柱床分层，如有气泡及柱床分层应重装。装完后，用玻棒轻轻搅动柱床表面层，待凝胶自然沉降形成平面后，在凝胶表面上放一片滤纸或尼龙滤布，以防将来在加样时凝胶被冲起。用洗脱液洗涤柱床半小时，使柱床稳定。

2. 装柱效果检查及外水体积测定 将层析柱出口打开，使柱面上溶液流出，直至床面与液面刚好平齐为止。关闭出口，用滴管于床面中心滴加 10 滴 0.5% 蓝色葡聚糖溶液，切勿搅动柱床表面。打开出口，使蓝色葡聚糖溶液进入柱床，且直至柱面与液面平齐，关闭出口。同上法加入 20 滴洗脱液，该洗脱液完全进柱后，再用洗脱液进行洗脱。蓝色葡聚糖进柱后，立即开始收集，同时观测蓝色葡聚糖在柱中的行为，若蓝色谱带较集中，表明装柱效果良好，待色带流至柱底时，开始按每管 5 滴收集。待蓝色液全部流出后，将从蓝色葡聚糖进柱开始所收集的流出液，直至颜色最深的一管的总体积相加，量出体积，即为外水体积。

3. 标准曲线的制定：

蛋白质标准样品混合液：分别称取 2.5 ~ 3.0mg 牛血浆清蛋白（分子量 67000Da），鸡卵清蛋白（分子量 43000Da），胰凝乳蛋白酶原 A（分子量 25000Da），结晶牛胰岛素（pH 2 ~ 6 时为二聚体，分子量 12000Da）共同溶于 1 ~ 1.5ml 0.025% 氯化钾 - 0.2mol/L 乙酸溶液中。

上柱和洗脱：将标准样品混合液上柱，然后用 0.025% 氯化钾 - 0.2mol/L 乙酸溶液洗脱。流速 3ml/10min，3ml/管，用部分收集器收集，紫外检测仪 280nm 处检测。或收集后用紫外分光光度计于 280nm 处测定每管 OD 值，以管号（或洗脱体积）为横坐标，OD 值为纵坐标绘出洗脱曲线。

根据洗脱峰位置量出每种蛋白质的洗脱体积（V_e）。然后以蛋白质分子量的对数（lgM）为横坐标，V_e 为纵坐标，作出标准曲线。为了结果可靠，应以同样条件重复 1 ~ 2 次，取 V_e 的平均值作图。

4. 未知样品分子量的测定 完全按照标准曲线的条件操作，根据紫外检测的洗脱峰位置，量出洗脱体积。重复测定 1 ~ 2 次，取其平均值，由标准曲线可查得样品的分子量。

【思考题】

扫码“练一练”

1. 本实验中，选择下列哪种凝胶测定效果更好些（Sephadex G - 75，SephadexG - 100），为什么？
2. 本实验中采用的是葡聚糖凝胶，若采用生物胶，哪种型号合适？

EXPERIMENT 12 Determination of molecular weights of proteins by gel chromatography

【Purpose】

1. To learn the principle of protein molecular weight determination by gel chromatography.
2. To understand the general operation of gel chromatography.

【Principle】

The separation of gel chromatography is done in a column containing a filler of porous substances(crosslinked polystyrene, porous glass, porous silica Gel, crosslinked glucan, etc.). The total volume of the column is V_A, which includes the frame volume of the packing V_{GM}, the pore volume of the packing V_i and the volume V_0 between the packing particles.

$$V_A = V_i + V_0 + V_{GM}$$

The column space includes the pore volume of the packing V_i and the volume V_0 between the packing particles. The solvent in pore volume V_i is fixed phase, and the solvent in intergranular volume V_0 is mobile phase.

The packed particles contain many pores of different sizes. These pores are large for solvent molecules to be diffused in and out freely. If the solute molecules are of the right size, they can be diffused into the pores to varying degrees. The largc solute molecules can only take up a few pores, while the smaller solute molecules can occupy several smaller holes in addition to the large ones. So, the pore volume that can be occupied increases rapidly with the decrease of solute molecular size. When a polymer solution with a certain molecular weight distribution passes through the column, the smaller molecules stay in the column longer than the larger ones, so the components of the sample are separated according to the order of the molecular size, and the largest molecules are leached out first.

The quantitative relationship is:

$$V_e = V_0 + V_{i,ace}$$

Where V_e is the leached volume, V_0 is the intergranular volume, $V_{i,ace}$ is the fraction of the pore volume that a solute molecule of some size can penetrate. V_i, ace is part of the total pore volume V_i, and it is a function of the solute molecular weight. Its ratio to V_i is Kd, which is the partition coefficient.

$$\mathrm{Kd} = V_{i,ace}/V_i$$

From the above formula:

$$V_e = V_0 + \mathrm{Kd}\ V_i$$

Determination of molecular weights of biological macromolecules by gel filtration chromatography is easy to operate, simple to instrument, less consumption of samples, and can be recycled. The elution volume V_e decreases with the increase of molecular weight, as long as the separation range of gel is between infiltration limit and exclusion limit. For a specific system(gel column), the relationship between the elution volume and molecular weight of the substance to be measured conforms to the formula:

$$V_e = -\mathrm{Klog}M + C$$

Where K and C are constants, respectively representing the slope and extrapolation intercept of the line equation.

The relationship between the elution volume and the molecular weight as can be seen from the figure, the elution volume of the substance is inversely related to the molecular weight.

Gel chromatography is a simple method to determine the molecular weight of proteins. Sometimes a pure substance is not required and just a crude product can be used. For example, in a crude en-

zyme preparation, in order to determine the molecular weight of an enzyme component, the molecular weight of the eluent can be determined from the standard curve as long as the eluent with the largest activity of the enzyme is determined and the elution volume is determined. Gel chromatography also has some limitations in the determination of molecular weights. In the range of pH 6 – 8, the linear relationship is good, but at the extreme pH, the general protein may be deviated by denaturation. Glycoproteins with a sugar content of more than 5% are measured at larger molecular weights than the real molecular weight, while ferritin, in contrast, is measured at smaller molecular weights. Some enzyme substrates are sugars, such as amylase and lysozyme, etc. , which will form complexes with dextran gel, and this complex is similar to the enzyme – substrate complex, so it will have abnormally in the analysis of dextran gel. The results obtained by gel chromatography should be compared with those obtained by other methods.

【Materials】

1. Apparatus

(1) chromatography column: 1 column tube 1.2cm × 100cm

(2) nucleic acid protein detector (or ultraviolet spectrophotometer)

(3) partial collector

(4) the bottle mouth

(5) the dropper

(6) test tube (1.5cm × 10cm)

(7) test tube rack

(8) beaker

2. Reagents

(1) standard proteins: bovine plasma albumin, chicken egg albumin, chymotrypsin progenitor A, bovine insulin, etc. , chromatographically pure.

(2) blue dextran – 2000

(3) 0.025mol/L potassium chloride – 0.2mol/L acetic acid solution

(4) Sephadex G – 100

(5) protein samples

【Procedures】

1. Gel preparation and column packing

Add 8g Sephadex G – 100 to 200ml distilled water to dissolve and swell in the boiling water bath for 5 hours. The gel water and fine particles are removed by tilting method and washed with distilled water repeatedly until there were no fine particles. Before packing the column, place the treated gel in a vacuum dryer to vacuum out the air in the gel. Take a cleaned chromatography column of 1.2cm × 100cm and fill with a little glass wool at the bottom. Add 20ml eluent (use 0.025mol/L potassium chloride – 0.2mol/L acetic acid solution in this experiment). Close the column opening. Stir the swollen gel well and add it to the column. When the height of gel deposition at the bottom of the column is 2cm, open the column opening. Continue to pack the column until the height of the column bed is about 95cm and then close the opening. It is forbidden to produce bubbles and stratification of column bed during column loading. If there are bubbles and stratification of column bed, it should be reinstalled. After loading, gently stir the surface layer of the column bed with a glass stick. After the gel naturally settles to form a plane, put a filter paper or nylon press cloth on the surface of the gel to prevent the gel from being washed up in the future. Wash the column bed with eluent for half an hour to make the column bed stable.

2. Inspection of column packing effect and determination of external water volume

Open the column outlet and allow the solution to flow out of the column until the bed surface is just flush with the liquid surface. Close the outlet and add 10 drops of 0.5% blue dextran solution to the center of the bed surface with a dropper. Do not disturb the bed surface. Open the outlet to allow the blue dextran solution to enter the column bed and close the outlet until the cylinder is flush with the liquid surface. 20 drops of the eluate are added in the same manner as above. After the eluate is completely charged into the column, elute with an eluent. Start to collect the eluent immediately after the blue dextran enter the column. Meanwhile, observe the behavior of blue dextran in the column. If the blue chromatographic band is more concentrated, it indicates that the column packing effect is good. When the band flows to the bottom of the column, start collecting 5 drops per tube. After all the blue liquid is discharged, the total volume of the outflow collected from the blue dextran inlet column until the total volume of the darkest tube is added to measure the volume, namely the volume of external water.

3. Standard curve determination

Protein standard sample mixture: respectively dissolve 2.5 – 3.0mg bovine plasma albumin (molecular weight 67000), chicken egg albumin (molecular weight 43000), chymotrypsin progenitor A (molecular weight 25000), and crystalline bovine insulin (molecular weight 12000) in 1 – 1.5ml 0.025% potassium chloride with 0.2mol/L acetic acid solution.

Loading on the column and elution: load the standard sample mixture on the column and then elute with 0.025% potassium chloride – 0.2mol/L acetic acid solution. The flow rate: 3ml/10min, 3ml/tube. Collect by partial collectors and detect the tubes at 280nm with ultraviolet detector. After collection, measure the OD value of each tube at 280nm with an ultraviolet spectrophotometer. Draw the elution curve with the tube number (or elution volume) as the horizontal coordinate and the OD value as the vertical coordinate. Measure the elution volume (V_e) of each protein according to the elution peak position. Then, use the logarithm (lgM) of protein molecular weight as the abscissa and V_e as the ordinate to draw the standard curve. In order to obtain reliable results, the same conditions should be repeated for 1 – 2 times, and the average value of V_e should be taken to plot.

4. Determination of the molecular weight of unknown sample

Measure the elution volume according to the position of the elution peak detected by the ultraviolet spectrophotometer as the method of the standard curve determination. Repeat determination 1 – 2 times and take its average value. The molecular weight of the sample can be found from the standard curve.

【Questions】

1. In this experiment, which of the following gels should be selected for better determination effect (Sephadex G – 75, Sephadex G – 100), and why?

2. Dextran gel is used in this experiment. If Bio - gel is used, which model is suitable?

扫码“学一学”

实验十三　亲和色谱法制备胰蛋白酶抑制剂

【实验目的】

1. 了解　亲和层析的基本步骤和洗涤、洗脱的基本规律。

2. 学习利用固定化胰蛋白酶作亲和吸附剂制取相应的酶抑制剂的方法。

【实验原理】

胰蛋白酶抑制剂广泛存在于豆类、谷类、油料作物等植物中。胰蛋白酶抑制剂在这些作物在各部位均有分布，但主要存在于作物的种子中。在种子内，胰蛋白酶抑制剂主要分布于蛋白质含量丰富的组织或器官，定位于蛋白体、液泡或存在于细胞液中。例如，大豆和绿豆种子中胰蛋白酶抑制剂的含量可达总蛋白的6%～8%。在不同作物种子中，胰蛋白酶抑制剂铁活性各不相同，通常以大豆中胰蛋白酶抑制活性最高。

在禽类的蛋清中存在两种蛋白酶抑制剂——卵类黏蛋白和卵清蛋白酶抑制剂。卵类黏蛋白可抑制猪、羊和牛的胰蛋白酶的活性，但对人的胰蛋白酶活性无抑制作用。卵清蛋白酶抑制剂可抑制牛和羊的胰蛋白酶及糜蛋白酶的活性。它们的存在可以防止蛋白质的分解，阻止细菌在蛋中繁殖，因而具有保护蛋黄和卵胚的作用。目前常用的蛋白酶抑制剂制备方法有盐析法、有机溶剂沉淀、离子交换法及亲和层析法等，其中亲和层析法纯化倍数最高。

亲和层析（Affinity chromatography）又称为生物专一性吸附层析，是根据生物大分子与一些相应物质进行专一性结合的特征设计的。这些能专一结合的物质有抗原和抗体、酶和底物或抑制剂、核酸互补链、药物与受体等。将其中一种物质（配基）结合于适当的固体支持物（载体）上，通过层析可以把微量的对应物质从大量的杂质中分离出来。亲和层析目前已发展成一种制备和纯化多种生物活性物质的简便而有效的技术。

本实验使用琼脂糖凝胶－胰蛋白酶亲和吸附剂，在一定的pH和离子强度下，通过亲和层析将鸡蛋清中的卵类黏蛋白（一种天然的胰蛋白酶抑制剂）吸附于柱上，再改变条件将抑制剂洗脱下来，达到提取、纯化的目的。

【实验材料】

1. 器材

（1）烧杯	2000ml，500ml，250ml，125ml	各1只
（2）磁力搅拌器		1台
（3）玻璃漏斗	4～5cm	1只
（4）砂芯漏斗	9cm或10cm	1只
（5）吸滤瓶	1000～2000ml	1只
（6）量筒	100ml	1只
（7）温度计		1只

（8）层析柱	1.5cm×20cm，	1根
（9）玻璃棒		6根
（10）小试管及试管架		
（11）吸量管	0.1ml，1ml，2ml	各2支
（12）停表		1只
（13）皮头滴管		1支
（14）751紫外分光光度计		1台
（15）恒温水浴		1台
（16）恒流泵		1台
（17）自动收集器		1台

2. 试剂

（1）0.05mol/L pH 7.8磷酸缓冲液　$Na_2HPO_4 \cdot 12H_2O$（分子量358.16）16.476g和0.625g $NaH_2PO_4 \cdot 2H_2O$（MW 156.03），溶于水，定容至1000ml，pH计校正。

（2）0.06% BANA乙醇溶液　取BANA 60mg，95%乙醇20ml，使溶解，用pH 7.8、0.2mol/L磷酸缓冲液定容至100ml。

（3）0.05%萘基乙二胺盐酸盐（NEDA）乙醇溶液

（4）0.1mol/L pH 7.5 Tris－HCl缓冲液（内含0.01mol/L $CaCl_2$0.5mol/L NaCl）　取三羟甲基氨基甲烷12.12g溶于水，分别加入$CaCl_2$ 1.1g、NaCl 29.2g和1mol/L HCl 40ml，定容至1000ml，pH计校正。

（5）0.5mol/L pH 2.5甘氨酸缓冲液　取3.75g甘氨酸溶于水，加1mol/L HCl溶液28.3ml，用水稀释至1000ml，pH计校正。

（6）0.05% NEDA乙醇溶液

（7）0.1%亚硝酸钠溶液

（8）标准酶液（20μg/ml）　称2mg胰蛋白酶结晶，用pH 7.8磷酸缓冲液定容至100ml。

（9）鸡蛋1枚

（10）大豆胰蛋白酶抑制剂

（11）0.5%氨基磺酸铵溶液

【实验方法】

1. 亲和　将制得的固定化胰蛋白酶（制备方法见实验五）先在室温放置30分钟，抽气后装柱，用50ml pH 7.5的Tris－HCl缓冲液过柱，以平衡凝胶，同时稳定柱床，流速2ml/min，至流出液达pH 7左右。取已打匀的新鲜鸡蛋清2ml，用上述缓冲液10ml稀释，用电磁搅拌机打匀，过滤后上柱吸附，流速小于3滴/分钟。然后用相同的缓冲液洗柱，流速小于2ml/min，直至流出液在紫外分光光度计280nm处读数不大于0.02（约洗100ml）。再用pH 4.8醋酸缓冲液15ml洗柱，流速1ml/min，以除去吸附力较强的杂蛋白。

2. 洗脱　以0.05mol/L pH 2.5的甘氨酸缓冲液洗脱，流速小于5滴/分钟。用小试管收集洗脱液，每管3ml，当洗脱液体积累计达25ml以上时，以紫外分光光度计测定各管蛋白质含量（OD_{280nm}）。

3. 抑制剂活力测定　将蛋白浓度高的2～3管洗脱液合并，测A_{280}。用蒸馏水稀释10

倍，测其对胰蛋白酶的抑制活力。将大豆胰蛋白酶抑制剂2mg溶于40ml pH 7.8磷酸缓冲液作对照。蛋清溶液：鲜蛋清1ml，用生理盐水稀释20倍，电磁搅拌器打匀后，再用pH 7.8磷酸缓冲液稀释20倍。测A_{280}及胰蛋白酶抑制剂活力。取7支试管，分别按下表操作。

步骤 \ 管号	0	1	2	3	4	5	6
标准溶液，20μg/ml		0.25	0.25	0.25	0.25	0.25	0.25
标准抑制剂，50μg/ml			0.25				
蛋清，1/400				0.125	0.125		
洗脱收集液，1/10						0.125	0.25
PH 7.8 PB，ml	0.5	0.25		0.125		0.125	
2mol/L HCl，ml	0.5						
底物（BANA），ml	0.5	0.5	0.5	0.5	0.5	0.5	0.5
37℃，准确反应15分钟							
2mol/L HCl，ml	0	0.5	0.5	0.5	0.5	0.5	0.5
0.1% $NaNO_2$，ml	1	1	1	1	1	1	1
摇匀，放置3分钟以上							
0.5%氨基磺酸铵	1	1	1	1	1	1	1
摇匀，放置2分钟以上							
NEDA，ml	2	2	2	2	2	1	1
摇匀，放置30分钟							
A_{280}							
剩余活力 U/mg					1	1	1
抑制效价							

4. 结果处理

根据下面公式计算各抑制效价：

$$P_{抑} = (P_1 - P_2) \times \frac{W_1}{W_2} \text{ U/mg}$$

式中，P_1为标准胰蛋白酶效价；P_2为标准胰白酶经抑制后的剩余效价；W_1、W_2为反应中标准酶及抑制剂的重量，mg。

要求分别计算：

（1）鸡蛋清抑制效价

（2）亲和后收集液抑制效价

（3）卵类黏蛋白亲和纯化倍数

（4）抑制剂活力回收率

【思考题】

1. 亲和层析时为何要将蛋清稀释，这样是否会影响抑制剂的收率？
2. 亲和层析的洗脱除用低pH洗脱液外，还有哪些洗脱方法？

EXPERIMENT 13 Preparation of Trypsin inhibitor by affinity chromatography

【Purpose】

1. To learn to use immobilized trypsin as an affinity adsorbent to prepare the corresponding enzyme inhibitor.

2. To know the basic steps of affinity chromatography and the basic rules of washing and elution.

【Principle】

Trypsin Inhibitor is widely found in plants such as legumes, cereals, and oil crops. Trypsin inhibitors are distributed in all parts of these crops, but mainly in the seeds of crops. In seeds, trypsin inhibitors are mainly distributed in tissues or organs rich in protein, localized in protein bodies, vacuoles or present in cell fluids. For example, the content of trypsin inhibitor in soybean and mung bean seeds can reach 6% to 8% of total protein. The trypsin inhibitor iron activity varies among different crop seeds, and the trypsin inhibitory activity is usually highest in soybean.

There are two protease inhibitors in the egg white of poultry – ovomucoid and ovalbumin inhibitors. Ovomucoid inhibits the activity of trypsin in pigs, sheep and cattle, but has no inhibitory effect on human trypsin activity. Ovalbumin inhibitors inhibit the activity of trypsin and chymotrypsin in cattle and sheep. Their presence prevents the breakdown of proteins and prevents bacteria from multiplying in eggs, thus protecting egg yolks and egg embryos. At present, the commonly used protease inhibitor preparation methods include salting out, organic solvent precipitation, ion exchange and affinity chromatography, among which the affinity chromatography has the highest purification ratio.

Affinity chromatography is also calledbiospecific adsorption chromatography. It is designed according to the characteristics of the specific combination of biological macromolecules and some corresponding substances. These specifically combined substances are: antigens and antibodies, enzymes and substrates or inhibitors, complementary strands of nucleic acids, drugs and receptors. One of the substances (ligands) is bound to a suitable solid support (carrier), and a trace amount of the corresponding substance can be separated from a large amount of impurities by chromatography. Affinity chromatography has now evolved into a simple and efficient technique for the preparation and purification of a wide variety of biologically active substances.

In this experiment, agarose gel – trypsin affinity adsorbent was used to adsorb the ovomucoid in egg white on the column by affinity chromatography at a certain pH and ionic strength, and then changing the conditions to achieve the purpose of extraction and purification.

【Materials】

1. Equipment

(1) Beaker

(2) Magnetic stirrer

(3) Glass funnel

(4) Sand core funnel

(5) Suction filter bottle

(6) Measuring cylinder

(7) Thermometer

(8) Column

(9) Glass rod

(10) Small test tube and test tube rack

(11) Pipette

(12) Stopwatch

(13) Leather dropper

(14) 751 UV spectrophotometer

(15) Constant temperature water bath

(16) Constant current pump

(17) Automatic collector

2. Reagent

(1) 0.05mol/L pH 7.8 phosphate buffer

$Na_2HPO_4 \cdot 12H_2O$ (MW 358.16) 16.476g and 0.625g $NaH_2PO_4 \cdot 2H_2O$ (MW 156.03), dissolved in water, adjusted to 1000ml, pH meter calibration.

(2) 0.06% BANA ethanol solution

Make 60mg BANA and 20ml 95% ethanol dissolve, and dilute to 100ml with pH 7.8, 0.2mol/L phosphate buffer.

(3) 0.05% naphthylethylenediamine hydrochloride (NEDA) ethanol solution

(4) 0.1mol/L pH 7.5 Tris – HCl buffer (containing0.01mol/L $CaCl_2$ 0.5mol/L NaCl)

Take 12.12g of trihydroxymethyl aminomethane dissolved in water, add $CaCl_2$ 1.1g, NaCl 29.2g and 1mol/L HCl 40ml respectively, then make up to 1000ml, pH meter calibration.

(5) 0.5mol/L pH 2.5glycine buffer

3.75g of glycine dissolved in water, add 1mol/L HCl solution 28.3ml, dilute with water to 1000ml, pH meter calibration.

(6) 0.05% NEDA ethanol solution

(7) 0.1% sodium nitrite solution

(8) Standard enzyme solution (20μg/ml)

2mg trypsin crystals, make up to 100ml with pH 7.8 phosphate buffer.

(9) Egg

(10) Soybean trypsin inhibitor

(11) 0.5% ammonium sulfamate solution

【Procedures】

1. Affinity

First place the prepared immobilized trypsin at room temperature for 30 minutes, evacuate and then load with column. The column is passed through 50ml pH 7. 5 Tris – HCl buffer to equilibrate the gel while stabilizing the column bed at a flow rate of 2ml/min until the effluent reached pH 7. Take 2ml of beaten fresh egg white, then dilute it with 10ml of the above buffer, mix with a magnetic stirrer, filter and adsorb on the upper column, the flow rate should be less than 3 drops/min. Then wash the column with the same buffer at a flow rate of less than 2ml/min until the effluent is read no more than 0. 02(about 100ml) at 280nm in the UV spectrophotometer. Then the column is washed with 15ml of pH 4. 8 acetate buffer at a flow rate of 1ml/min to remove the heterologous protein with strong adsorption.

2. Elution

Elutionis carried out with a 0. 05mol/L pH 2. 5glycine buffer at a flow rate of less than 5 drops/min. Collect the eluate was collected in a small tube, 3ml per tube, and when the elution liquid accumulation amount is 25ml or more, the protein content(OD_{280nm}) of each tube is measured by an ultraviolet spectrophotometer.

3. Inhibitor activity assay

Combine two to three tube eluates with high protein concentrations and measure A_{280}. Dilute 10 times with distilled water to measure its inhibitory activity of trypsin. Dissolve 2mg soybean trypsin inhibitor in 40ml pH 7. 8 phosphate buffer as a control. Egg white solution: 1ml of fresh egg white, dilute 20 times with physiological saline, mixed well with a magnetic stirrer, and then diluted 20 times with pH 7. 8 phosphate buffer. Measure A_{280} and trypsin inhibitor activity. Take 7 tubes and operate according to the table below.

Steps \ No.	0	1	2	3	4	5	6
Standard solution, 20μg/ml		0. 25	0. 25	0. 25	0. 25	0. 25	0. 25
Standard inhibitor, 50μg/ml			0. 25				
Egg white, 1/400				0. 125	0. 125		
Elution collector, 1/10						0. 125	0. 25
PH 7. 8 PB, ml	0. 5	0. 25		0. 125		0. 125	
2mol/L HCl, ml	0. 5						
Substrate(BANA), ml	0. 5	0. 5	0. 5	0. 5	0. 5	0. 5	0. 5
37℃, React accurately for 15 minutes							
2mol/L HCl, ml	0	0. 5	0. 5	0. 5	0. 5	0. 5	0. 5
0. 1% $NaNO_2$, ml	1	1	1	1	1	1	1
Shake well and place for more than 3 minutes							
0. 5% Ammonium sulfamate	1	1	1	1	1	1	1
Shake well and place for more than 2 minutes							
NEDA, ml	2	2	2	2	2	1	1
Shake well, place for 30 minutes							
A_{280}							
Remainingactivity(U/mg)					1	1	1
Inhibition titer							

4. Result processing:

Calculate the inhibitor titers according to the following formula:

$$P_{抑} = (P_1 - P_2) \times \frac{W_1}{W_2} U/mg$$

In the formula:

P_1——Standard trypsin titer

P_2——The remaining titer after the inhibition of standard trypsin

W_1、W_2——The weight of the standard enzyme and inhibitor in the reaction, mg

Required to calculate separately:

①Egg white inhibits titer

②Collection liquid inhibition titer after affinity

③Ovum mucin affinity purification multiple

④Inhibitor activity recovery

【Questions】

1. Why should the egg white be diluted during affinity chromatography? Will this affect the yield of the inhibitor?

2. What is the elution method for the affinity chromatography except for the low pH eluent?

实验十四　超滤法制备胸腺肽

扫码“学一学”

扫码“看一看”

【实验目的】

1. **掌握**　超滤法的原理。

2. 学习用超滤法制备胸腺肽的方法。

【实验原理】

超滤技术（Ultrafiltration technology）是最近几十年迅速发展起来的一项分子级薄膜分离手段。它以特殊的超滤膜为分离介质，以膜两侧的压力差为推动力，将不同分子量的物质进行选择性分离。超滤膜在生物制药中可用来分离蛋白质、核酸、多糖、多肽、抗生素、病毒等。超滤的优点包括没有相转移，无需添加任何强烈化学物质，可以在低温下操作，过滤速率较快，便于做无菌处理等。所有这些都能使分离操作简化，避免了生物活性物质的活力损失和变性。

鉴于超滤技术有以上诸多优点，故常被用作：大分子物质的脱盐和浓缩，以及大分子物质溶剂系统的交换平衡；小分子物质的纯化；大分子物质的分级分离；生化制剂或其他制剂的去热原处理。超滤技术已成为制药工业、食品工业、电子工业以及环境保护诸领域中不可缺少的有力工具。

胸腺肽（Thymus peptides）是从冷冻的小牛（或猪、羊）胸腺中，经提取、部分热变性、超滤等工艺过程制备出的一种具有高活力的混合肽类药物制剂。十二烷基硫酸钠聚丙

烯酰胺凝胶电泳（SDS－PAGE）分析表明，胸腺肽主要是分子量9600Da和7000Da左右的两类蛋白质或肽类，氨基酸组成达15种，且必需氨基酸含量高，还含有RNA 0.2～0.3mg/ml，DNA 0.12～0.18mg/ml；对热较稳定，80℃加热生物活性不降低，但蛋白水解酶作用会使其生物活性消失。

胸腺肽可调节细胞免疫功能，有较好的抗衰老和抗病毒作用，适用于原发和继发性免疫缺陷病以及因免疫功能失调所引起的疾病，对肿瘤有很好的辅助治疗效果，也用于再生障碍性贫血、急慢性病毒性肝炎等。无过敏反应和不良的副作用。

本实验采用小牛胸腺为原料，经过热变性、过滤等操作，最后用超滤技术制备胸腺肽，并用SDS－PAGE检测胸腺肽的纯度。

【实验材料】

1. 材料

小牛胸腺

2. 器材

（1）绞肉机		1台
（2）组织捣碎机		1台
（3）剪刀		1只
（4）烧杯	500ml，250ml，200ml	各1个
（5）超滤器		1台
（6）布氏漏斗		1只
（7）恒温水浴		1台
（8）低温冰箱		1台
（9）冻干机		1台
（10）灭菌锅		1台

【实验方法】

1. 预处理及细胞破碎 取－20℃冷藏小牛胸腺100g，用无菌剪刀剪去脂肪、筋膜等非胸腺组织，再用冷无菌蒸馏水冲洗，置于灭菌绞肉机中绞碎。

2. 制匀浆、提取 将绞碎胸腺与冷蒸馏水按1∶1的比例混合，置于10000r/min的高速组织捣碎机中捣碎1分钟，制成胸腺匀浆。浸渍提取，温度应在10℃以下，并放置－20℃冰冻贮藏48小时。

3. 部分热变性、离心、过滤 将冻结的胸腺匀浆融化后，置水浴上搅拌加温至80℃，保持5分钟，迅速降温，放置－20℃以下冷藏2～3天。然后取出融化，以5000r/min离心40分钟，温度2℃，收集上清液，除去沉渣，用滤纸浆或微孔滤膜（水相，0.22μm）减压抽滤，得澄清滤液。

4. 超滤、冻干 将滤液用分子量截流值为1万Da以下的超滤膜进行超滤，收取分子量1万Da以下的活性多肽，得精制液，冻干。

5. 纯度测定 用SDS－PAGE电泳检测（参考实验九）。

扫码“练一练”

【思考题】

1. 超滤技术的特点？说明本实验使用超滤的理由？

2. 胸腺肽临床应用有哪些?

EXPERIMENT 14 Preparation of thymus peptides by ultrafiltration

【Purpose】

1. To master the principle of ultrafiltration
2. To learn how to prepare thymosin by ultrafiltration

【Principle】

Ultrafiltration technology is a molecular - scale membrane separation method that has developed rapidly in recent decades. This technology can separate substances of different molecular weights selectively by the special ultrafiltration membrane and the pressure difference on both sides of the membrane. Ultrafiltration can be used to separate proteins, nucleic acids, polysaccharides, peptides, antibiotics, viruses, and other macromolecules in biopharmaceuticals. The advantages of ultrafiltration include no phase transfer, no need to add any strong chemicals, can be operated at low temperatures. Besides, the filtration is fast and the aseptic processing is convenient according to the technology. The separation operation is easy, and loss of vitality and denaturation of the bioactive materials could be avoided.

In view of those advantages of ultrafiltration technology, it is often used as: desalting and concentrating the macromolecular substances, exchanging equilibrium of solvent systems of macromolecular substances; purifying of small molecular substances; fractionating of macromolecular substances; depyrogenating of biochemical preparations. Ultrafiltration technology has become an indispensable tool in the pharmaceutical, food, electronics, and environmental protection fields.

Thymus peptides are high - activity mixed peptide pharmaceutical preparations prepared from the thymus of frozen calves (or pigs, sheep) by extraction, partial heat denaturation, ultrafiltration, and other processes. SDS - PAGE analysis showed that thymosin contained mainly two kinds of proteins or peptides with molecular weights of 9600Da and 7000Da, respectively. Besides, the amino acids composition of thymosin is up to 15 and the content of essential amino acids is high. Thymus peptides also contain 0.2 - 0.3mg/ml of RNA, 0.12 - 0.18mg/ml of DNA and is stable to heat. The biological activity of thymus peptides does not decrease at 80℃, but the action of proteolytic enzymes will make its biological activity disappear.

Thymosin can regulate cellular immune function, has good anti - aging and anti - viral effects, is suitable for primary and secondary immunodeficiency diseases and diseases caused by immune dysfunction, has a good auxiliary therapeutic effect on tumors and can be used for aplastic anemia, acute and chronic viral hepatitis. Thymosin has no allergic reactions and adverse side effects.

In this experiment, the calf thymus is used as raw material, and subjected to thermal denaturation, filtration, and finally thymosin will be prepared by ultrafiltration. The purity of the thymosin is measured by SDS - PAGE.

【Materials】

1. Calf thymus

2. Equipment

(1) Meat grinder

(2) Tissue masher

(3) scissors

(4) beakers

(5) Ultrafilter

(6) Buchner funnel

(7) Water bath kettle

(8) Low temperature refrigerator

(9) Freeze dryer

(10) Sterilizer Pot

【Procedures】

1. Tissue pretreatment

Take –20℃ chilled calf thymus 100g, use the sterile scissor to cut off non – thymus tissue such as fat, fascia, and then place the left tissue in the sterile meat grinder to grind after rinsing with cold sterile distilled water.

2. Tissue homogenization

Mix the ground thymus with cold distilled water in a ratio of 1 : 1, and then mash them in a high – speed tissue masher at 10000r/min for 1 minute to prepare thymus homogenate. Dip extraction under the temperature that should be below 10℃, and then store them at –20℃ for 48 hours.

3. Partial thermal denaturation, centrifugation, filtration

Melt the frozen thymus homogenate, stir, and heat the homogenate to 80℃ in a water bath for 5 minutes. Cool rapidly, and store the sample under –20℃ for 2 to 3 days. Then, the homogenate will be taken out, thawed, and centrifuged at 5000r/min for 40 minutes at a temperature of 2℃. Collect the supernatant to remove the sediment, and filter with filter paper or microporous membrane (aqueous phase, 0. 22μm) under reduced pressure to obtain a clear filtrate.

4. Ultrafiltration and lyophilization

Ultrafilter the filtrate with an ultrafiltration membrane with a molecular weight cutoff of 10000Da or less, collect the bioactive polypeptide with a molecular weight of 10000Da or less, lyophilize.

5. Determination of purity

SDS – PAGE (referring experiment 9).

【Questions】

1. What are the characteristics of ultrafiltration technology? Explain why ultrafiltration can be used.

2. What are the clinical applications of thymosin?

第二部分　生物药物制备综合性实验

第一节　天然生物药物的制备

实验十五　固定化细胞法生产L－天冬氨酸和L－丙氨酸

扫码“学一学”

【实验目的】

1. **掌握**　固定化细胞的技术。
2. **了解**　酶法生产这两种氨基酸的原理和细胞固定化的原理及优点。
3. 学习L－天冬氨酸和L－丙氨酸的制备方法。

【实验原理】

固定化细胞（Immobilized cell）技术，就是利用物理或化学手段将游离的微生物细胞、动物细胞，定位于限定的空间领域，并使其保持活性且能反复利用的一项技术。

固定化细胞的制备方法主要有以下几种：吸附法、共价交联法、絮凝法、包埋法。吸附法是细胞通过静电相互作用（范德华力、离子键和氢键）吸附到支持物表面，此法简单价廉，但经常发现有细胞泄露；交联可以是细胞通过离子相互作用或共价连接到一个表面，也可以是细胞与细胞之间用过天然或化学试剂诱导产生连接；絮凝法是利用某些微生物细胞具有絮凝形成颗粒的能力而对细胞进行固定化的方法；包埋法是近年来发展迅速的一种新兴固定化细胞技术，它是将细胞捕获在一个保护性基质结构或胶囊中，减少了细胞泄露，因此具有操作简单、对细胞活性影响较小、效率高等特点，是目前细胞固定化研究和应用最广泛的方法之一。

L－天冬氨酸（L－Aspartic acid）是天然存在的重要氨基酸，在食品、医药、日用化工、纺织等行业有着广泛的应用。L－天门冬氨酸钾、镁盐在细胞代谢中起着重要作用，是钾、镁的有效补充剂，适用于各种心脏病。在食品方面，L－天冬氨酸作为食品添加剂可改善食品风味，L－天冬氨酸与D－丙氨酸可合成L－天冬酰－D－丙氨酸甲酯（其甜度为蔗糖的150倍），是一种新型甜味剂。在化工方面，L－天冬氨酸还可以作为制造合成树脂的原料，其衍生物还可以合成量型表面活性剂以及制造化妆品。

工业上生产L－天冬氨酸以往采用化学合成法和微生物发酵法。随着固定化酶和固定化细胞技术的不断完善和发展，人们开始改用含有天冬氨酸酶的固定化细胞直接将延胡索酸铵转化成L－天冬氨酸来实现工业化生产。

L－丙氨酸（L－Alanine）是一种脂肪族的非极性氨基酸，是丙酮酸代谢体系的非必需氨基酸，在血液中含量最多，与糖代谢密切相关，使转氨反应中重要的氨基供体，具有重要的生理功能；同时又是一种重要的氨基酸类药物，为多种复方氨基酸输液的重要组成成分，并可作多种医药中间体。

本实验用卡拉胶包埋天冬氨酸酶产生菌和L－天冬氨酸β－脱羧酶产生菌制成固定化细

胞，利用生物反应器，可连续生产 L－天冬氨酸和 L－丙氨酸。

$$\underset{\text{延胡索酸（富马酸）}}{HOOC-CH=CH-COOH}+NH_4^+ \xrightleftharpoons{\text{天冬氨酸酶}} \underset{\text{L－天冬氨酸}}{HOOC-CH_2-CHNH_2-COOH}$$

$$\underset{\text{L－天冬氨酸}}{HOOC-CH_2-CHNH_2-COOH} \xrightleftharpoons{\text{L－天冬氨酸 β－脱羧酸}} \underset{\text{L－丙氨酸}}{CH_3CH_2-CHNH_2-COOH}$$

【实验材料】

1. 器材

（1）水浴振摇培养箱		1 台
（2）磁力搅拌器		1 台
（3）HPLC		1 台
（4）三角烧瓶	250ml	5 个
（5）抽滤瓶		1 个
（6）离心机		1 台
（7）布式漏斗		2 只
（8）恒温水浴锅		1 台
（9）酶柱		1 根
（10）烧杯	1000ml	1 个
（11）烧杯	500ml	2 个
（12）烧杯	250ml	2 个
（13）超级恒温水浴		1 台
（14）恒流泵		1 台
（15）高压灭菌锅		1 个
（16）锋利小刀		1 把
（17）试管		12 支

2. 试剂

（1）菌种

（2）玉米浆

（3）延胡索酸铵

（4）PLP（5－磷酸吡哆醛）

（5）牛肉浸膏

（6）KH_2PO_4

（7）$MgSO_4 \cdot 7H_2O$

（8）浓氨水

（9）氯化钾

（10）氯化镁

（11）浓硫酸

（12）95% 乙醇

（13）L－天冬氨酸标准品

（14）延胡索酸标准品

（15）L－丙氨酸标准品

【实验方法】

1. 细菌的培养及菌体制备 接种1ml对数生长期天冬氨酸酶产生菌种至250ml三角烧瓶中，内含50ml已经灭菌的培养基（1%延胡索酸铵，2%玉米浆，2%牛肉浸膏，0.5% KH_2PO_4，0.05% $MgSO_4 \cdot 7H_2O$，pH 7.0），37℃振摇培养24小时，培养液离心（3000r/min，20分钟）倾出上清液，再用生理盐水洗涤1次，离心收集菌体，置冰箱备用。

接种1ml对数生长期L－天冬氨酸β－脱羧酶产生菌种至250ml三角烧瓶中，内含50ml已灭菌的培养基（1.5% L－谷氨酸钠，0.5%富马酸铵，1%富马酸钠，3.0%玉米浆，2%蛋白胨，0.05% KH_2PO_4，0.01% $MgSO_4 \cdot 7H_2O$，pH 7.0），37℃振摇培养24小时，培养液离心（3000r/min，20分钟）倾出上清液，再用生理盐水洗涤一次，离心收集菌体，置冰箱备用。

2. 固定化细胞的制备

（1）天冬氨酸酶产生菌的固定化 将4g湿菌体悬浮于4ml生理盐水中，置45℃恒温水浴保温，0.8g卡拉胶于17ml生理盐水中，加热至70～80℃使之溶解，再降温至45℃，两者于45℃混匀，混合物置4℃冰箱放置30分钟，将凝胶在100ml 0.3mol/L KCl溶液中浸泡4小时，然后切成3mm×3mm×3mm的立方体颗粒。经生理盐水洗涤后，将固定化细胞置于1mol/L延胡索酸铵中，于37℃活化24小时，即可使用。

（2）L－天冬氨酸β－脱羧酶产生菌的固定化 将5g湿菌体悬浮于5ml生理盐水中，置45℃恒温水浴保温，0.8g卡拉胶于17ml生理盐水中，加热至70～80℃使之溶解，降温至45℃，两者于45℃混匀，混合物置4℃冰箱放置30分钟，将凝胶在100ml 0.3mol/L KCl溶液中浸泡4小时，然后切成3mm×3mm×3mm的立方体颗粒。用0.1mol/L甘氨酸充分洗涤，在1mol/L天冬氨酸铵（含0.1mol/LPLP）底物中，于37℃活化24小时，即可使用。

3. 酶活力测定

（1）湿菌体天冬氨酸酶活力的测定 取0.5g湿细胞悬浮于2ml蒸馏水中，加入30ml 1.0mol/L延胡索酸铵，内含1mmol/L $MgCl_2$，1% Triton pH 9.0，37℃搅拌反应30分钟，煮沸终止反应，1200r/min，5分钟离心后，稀释反应上清液于240nm处测定吸收值，按延胡索酸残余量计算酶活力，一个酶活力单位定义为在测定条件下，每小时每克细胞转化生成1μmol天冬氨酸所需的酶量。

（2）固定化细胞的天冬氨酸酶活力的测定 取相当于0.5g天然细胞的固定化细胞，置于30ml 1.0mol/L延胡索酸铵，内含1mmol/L $MgCl_2$，37℃搅拌反应30分钟，迅速过滤除去固定化细胞，分离反应液，经稀释后于240nm处测定延胡索酸残余量，并计算每克固定化细胞的酶表现活力。总活力用每克固定化细胞，单位小时内所产生的L－天冬氨酸的微摩尔数表示（μmol/h·g·cell）。

（3）延胡索酸（即富马酸）标准曲线的绘制 延胡索酸在240nm处有特征峰吸收，在一定范围内与延胡索酸含量成线性关系，且反应系统中无干扰。

标准曲线的绘制

试管号	1	2	3	4	5	6
延胡索酸标准液（50μg/ml）	0	1	2	3	4	5
蒸馏水	5	4	3	2	1	0
$OD_{240\ nm}$						

根据测定的吸收值，作出标准曲线或回归方程。

（4）湿菌体 L－天冬氨酸 β－脱羧酶产生菌活力的测定　取 1g 湿菌体悬浮于 10ml 生理盐水中，取 0.5ml 悬浮液，加入含 0.1mmol/L PLP 的 1mol/L 天冬氨酸铵的底物 2ml（pH 6.0）于 37℃反应 30 分钟，煮沸终止反应。生成的 L－丙氨酸用纸层析法定量测定。

展开剂选用正丁醇：醋酸：水＝4：1：1，显色剂用 0.5% 茚三酮溶液，烘干后，将显色斑点剪下用 0.1% $CuSO_4 \cdot 5H_2O$：75% 乙醇＝2：38 洗脱后，于 520nm 处比色，再从标准曲线上读得 L－丙氨酸的含量。L－天冬氨酸 β－脱羧酶产生菌活力的定义为：每克细胞每小时生成 1μmol L－丙氨酸为 1 个酶活力单位（U）。

（5）固定化细胞的 L－天冬氨酸 β－脱羧酶活力的测定　取 6g 固定化细胞（相当于 1g 的湿细胞）加入 20ml 生理盐水，于 37℃保温，加入含 0.1mmol/L PLP 的 1mol/L L－天冬氨酸铵 4ml，pH 6.0，于 37℃反应 30 分钟，迅速过滤除去固定化细胞，分离反应液，生成的 L－丙氨酸用纸层析法定量测定。

（6）L－丙氨酸标准曲线的绘制　按（5）法用纸层析法定量测定 L－丙氨酸的含量与显色斑点脱色后在 520nm 处比色的关系。

标准曲线的绘制

试管号	1	2	3	4	5	6
L－丙氨酸标准液（40μg/ml）	0	1	2	3	4	5
蒸馏水	5	4	3	2	1	0
$OD_{520\ nm}$						

根据测定的吸收值，作出标准曲线或回归方程。

4. L－丙氨酸的制备　分别将 6g 天冬氨酸酶和 L－天冬氨酸 β－脱羧酶的固定化细胞装入带夹套的固定化生物反应器内（Φ1.5cm×30cm），1.0mol/L 延胡索酸铵（含 1mmol/L $MgCl_2$，pH 9.0）底物，以恒流速度通过天冬氨酸酶固定化细胞柱，SV＝0.87/h，然后将流出液通过 L－天冬氨酸 β－脱羧酶固定化细胞柱，并加 0.1mmol/L PLP 和 28% 的氨水，使 pH 6.0，流出液用纸层析法计算 L－丙氨酸的量，并计算转化率（L－丙氨酸转化率＝L－丙氨酸浓度/延胡索酸铵浓度×100%）。

【思考题】

扫码“练一练”

1. 分别求出湿菌体天冬氨酸酶活力、固定化细胞的天冬氨酸酶活力及湿菌体 L－天冬氨酸 β－脱羧酶产生菌活力、固定化细胞的 L－天冬氨酸 β－脱羧酶活力。
2. 分别求出两种菌体包埋后的活力回收率。
3. 分别求出 L－天冬氨酸和 L－丙氨酸的转化率。

Part 2 Preparation experiments of biopharmaceutical drugs

Section 1 Preparation of natural biopharmaceutical drugs

EXPERIMENT 15 Production of L - Aspartic acid and L - Alanine by immobilized cell method

【Purpose】

1. To master the technology of immobilized cells.

2. To know the principles of enzymatic production of these two amino acids and the principles and advantages of cell immobilization.

3. To learn the preparation method of L - aspartic acid and L - alanine.

【Principle】

Immobilized cell technology is a technique that uses physical or chemical means to localize free microbial cells and animal cells in a defined spatial domain, and to keep it active and reusable.

There are mainly 4 methods for preparing immobilized cells: adsorption method, covalent cross - linking method, flocculation method, and embedding method. The adsorption method is that the cells are adsorbed to the surface of the support by electrostatic interaction (van der Waals force, ionic bond and hydrogen bond), which is simple and inexpensive, but cell leakage is often found. Cross - linking may be through cell interaction or covalent attachment to a surface, cells are induced to be connected by natural or chemical agents. Flocculation is a method of immobilizing cells by utilizing the ability of certain microbial cells to flocculate to form particles. The embedding method is an emerging immobilized cell technology that has been rapidly developed in recent years. It captures cells in a protective matrix. In the structure or capsule, the cell leakage is reduced, so that it has the characteristics of simple operation, less influence on cell activity, high efficiency. The embedding method is one of the most extensive methods for cell immobilization research and application.

L - Aspartic acid is an important amino acid that exists naturally and has a wide range of applications in food, pharmaceutical, household chemicals, textile and other industries. Potassium and magnesium salts of L - Aspartate play an important role in cell metabolism and are effective supplements for potassium and magnesium, which are suitable for various heart diseases. In food, L - Aspartic acid as a food additive can improve the flavor of food, L - aspartic acid and D - alanine can synthesize L - Aspartyl - D - alanine methylester (its sweetness is 150 times of sucrose) is a new type of sweetener. In the chemical industry, L - Aspartic acid can also be used as a raw material for the manufacture of synthetic resins, and derivatives can also synthesize a type of surfactant and man-

ufacture a cosmetic.

Industrial production of L – Aspartic acid has previously been carried out by chemical synthesis and microbial fermentation. With the continuous improvement and development of immobilized enzyme and immobilized cell technology, people began to use the immobilized cellscontaining aspartase to directly convert ammonium fumarate into L – Aspartic acid to achieve industrial production.

L – Alanine is an aliphatic non – polar amino acid. It is a non – essential amino acid in the pyruvate metabolism system. It is the most abundant in the blood and is closely related to sugar metabolism, making it important in the transamination reaction. Amino donor has important physiological functions; at the same time, it is an important amino acid drug, an important component of a variety of compound amino acid infusions and can be used as a variety of pharmaceutical intermediates.

In this experiment, carrageenan – incorporated aspartase – producing bacteria and L – Aspartate β – decarboxylase – producing bacteria will be used to prepare immobilized cells, and L – Aspartic acid and L – Alanine could be continuously produced by using a bioreactor.

$$\underset{\text{Fumaric acid}}{HOOC-CH=CH-COOH} + NH_4^+ \xrightleftharpoons{\text{Aspartase}} \underset{\text{L – Aspartic acid}}{HOOC-CH_2-CHNH_2-COOH}$$

$$\underset{\text{L – Aspartic acid}}{HOOC-CH_2-CHNH_2-COOH} \xrightleftharpoons{\text{L – Aspartate}\beta\text{ – decorboxylase}} \underset{\text{L – Alanine}}{CH_3CH_2CHNH_2-COOH}$$

【Materials】

1. Apparatus

(1) Water bath shaking incubator

(2) Magnetic stirrer

(3) HPLC

(4) Triangular flask (250ml)

(5) suction filter bottle

(6) Centrifuge

(7) Cloth funnel

(8) Constant temperature water bath

(9) Enzyme column

(10) Beaker (1000ml)

(11) Beaker (500ml)

(12) Beaker (250ml)

(13) Super constant temperature water bath

(14) Constant current pump

(15) Autoclave

(16) sharp knife

(17) test tube

2. Reagents

(1) strain

(2) corn syrup

(3) Ammonium fumarate

(4) PLP(pyridoxal 5 – phosphate)

(5) Beef extract

(6) KH_2PO_4

(7) $MgSO_4 \cdot 7H_2O$

(8) concentrated ammonia

(9) Potassium chloride

(10) Magnesium chloride

(11) concentrated sulfuric acid

(12) 95% ethanol

(13) L – aspartic acid standard

(14) fumaric acid standard

(15) L – alanine standard

【Procedures】

1. Bacterial culture and bacterial preparation

Inoculate 1ml logarithmic growth phase aspartase producing strain into a 250ml Erlenmeyer flask containing 50ml of sterilized medium(1% ammonium fumarate, 2% corn syrup, 2% beef extract, 0.5% KH_2PO_4, 0.05% $MgSO_4 \cdot 7H_2O$, pH 7.0), shake at 37℃ for 24 hours, centrifuge the culture solution(3000 r/min, 20 minutes), pour the supernatant, wash once with physiological saline, collect the cells by centrifugation, and place in the refrigerator.

Inoculate 1ml logarithmic growth phase L – aspartate β – decarboxylase producing strain into a 250ml erlenmeyer flask containing 50ml of sterilized medium(1.5% L – glutamate, 0.5% ammonium fumarate, 1% sodium fumarate, 3.0% corn syrup, 2% peptone, 0.05% KH_2PO_4, 0.01% $MgSO_4 \cdot 7H_2O$, pH 7.0), shake at 37℃ for 24 hours, centrifuge(3000 r/min, 20 minutes). Decant the supernatant, wash once with physiological saline, collect the cells by centrifugation and place in a refrigerator for use.

2. Immobilized cell preparation

(1) Immobilization of aspartase producing bacteria

Suspend 4g of wet cells in 4ml of physiological saline, keep in a constant temperature water bath at 45℃, 0.8g carrageenan in 17ml normal saline, heat to 70 – 80℃ to dissolve, and then cool to 45℃, both at 45℃ After mixing, place the mixture in a refrigerator at 4℃ for 30 minutes, and immerse the gel in 100ml of 0.3mol/L KCl solution for 4 hours, and then cut into cube pieces of 3mm × 3mm × 3mm. After washing with physiological saline, place the immobilized cells in 1mol/L ammonium fumarate and activated at 37℃ for 24 hours to be used.

(2) Immobilization of L – aspartate β – decarboxylase producing bacteria

Suspend 5g of wet cells in 5ml of physiological saline, keep in a constant temperature water bath at 45℃, 0.8g carrageenan in 17ml normal saline, heat to 70 – 80℃ to dissolve, cool to 45℃, the two mixed at 45℃. Place the mixture in a refrigerator at 4℃ for 30min, immerse the gel in 100ml of 0.3mol/L KCl solution for 4 hours, and then cut into cubes of 3mm × 3mm × 3mm. Wash

thoroughly with 0. 1mol/L glycine, and activated at 37℃ for 24 hours in 1mol/L ammonium aspartate (containing 0. 1mol/L PLP) substrate.

3. Enzyme activity assay

(1) Determination of wet cell aspartase activity

In 2ml of distilled water, suspend 0. 5g wet cells, add 30ml 1. 0mol/L ammonium fumarate, containing 1mmol/L $MgCl_2$, 1% Triton pH 9. 0, stirred at 37℃ for 30 minutes, boil to terminate the reaction, 1200r/min, 5 minutes after centrifugation, dilut the reaction supernatant, measure for absorbance at λ240nm, and calculate the enzyme activity using the residual amount of fumaric acid. One unit of enzyme activity is defined as the amount of enzyme required to produce 1μmol of aspartic acid per gram of cells per hour under the assay conditions.

(2) Determination of aspartase activity of immobilized cells

Place the immobilized cells corresponding to 0. 5g of natural cells in 30ml of 1. 0mol/L ammonium fumarate, containing 1mmol/L $MgCl_2$, and stirred at 37℃ for 30 minutes. Quickly remove the immobilized cells by filtration, separate and dilute the reaction solution. Measure the residual amount of fumaric acid at λ240nm and calculate the enzyme performance vigor per gram of the immobilized cells. The total viability is expressed in micromoles of L - aspartic acid produced per unit of immobilized cells per unit hour (μmol/h · g · cell).

(3) Drawing of the standard curve of fumaric acid

The fumaric acid has a characteristic peak absorption at λ240nm, which is linear with the fumaric acid content within a certain range, and there is no interference in the reaction system.

Drawing of standard curve

Test tube number	1	2	3	4	5	6
Fumaric acid standard solution (50μg/ml)	0	1	2	3	4	5
Distilled water	5	4	3	2	1	0
OD_{240nm}						

A standard curve or regression equation is made based on the measured absorbance values.

(4) Determination of viable activity of L - aspartate β - decarboxylase producing bacteria in wet cells

In 10ml of physiological saline suspend 1g of wet cells, take 0. 5ml of the suspension, and add 2ml (pH 6. 0) of a substrate of 1mol/L ammonium aspartate containing 0. 1mmol/L of PLP to react at 37℃ for 30 minutes, and boil. Stop the reaction. The resulting L - alanine is quantitatively determined by paper chromatography.

The developing agent is n - butanol : acetic acid : water = 4 : 1 : 1, and the developer is 0. 5% ninhydrin solution. After drying, the color spots are cuthed, after elution with 0. 1% $CuSO_4 \cdot 5H_2O$: 75% ethanol = 2 : 38, the color is measured at λ520nm, and read the content of L - alanine from the standard curve. The activity of L - aspartate β - decarboxylase - producing bacteria is defined as: 1μmol L - alanine per gram of cells per hour is an enzyme activity unit (U).

(5) Determination of L - aspartate β - decarboxylase activity in immobilized cells

Take 6g of immobilized cells (corresponding to 1g of wet cells), add 20ml of normal saline, incubate at 37℃, add 4ml of 1mol/L L – aspartic acid ammonium containing 0.1mmol/L PLP, pH 6.0, at 37℃. After reacting for 30 minutes, quickly remove the immobilized cells by filtration, prepare the reaction solution, and determine the resulting L – alanine quantitatively by paper chromatography.

(6) Drawing of L – alanine standard curve

According to the method (5), the relationship between the content of L – alanine and the coloration at λ520nm after decolorization of the colored spots is quantitatively determined by paper chromatography.

Drawing of standard curve

Test tube number	1	2	3	4	5	6
L – alanine standard solution (40μg/ml)	0	1	2	3	4	5
Distilled water	5	4	3	2	1	0
$OD_{520\ nm}$						

A standard curve or regression equation is made based on the measured absorbance values.

4. Preparation of L – alanine

Place 6g the immobilized cells containing aspartase and L – aspartate β – decarboxylase in a jacketed immobilized bioreactor (Φ1.5cm × 30cm), 1.0mol/L of ammonium fumarate (A substrate containing 1mmol/L $MgCl_2$, pH 9.0), immobilize on a column of cells by aspartase at a constant flow rate, SV = 0.87/h, and then pass the effluent through L – aspartate β – decarboxylase. The cell column is immobilized. Add 0.1mmol/L PLP and 28% ammonia water to make pH 6.0. The amount of L – alanine is calculated by paper chromatography, and the conversion rate is calculated (L – alanine conversion). Rate = L – alanine concentration/ammonium fumarate concentration × 100%).

【Questions】

1. Calculate the wet cell aspartase activity, the aspartase activity of the immobilized cells, the L – aspartate β – decarboxylase – producing bacteria activity of the wet cells, and the beta – decarboxylase activity of the immobilized cells.
2. Determine the viability recovery rate of the two cells after embedding.
3. The conversion rates of L – aspartic acid and L – alanine were determined separately.

扫码“学一学”

实验十六 胰岛素的制备及含量测定

【实验目的】

扫码“看一看”

1. 掌握 胰岛素的酶联免疫法检测方法。
2. 了解 胰岛素的制备方法。

【实验原理】

胰岛素（Insulin）是体内一类重要的多肽激素，是由胰岛 β 细胞受内源性或外源性物质如葡萄糖、乳糖、核糖、精氨酸等的刺激而分泌的一种蛋白质激素，是机体内唯一降低

血糖的激素，同时促进糖原、脂肪、蛋白质合成。胰岛素在水、乙醇、氯仿或乙醚中几乎不溶；在矿酸（无机酸）或氢氧化碱溶液中易溶。同时它也是临床上治疗糖尿病的一线药物。

酶联免疫法（Enzyme－linked immunosorbent assays，ELISA）的原理：被测药物胰岛素先与固定相上的抗体形成复合物，再与第二抗体结合，第二抗体是用可以与底物发生显色反应的酶来标记，根据酶催化反应产物与胰岛素之间量的比例关系来定量胰岛素。ELISA已成为胰岛素测定的主要方法，该方法具有使用寿命长、操作简便快速、重复性好、无辐射源的优点，并且已有不少实验证明，它与生物检定法具有一定的量效关系及相关性，说明它可部分地反映药物的生物活性。

本实验以新鲜猪胰脏为原料，采用酸－醇提取减压浓缩法提取胰岛素粗品，再经酶联免疫法检测胰岛素含量。

【实验材料】

1. 器材

（1）组织捣碎机		1台
（2）电动搅拌器		1台
（3）旋转蒸发仪		1台
（4）酶标仪		1台
（5）剪刀		1把
（6）烧杯	800ml	1只
	400ml	1只
（7）量筒	500ml	1只
	100ml	1只
（8）真空干燥器及真空泵		1套
（9）纱布		1卷
（10）离心机		1台

2. 试剂

（1）86%乙醇溶液

（2）5%草酸溶液

（3）6mol/L的硫酸溶液

（4）4mol/L的氨水溶液

（5）氯化钠固体

（6）4mol/L盐酸溶液

（7）猪胰岛素ELISA检测试剂盒

【实验方法】

1. 预处理与提取　取冻胰脏，剪去脂肪，切成小块，称取100g，用组织捣碎机刨碎，加86%乙醇100ml和5%草酸100ml再加硫酸调pH（2.2～3.0），在10～15℃搅拌提取2小时，用纱布过滤，留滤液。

2. 蛋白沉淀　在10～15℃下不断搅拌提取液并加入浓氨水调pH 8.0～8.4后，立即在

4000r/min 离心 10 分钟，除去碱性蛋白，取上清液，用硫酸酸化至 pH 3. 6 ~ 3. 8，5℃静置 2 小时，使酸性蛋白充分沉淀。

3. 浓缩 将溶液用离心机 4000r/min 离心 10 分钟，取上清液加入到蒸馏瓶中（350ml/瓶），35℃左右在旋转蒸发仪中除去乙醇（时间约为 1 个小时），浓缩液体积约为原体积的 1/2。

4. 成品收集 浓缩液用 4mol/L 的 HCl 调至 pH 2. 3 ~ 2. 5，并于 20 ~ 25℃搅拌下加入 23%（W/V）（kg/L）固体氯化钠，室温静置 12 分钟。析出物为胰岛素粗品（含水量约为 40%）。4000r/min 离心 10 分钟收集胰岛素粗品，用 10 倍体积的冷丙酮脱水，再放于真空干燥箱中用 P_2O_5 干燥得胰岛素白色结晶。

5. 含量测定 参照 ELISA 试剂盒使用说明书操作

扫码“练一练”

【思考题】

1. 测定胰岛素活性的方法有哪些?
2. 影响胰岛素活性的因素有哪些?

EXPERIMENT 16 Preparation of insulin from pancreas and content determination

【Purpose】

1. To master the principle of ELISA.
2. To know the principle of preparation of Insulin from pancreas.

【Principle】

Insulin is an important polypeptide hormone in the body. It is a protein hormone secreted by islet β cells stimulated by endogenous or exogenous substances such as glucose, lactose, ribose, arginine, etc. It is the only hormone in the body to reduce blood glucose and promote the synthesis of glycogen, fat and protein. Insulin is almost insoluble in water, ethanol, chloroform or ether. Soluble in mineral acid(inorganic acid) or alkaline hydroxide solution. At the same time, it is also a first - line drug for clinical treatment of diabetes.

Enzyme - linked immunosorbent assays(ELISA) principle: the tested insulin first forms a complex with the antibody in the fixed phase, and then binds with the second antibody. The second antibody is labeled with enzymes that can react with the surface to color, and the insulin is quartified according to the proportion between the enzyme catalyzed reaction product and the amount of insulin. ELISA has become the main method for the determination of insulin, which has the advantages of long service life, simple operation, good repeatability and no radiation. It has been proved by many experiments that it has a dose effect relationship with biological assay. It shows that it can pantically reflect the biological activity of drugs.

In this study, fresh pig pancreas will be used as raw material to extract crude insulin by acid-alcohol extration and reduced pressure concentration, and then content determination by ELISA.

【Materials】

1. Apparatus

(1) Organization stamp machine

(2) Electric blender

(3) Rotary evaporation apparatus

(4) Microplate reader

(5) Scissors

(6) Beaker

(7) Measuring cylinder

(8) Vacuum drier and vacuum pump

(9) Gauze volume

(10) Centrifuge

2. Reagents

(1) 86% ethanol solution

(2) 5% oxalic acid solution

(3) 6mol/L sulfuric acid solution

(4) 4mol/L ammonia solution

(5) Sodium chloride solid

(6) 4mol/L HCl solution

(7) Pig insulin ELISA kit

【Procedures】

1. The preprocessing and extraction

Remove fat of frozen pancreatic and cut into small pieces, then take 100g. Mash it withorganization stamp machine, then add 100ml 86% ethanol and 100ml 86% oxalic acid, use sulfuric acid to adjust pH(2.2 ~ 3.0). After stirring extraction 2 hours in 10 to 15℃, remove precipitation by gauze filtration.

2. Protein precipitation

Adjust pH to 8.0 – 8.4 under stirring constantly and then centrifuge at 4000r/min for 10 minutes to immediately remove the alkaline protein. Take supernatant, acidification with sulfuric acid to pH 3.6 ~ 3.8, and then place at 5℃ for 2 hours, make the acidity protein precipitation adequately.

3. The enrichment

Centrifuge solution at 4000r/min for 10 minutes and take the supernatant in a distillation bottle (350ml). Remove ethanol by the rotary evaporation apparatus at about 35℃ (time is about 1 hour), concentrate volume is about 1/2 of the original volume.

4. The finished product collection

Concentrate is adjusted to pH 2.3 ~ 2.5 in 4mol/L HCl, and then add 23% (W/V) of solid sodium chloride under stir in 20 to 25℃. Place it at room temperature overnight. Deposition is insulin products (water content about 40%). Collect raw insulin by centrifuge at 4000r/min for 10 minutes

and dehydrate with cold acetone. Put in the vacuum drying oven to dry white crystalline insulin.

5. Content determination

According to ELISA kit manual operation.

【Questions】

1. How many ways to determine the activity of insulin?
2. What are the factors that affect insulin activity?

扫码"学一学"

实验十七　酵母 RNA 的制备和单核苷酸的离子交换柱色谱分析

【实验目的】

1. **掌握**　离子交换法分离单核苷酸的原理和方法
2. 学习以酵母为原料采用稀碱法制备 RNA 的原理和方法

【实验原理】

从微生物中提取 RNA 是工业上最实际有效的方法。一些最常见的菌体，如啤酒酵母、纸浆酵母、石油酵母、面包酵母、白地霉等均含有丰富的核酸资源。工业上制备 RNA 一般选用成本较低、适于大规模操作的稀碱法和浓盐法。稀碱法是用氢氧化钠溶液，使细胞壁变性裂解；浓盐法是用高浓度盐溶液处理，同时加热，以改变细胞壁的通透性，使核酸从细胞内释放出来。如用稀碱法，需用酸中和，然后除去菌体碎片，将 pH 调至 RNA 的等电点（pH 2.5），使 RNA 沉淀下来，该法的缺点是制备的 RNA 分子量较低。

核酸经酸、碱或酶水解可以产生各种核苷酸，核苷酸的可解离基团是第一磷酸基、含氮环上的 $-NH_2$和第二个磷酸基等，它们的解离常数（pKa）和由此得到的等电点差异是进行离子交换层析分离的基础。四种单核苷酸的解离常数和等电点见下表。

表 17-1　四种单核苷酸的解离常数（pKa）和等电点（pI）

核苷酸	第一磷酸基 (pKa 1)	含氮环 (pKa 2)	第二磷酸基 (pKa 3)	等电点 (pI)
腺苷酸（AMP）	0.9	3.7	6.2	2.35
鸟苷酸（GMP）	0.7	2.4	6.1	1.55
胞苷酸（CMP）	0.8	4.5	6.3	2.65
尿苷酸（UMP）	1.0	—	6.4	—

在一定条件下，离子交换树脂对不同单核苷酸的吸附能力是不同的，因此，选择适当类型的离子交换树脂，控制吸附及洗脱的条件便可分离各种单核苷酸。为了增加单核苷酸在离子交换树脂上的吸附能力，需要控制条件使单核苷酸带上大量相应电荷，这主要是通过调节 pH 值，使单核苷酸的一些可解离基团（磷酸基、氨基、烯醇基）解离，同时应减少上柱溶液中除单核苷酸外的其他离子强度。而洗脱时则相反，应使被吸附的单核苷酸的相应电荷降低，通常是增加洗脱液中竞争性的离子强度，必要时提高温度使离子交换树脂对单核苷酸的非极性吸附作用减弱。

RNA 可被碱水解成2’-或3’-核苷酸。可利用阳离子交换树脂（聚苯乙烯-二乙烯苯，磺酸型）或阴离子交换树脂（聚苯乙烯-二乙烯苯、季铵碱型）分离核苷酸。本实验利用强碱型阴离子交换树脂（强碱型201×8、强碱型201×7、国产717、Dowexl、Amberite IRA-400 或 Zerolit FF）将各类核苷酸分开，测定核苷酸的紫外吸收光谱的比值：OD_{250}/OD_{260}、OD_{280}/OD_{260}、OD_{290}/OD_{260}，对照标准比值表，可以确定其为何种核苷酸。

【实验材料】

1. 器材

（1）烘箱	1台
（2）电动搅拌器	1台
（3）桶或缸	1个
（4）离心机	1台
（5）滤布	若干
（6）玻璃层析柱 1.1cm×20cm	1根
（7）部分收集器	1台
（8）紫外分光光度计	1台
（9）恒温水浴	1台

2. 试剂

（1）6mol/L 盐酸。

（2）95%乙醇：乙醇与水按体积比 95：5 进行混合。

（3）1mol/L 甲酸：量取 21.4ml 88% 甲酸定容至 500ml。

（4）1mol/L 甲酸钠溶液：称取 32.15g 甲酸钠用水溶液定容至 500ml。

（5）0.02mol/L 甲酸：量取 10ml 1mol/L 甲酸定容至 500ml。

（6）0.15mol/L 甲酸：量取 75ml 1mol/L 甲酸定容至 500ml。

（7）0.01mol/L 甲酸-0.05mol/L 甲酸钠溶液（pH 4.44）：量取 5ml 1mol/L 甲酸，25ml 1mol/L 甲酸钠溶液定容 500ml。

（8）0.01mol/L 甲酸-0.1mol/L 甲酸钠溶液（pH 3.74）：量取 50ml 1mol/L 甲酸，50ml 1mol/L 甲酸钠溶液定容至 500ml。

（9）0.3mol/L 氢氧化钾溶液：称取 1.68g 氢氧化钾用水溶液定容至 100ml。

（10）2mol/L 过氯酸：量取 17ml 70%～72% 过氯酸定容至 100ml。

（11）2mol/L 氢氧化钠溶液：称取 NaOH 8.0g 用水定容至 100ml。

（12）0.5mol/L 氢氧化钠溶液：称取 NaOH 2.0g 用水定容至 100ml。

（13）新鲜啤酒酵母。

【实验方法】

1. RNA 的制备　取啤酒酵母于 2～3℃条件下每天用水洗一次，共洗 3～4 次，最后沉淀 1～2 天。倾去上清液，得酵母泥。取酵母泥 10g，置培养皿内，于 105℃烘箱内烤干，测定其含水量。这样的酵母泥一般含干酵母 15% 左右。

加水将酵母泥调成 5%～10% 干重的菌悬液，再加入浓的氢氧化钠溶液，使氢氧化钠的最终浓度（即提取时的浓度）为 1%。在 20℃条件下电动搅拌 30～45 分钟（如室温在 20℃

以下，可以适当延长时间）。然后用6mol/L盐酸中和至pH 7.0，再搅拌10分钟，直火或用蒸汽迅速加热到90℃在该温度下保持10分钟，迅速下降到10℃以下，与低温处静置3～6天，沉出蛋白质和酵母残渣。虹吸取出上清液。向下层混浊液中加入1/3体积的水，搅匀，静置过夜。于3000r/min离心20分钟左右，取出上清液。将上述二次上清液（含RNA）合并。

用6mol/L盐酸调至pH 2.5（RNA的等电点），低温放置过夜。离心收集RNA沉淀，用少量乙醇洗二次，干燥。所得的RNA制品，色微黄，产率为4%～5%，RNA含量可达60%～70%。

2. 单核苷酸的分离

（1）样品处理　取20mg上述RNA制品，溶于2ml 0.3mol/L氢氧化钾溶液中，于37℃水解20小时，RNA在碱作用下水解成单核苷酸，水解完成后，用2mol/L过氯酸溶液调至pH 2.0以下，以4000r/min离心10分钟，取上清液，用2mol/L氢氧化钠溶液调至pH 8.0，并用紫外分光光度计准确测得含量后待用。

（2）离子交换柱的安装　取内径1.1～1.2cm的层析柱，将处理好的强碱型阴离子交换树脂悬浮液一次倒入玻璃柱内，使树脂自由沉降至柱下部，用一小片圆滤纸盖在树脂面上。缓慢放出液体，使液面降至滤纸片下树脂面上（注意在整个操作过程中防止液面低于树脂，当液面低于树脂表面时空气将进入，在树脂内形成气泡，妨碍层析结果）。经沉积后离子交换树脂柱床高7～8cm。

（3）加样　将RNA水解液小心地用滴管加到离子交换树脂柱上，待样品液面降低到滤纸片内时，用50ml蒸馏水淋洗树脂柱。碱基、核苷及其他不被阴离子交换树脂吸附的杂质均被洗出。

（4）核苷酸混合物的洗脱　收集蒸馏水洗脱液，在紫外分光光度计上测260nm处光密度，待洗脱液不含紫外吸收物质（光密度值低于0.02）时，可用甲酸及甲酸钠溶液进行洗脱。

依次用下列洗脱液分段洗脱：500ml 0.02mol/L甲酸；500ml 0.15mol/L甲酸；500ml 0.1mol/L甲酸－0.05mol/L甲酸钠溶液（pH 4.44）；最后用500ml 0.1mol/L甲酸－0.1mol/L甲酸钠溶液（pH 3.74）。用部分收集器收集流出液，控制流速8ml/10min，8ml/管。

（5）由层析柱所得各部分洗脱液的分析　以相应浓度的甲酸或甲酸钠溶液作为空白对照，用紫外分光光度计测各管溶液在260nm波长处光密度值，以洗脱液体积（或管数）为横坐标，光密度值为纵坐标作图，分析各部分波峰位置。

根据各部分核苷酸在不同波长时光密度的比值（OD_{250}/OD_{260}、OD_{280}/OD_{260}、OD_{290}/OD_{260}），对照标准比值（表17－2）以及洗脱时的相应位置确定其为何种核苷酸。由洗脱液的体积和它们在紫外部分的光密度值，计算各种核苷酸的含量。

附：强碱型阴离子交换树脂201×8（聚丙乙烯－二乙烯苯，三甲胺季铵碱型，全交换量大于3毫克当量/克干树脂，100～200目）的处理方法：用水浸泡并利用浮选法除去细小颗粒，使用时先用0.5mol/L氢氧化钠溶液浸泡1小时，以除去碱性溶液杂质然后用无离子水洗至中性。再用1mol/L盐酸浸泡0.5小时，除去酸溶性杂质，再用无离子水洗至中性。然后用1mol/L甲酸钠溶液浸泡，使树脂转变成甲酸型。将树脂装入柱内，继续用1mol/L甲酸钠溶液洗，直至流出液中不含氯离子（用1%硝酸银溶液检查）。最后用1mol/L甲酸洗，直至260nm处光密度值低于0.02，并用蒸馏水洗至接近中性，即可使用。

表 17－2　部分核苷酸的物理常数

核苷酸	分子量	异构体 \ pH	紫外吸收广谱性质							
			克分子消光系数		光密度比值					
			$\varepsilon 260 \times 10^{-3}$		250/260		280/260		290/260	
			2	7	2	7	2	7	2	7
AMP	347.2	2'	14.5	15.5	0.85	0.8	0.23	0.15	0.038	0.009
		3'	14.5	15.3	0.85	0.8	0.23	0.15	0.038	0.009
		5'	14.5	15.3	0.85	0.8	0.22	0.15	0.03	0.009
GMP	363.2	2'	12.3	12.0	0.90	1.15	0.68	0.68	0.48	0.285
		3'	12.3	12.0	0.90	1.15	0.68	0.68	0.48	0.285
		5'	11.6	11.7	1.22	1.15	0.68	0.68	0.40	0.28
CMP	323.2	2'	6.9	7.75	0.48	0.86	1.83	0.86	1.22	0.26
		3'	6.9	7.6	0.46	0.84	2.00	0.93	1.45	0.30
		5'	6.3	7.4	0.46	0.84	2.10	0.99	1.55	0.30
UMP	324.2	2'	9.9	9.9	0.79	0.85	0.30	0.25	0.03	0.02
		3'	9.9	9.9	0.74	0.83	0.33	0.25	0.03	0.02
		5'	9.9	9.9	0.74	0.73	0.38	0.40	0.03	0.03

【思考题】

1. 离子交换法分离单核苷酸的原理？
2. 离子交换时常用的洗脱方式有哪些？各有何特点？

扫码“练一练”

EXPERIMENT 17　Preparation of Yeast RNA and analysis of single nucleotide by ion exchange column chromatography

【Purpose】

1. To master the principle and method of separating single nucleotides by ion exchange
2. To learn the principle and method of preparing RNA by using dilute alkali method with yeast

【Principle】

Extracting RNA from microorganisms is the most effective method in the industry. Some common bacteria, such as brewer's yeast, pulp yeast, petroleum yeast, baker's yeast and Geotrichum candidum, are rich of nucleic acid. Dilute alkali method and concentrated salt method are usually used to produce RNA. Because these methods are low cost and suitable for large－scale preparation. The dilute alkali method uses a sodium hydroxide solution to denature and break down the cell wall. The concentrated salt method uses salt solution with high concentration, together with heating, to change the permeability of the cell wall and release the nucleic acid from the cells. If the dilute alkali method is used, it is necessary to neutralize with acid. The bacterial fragments are then removed, and the pH is adjusted to the isoelectric point of the RNA (pH 2.5) to precipitate the RNA. The disadvantage of this method is that the prepared RNA has a lower molecular weight.

Nucleic acid, hydrolyzed by acid, base or enzyme, can produce nucleotides. The ionizable groups of nucleotides are the first phosphate group, the $-NH_2$ on the nitrogen – containing ring, and the second phosphate group. The basis of ion exchange chromatography separation is their dissociation constant (pK) and the resulted difference of isoelectric point. The dissociation constant and isoelectric point of the four single nucleotides are listed in the following table (Tab. 17 – 1).

Tab. 17 – 1 Dissociation constants (pKa) and isoelectric points (pI) of four single nucleotides

Nucleotide	First phosphate group (pKa 1)	Nitrogen ring (pKa 2)	Second phosphate group (pKa 3)	isoelectric points (pI)
Adenylate (AMP)	0.9	3.7	6.2	2.35
Guanylate (GMP)	0.7	2.4	6.1	1.55
Cytidine (CMP)	0.8	4.5	6.3	2.65
Uridine (UMP)	1.0	—	6.4	—

Under certain conditions, the ability of the ion exchange resin to adsorb different single nucleotides is different. Therefore, nucleotides can be separated by selecting an appropriate type of ion exchange resin and controlling the conditions of adsorption and elution. In order to increase the adsorption capacity of single nucleotide, it is necessary to control the condition to increase charge of nucleotides. It can be achieved by adjusting the pH to make some ionizable groups of the single nucleotide (The phosphate group, the amino group, and the enol group) dissociated. And the ionic strength of the upper column solution other than the single nucleotide should be reduced. Conversely in elution, the corresponding charge of the adsorbed single nucleotide should be lowered, usually by increasing the competitive ionic strength in the eluent, and if necessary, increasing the temperature to decrease the non – polar absorption of the ion exchange resin to the nucleotides.

RNA can be hydrolyzed by base to 2' – or 3' – nucleotides which can be isolated by cation exchange resin (polystyrene – divinylbenzene, sulfonic acid type) or anion exchange resin (polystyrene – divinylbenzene, quaternary ammonium base type). In this experiment, a strong base type anion exchange resin (strong base type 201 × 8, strong base type 201 × 7, domestic 717, Dowexl, Amberite IRA – 400 or Zerolit FF) was used to isolate nucleotides. The ratio of the ultraviolet absorption spectrum (OD_{250}/OD_{260}, OD_{280}/OD_{260}, OD_{290}/OD_{260}), compared to standard ratio table (Tab. 17 – 2), can be used to identify the type of the nucleotide.

【Materials】

1. Apparatus

(1) Oven

(2) Electric mixer

(3) barrel or cylinder

(4) Centrifuge

(5) filter cloth

(6) Glass chromatography column (1.1cm × 20cm)

(7) Partial collector

(8) UV spectrophotometer

(9) water bath

2. Reagents

(1) 6mol/L hydrochloric acid

(2) 95% ethanol: 95ml ethanol + 5ml water.

(3) 1mol/L formic acid: 21.4ml of 88% formic acid to a volume of 500ml.

(4) 1mol/L sodium formate solution: dissolve 32.15g sodium formate with an aqueous solution to a volume of 500ml.

(5) 0.02mol/L formic acid: measure 10ml of 1mol/L formic acid to a volume of 500ml.

(6) 0.15mol/L formic acid: measure 75ml of 1mol/L formic acid to a volume of 500ml.

(7) 0.01mol/L formic acid - 0.05mol/L sodium formate solution (pH 4.44): measure 5ml 1mol/L formic acid, 25ml 1mol/L sodium formate solution to a constant volume of 500ml.

(8) 0.01mol/L formic acid - 0.1mol/L sodium formate solution (pH 3.74): measure 50ml of 1mol/L formic acid, 50ml of 1mol/L sodium formate solution to a volume of 500ml.

(9) 0.3mol/L potassium hydroxide solution: dissolve 1.68g of potassium hydroxide with an aqueous solution, adjust volume to 100ml.

(10) 2mol/L perchloric acid: measure 17ml of 70% ~ 72% perchloric acid, add water to 100ml.

(11) 2mol/L sodium hydroxide solution.

(13) 0.5mol/L sodium hydroxide solution.

(14) Fresh brewer's yeast.

【Procedures】

1. Preparation of RNA

Wash the brewer's yeast once a day at 2 - 3℃ for 3 - 4 times, and precipitate for 1 - 2 days. Remove the supernatant to obtain a yeast sludge. Weigh 10g yeast sludge, place in a petri dish, and bake in the oven at 105℃ to dry, measure the water content. Such yeast sludge generally contains about 15% dry yeast.

Add the water to adjust the yeast sludge to a 5% - 10% dry weight bacterial suspension, and then add a concentrated sodium hydroxide solution to make the final concentration of sodium hydroxide (ie, the concentration at the time of extraction) 1%. Electric stir at 20℃ for 30 - 45 minutes (if the room temperature is below 20℃, the time can be extended). Then neutralize to pH 7.0 with 6mol/L hydrochloric acid, stir for another 10 minutes, heat rapidly to 90℃ with steam, keep at this temperature for 10 minutes, rapidly drop below 10℃, and let stand for 3 - 6 days with low temperature, sinking out protein and yeast residues. Remove the supernatant by siphoning. Add 1/3 volume of water to the lower layer turbid solution, stir well, and let stand overnight. Centrifuge at 3000r/min for about 20min and remove the supernatant. Combine the above two supernatants (containing RNA).

Adjust to pH 2.5 (isoelectric point of RNA) with 6mol/L hydrochloric acid and leave at low temperature overnight. Centrifuge to obtain the RNA pellet, wash twice with a small amount of ethanol, and dry. The obtained RNA preparation has a yellowish color, a yield of 4% to 5%, and an

RNA content of 60% to 70%.

2. Single nucleotideisolation

2.1 Sample pretreatment

Dissolve 20mg the above RNA product in 2ml potassium hydroxide solution (0.3mol/L), hydrolyze at 37℃ for 20 hours, and hydrolyze the RNA into a single nucleotide under the action of alkali. After the hydrolysis, use 2mol/L perchloric acid solution. Adjust to pH 2.0 or below, centrifuge at 4000r/min for 10 minutes, take the supernatant, adjust to pH 8.0 with 2mol/L sodium hydroxide solution, and accurately measure the content with UV spectrophotometer before use.

2.2 Installation of ion exchange column

Pour the treated strong alkali – type anion exchange resin suspension into a glass column with an inner diameter of 1.1 to 1.2cm at a time. Settle the resin freely to the bottom of the column and place a small piece of circular filter paper on the resin surface. Slowly release the liquid to lower the liquid level to the resin surface under the filter paper (Note that prevent the liquid level lower than the resin during the entire operation. When the liquid level is lower than the surface of the resin, air will enter, forming bubbles in the resin, which hinders the chromatographic result). After deposition, the ion exchange resin column bed is 7 – 8cm high.

2.3 Sample loading

Add the RNA hydrolysate carefully to the ion exchange resin column with a dropper, and when the liquid level of the sample is lowered into the filter paper sheet, rinse the resin column with 50ml distilled water. Bases, nucleosides, and other impurities not absorbed by the anion exchange resin are washed out.

2.4 Elution of nucleotide mixture

Collect the distilled water eluate, measure the optical density at λ260nm on the ultraviolet spectrophotometer. Use the formic acid and sodium formate solution for elution when the eluent did not contain ultraviolet absorbing material (the optical density value was lower than 0.02).

Elute the fractions sequentially with the following eluent, 500ml 0.02mol/L formic acid, 500ml 0.15mol/L formic acid, 500ml 0.1mol/L formic acid – 0.05mol/L sodium formate solution (pH 4.44), finally, 500ml 0.1mol/L – formic acid – 0.1mol/L sodium formate solution (pH 3.74). Collect the effluent using a partial collector and control the flow rate at 8ml/10min, 8ml/tube.

2.5 Analyze each part of the eluate obtained from the column

Use the corresponding concentration of formic acid or sodium formate solution as a blank control. Measure the optical density of each tube solution at λ260nm by ultraviolet spectrophotometer. Plot the eluent volume (or tube number) on the abscissa and the optical density on the ordinate. Plot and analyze the peak position of each part.

According to the ratio of the optical density of each part of the nucleotide at different wavelengths (OD_{250}/OD_{260}, OD_{280}/OD_{260}, OD_{290}/OD_{260}), the control standard ratio, elution and the corresponding position determines which nucleotide it is. Calculate the contents of various nucleotides from the volume of the eluate and their optical density values in the ultraviolet portion.

Attachment: Strong base type anion exchange resin 201 × 8 (polypropylene – divinylbenzene, trimethylamine quaternary ammonium base type, total exchange capacity greater than 3 meq/g dry

resin, 100 – 200 mesh) treatment method: soak with water and The fine particles were removed by flotation, and were first soaked in a 0.5mol/L sodium hydroxide solution for 1 hour to remove impurities in the alkaline solution and then washed to neutral with ion – free water. After soaking for 1 hour with 1mol/L hydrochloric acid, the acid – soluble impurities were removed, and then washed with ion – free water until neutral. Then, it was immersed in a 1mol/L sodium formate solution to convert the resin into a formic acid type. The resin was placed in the column and washed with a 1mol/L sodium formate solution until the effluent contained no chloride ions (checked with a 1% silver nitrate solution). Finally, wash with 1mol/L formic acid until the optical density value at 260nm is lower than 0.02 and wash with distilled water until it is nearly neutral, and it can be used.

Tab. 17 – 2 Physical constants of some nucleotides

Nucleotide	Molecular weight	isomer	Ultraviolet absorption broad spectrum properties							
			Molecularextinction coefficient		Optical density ratio					
			$\varepsilon 260 \times 10^{-3}$		250/260		280/260		290/260	
		pH	2	7	2	7	2	7	2	7
AMP	347.2	2'	14.5	15.5	0.85	0.8	0.23	0.15	0.038	0.009
		3'	14.5	15.3	0.85	0.8	0.23	0.15	0.038	0.009
		5'	14.5	15.3	0.85	0.8	0.22	0.15	0.03	0.009
GMP	363.2	2'	12.3	12.0	0.90	1.15	0.68	0.68	0.48	0.285
		3'	12.3	12.0	0.90	1.15	0.68	0.68	0.48	0.285
		5'	11.6	11.7	1.22	1.15	0.68	0.68	0.40	0.28
CMP	323.2	2'	6.9	7.75	0.48	0.86	1.83	0.86	1.22	0.26
		3'	6.9	7.6	0.46	0.84	2.00	0.93	1.45	0.30
		5'	6.3	7.4	0.46	0.84	2.10	0.99	1.55	0.30
UMP	324.2	2'	9.9	9.9	0.79	0.85	0.30	0.25	0.03	0.02
		3'	9.9	9.9	0.74	0.83	0.33	0.25	0.03	0.02
		5'	9.9	9.9	0.74	0.73	0.38	0.40	0.03	0.03

【Questions】

1. What is the principle ofisolating single nucleotide by ion exchange?

2. What are the commonly used elution methods for ion exchange? What are their characteristics?

实验十八 胰弹性蛋白酶的制备及活力测定

扫码"学一学"

【实验目的】

1. 掌握 弹性蛋白酶活力测定的原理。

2. 熟悉 弹性蛋白酶制备的方法。

【实验原理】

弹性蛋白酶（Elastase E $C_{3.4.4.7}$）又称胰肽酶 E，是一种肽链内切酶，根据它水解弹性

扫码"看一看"

蛋白的专一性又称为弹性水解酶。弹性酶为白色针状结晶，是由 240 个氨基酸残基组成的单一肽链，分子量为 25900Da，等电点为 9.5。其最适 pH 随缓冲体系而略异，通常为 pH 7.4～10.3，在 0.1mol/L 碳酸缓冲液中为 8.8。

结晶弹性酶难溶于水，电泳纯的弹性酶易溶于水和稀盐酸溶液（可达 50mg/ml），在 pH 4.5 以下溶解度较小，增加 pH 可以增加溶解度。弹性酶在 pH 4.0～10.5，于 20℃较稳定，pH＜6.0 稳定性有所增加，冻干粉于 5℃可保存 6～12 个月。在 −10℃保存更为稳定。

弹性蛋白唯有弹性酶才能水解。弹性酶除能水解弹性蛋白外，还可水解血红蛋白、血纤维蛋白等。许多抑制剂能使弹性酶活力降低或消失，如 10^{-5}mol/L 硫酸铜、7×10^{-2}mol/L 氯化钠可抑制 50% 酶活力，氰化钠、硫酸铵、氯化钾、三氯化磷也有类似作用。上述抑制作用一般多为可逆。另外，大豆胰蛋白酶抑制剂、血清或肠内非透析物等也有抑制作用。其他如硫代苹果酸、巯基琥珀酸、二异丙基氟代磷酸等均能强力抑制酶活力。

弹性酶广泛存在于哺乳动物胰脏，弹性酶原合成于胰脏的腺泡组织（Micin），经胰蛋白酶或肠激酶激活后才成为活性酶。本实验以新鲜猪胰脏为原料进行提取，然后再用离子交换层析法进行纯化，得到弹性酶。所得产品采用比色法测其活力，以刚果红弹性蛋白为底物。

【实验材料】

1. 器材

名称	规格	数量
（1）组织捣碎机		1 台
（2）电动搅拌器		1 台
（3）布氏漏斗	10cm	1 只
（4）吸滤瓶	1000ml	1 只
（5）抽气泵		1 台
（6）剪刀		1 把
（7）烧杯	800ml	1 只
	400ml	1 只
（8）绸布	25cm×25cm	1 块
（9）量筒	500ml	1 只
	100ml	1 只
（10）玻璃棒（大）		2 根
（11）玻璃棒（小）		6 根
（12）试管	10ml	18 支
（13）吸量管	5ml	18 支
	1ml	2 支
（14）721 型分光光度计		1 台
（15）乳钵	10cm	1 只
（16）真空干燥器及真空泵		1 套

2. 试剂

（1）0.1M pH 4.5 醋酸缓冲液：$NaAc\cdot3H_2O$ 5.85g 与 36% 醋酸溶液 9.51ml（冰醋酸 3.42ml）溶于水，稀释至 1000ml，pH 计校正。

（2）1.0M pH 9.3 氯化铵缓冲液：26.8g 氯化铵溶于 500ml 水中，用浓氯水调整至 pH 9.3。

（3）pH 8.8 硼酸缓冲液：取 3.72g 硼酸和 13.43g 硼砂溶于水中，稀释至 1000ml，pH 计校正。

（4）pH 6.0 磷酸缓冲液：取 KH_2PO_4 6.071g，NaOH 0.215g 溶于 1000ml 水，pH 计校正。

（5）丙酮

（6）刚果红弹性蛋白

（7）弹性酶纯品

（8）Amberlite CG_{50}树脂

【实验方法】

1. 预处理及细胞破碎　取冻胰脏，剪去脂肪，切成小块。称取 100g 加入 50ml 醋酸缓冲液（内含 0.05M $CaCl_2$），用组织捣碎机搅碎，静置活化。

2. 提取　再加入 200ml pH 4.5 0.1M 醋酸缓冲液，25℃搅拌（机械搅拌）提取 1.5 小时。离心（3000r/min，15 分钟）除去上层油脂及沉淀。用绸布挤滤，保留滤液。

3. 树脂吸附　滤液中加 100ml 蒸馏水及 40g（抽干重）经过处理的 Amberlite CG_{50}树脂，于 20 ~ 25℃搅拌吸附 2.5 小时。倾去上层液体，树脂用蒸馏水洗涤。重复洗涤 5 ~ 6 次。

4. 解析　树脂中加 50ml 1M pH 9.3 NH_4Cl 液，搅拌洗脱 1 小时。洗脱过程中每隔 10 分钟测一次 pH，整个过程须保证 $5.2 < pH < 6.0$，否则用氨水调节。经布氏漏斗滤过的洗脱液至 pH 7 置冰箱或冰盐浴预冷 15 分钟。

5. 成品收集　在 −5℃条件下一边搅边加入 3 倍体积冷丙酮，继续搅 2 分钟。低温静置 20 分钟。离心（3000r/min，15 分钟）收集沉淀，将沉淀移入离心试管中，用 10 倍量冷丙酮分两次洗涤离心，再用 5 倍量乙醚洗一次，离心。置真空干燥器内用 P_2O_5，干燥得弹性酶粉，称重。

6. 活力测定　活力单位定义为在 pH 8.8，37℃条件下作用 20 分钟，水解 1.0mg 刚果红弹性蛋白的酶量定义为一个活力单位。

（1）标准曲线的制作　取 6 支试管按表 18 − 1 操作。

注意：①刚果红弹性蛋白称量须准确。②标准液中酶量是过量的，以保证水解完全。

表 18 − 1　标准曲线的制作

管　号	1	2	3	4	5	0
刚果红弹性蛋白（mg）	5	10	15	20	25	10
弹性酶液（ml）	5	5	5	5	5	0
pH 8.8 硼酸缓冲液（ml）	—	—	—	—	—	5
37℃水解 60 分钟（间歇搅拌 30 次）						
磷酸缓冲液（ml）	5	5	5	5	5	5
3000r/min × 10 分钟离心后取上清						
A_{495nm}值						

（2）样品测定　精确称取样品 5mg 左右（如效价高可适当减少）置乳钵中，加 5ml

（先加少量）pH 8.8 硼缓冲液研磨至完全溶解。吸取 1ml 于大试管中，用上述缓冲液配制每毫升 5～10 单位的待测液，取 3 支试管接表 18－2 操作。

表 18－2 样品测定

管　号	1	2	0
刚果红弹性蛋白（mg）	6	6	3
pH 8.8 硼酸缓冲液（ml）	4	4	5
待测酶液（ml）	1	1	0
37℃水解 20 分钟（间歇搅拌 20 次以上）			
磷酸缓冲液（ml）	5	5	5
3000r/min×10 分钟离心后取上清			
A_{495nm}			

取平均吸收值由标准曲线查得单位数，再由稀释倍数加以换算出弹性酶比活，并折算总收率。

附　树脂处理：取干 Amberlite CG_{50} 树脂，水漂洗后加 5 倍体积 1mol/L HCl 搅拌 2 小时，水洗至中性。再加 5 倍体积 1mol/L NaOH 搅拌 2 小时，水洗至中性，再用 pH 4.5 0.1mol/L 醋酸缓冲液平衡过夜。

扫码“练一练”

【思考题】

1. 弹性蛋白酶制备的原理？
2. 影响弹性蛋白酶收率的因素有哪些？

EXPERIMENT 18　Preparation and viability determination of pancreatic elastase

【Purpose】

1. To master the principle of determination of elastase viability.
2. To be familiar with the method of preparing elastase.

【Principle】

Elastase(E $C_{3.4.4.7}$), also known as trypsin E, is an endopeptidase that is also called elastic hydrolase according to its specificity for hydrolyzing elastin. Elastase is white needle crystal, a single peptide chain consisting of 240 amino acid residues with a molecular weight of 25900Da and pI is 9.5. The optimum pH is slightly different with the buffer system, usually pH 7.4 to 10.3, and 8.8 in 0.1mol/L carbonate buffer.

The crystalline elastase is hardly soluble in water. The electrophoresis－purified elastase is easily soluble in water and dilute hydrochloric acid solution(up to 50mg/ml). The solubility is lower at pH 4.5, and increasing the pH can increase the solubility. The elastase is stable at 20℃ at pH 4.0～10.5, and has an increased stability at pH < 6.0. The lyophilized powder can be stored at

5℃ for 6 to 12 months. It is more stable at −10℃.

Elastin can only be hydrolyzed by elastase. In addition to hydrolyzing elastin, elastase can hydrolyze hemoglobin, fibrin and so on. Many inhibitors can reduce or eliminate the activity of elastase. For example, 10^{-5}mol/L copper sulfate and 7×10^{-2}mol/L sodium chloride can inhibit 50% enzyme activity. Sodium cyanide, ammonium sulfate, potassium chloride and phosphorus trichloride also have Similar effect. The above inhibition is generally mostly reversible. In addition, soybean trypsin inhibitor, serum or intestinal non−dialyzed substances also have an inhibitory effect. Others such as thiomalic acid, decyl succinic acid, diisopropyl fluorophosphoric acid can strongly inhibit enzyme activity.

Elastase is widely present in the mammalian pancreas. Elastase is synthesized in the pancreatic acinar tissue (Micin), which is activated by trypsin or enterokinase. In this experiment, fresh pig pancreas was used as a raw material for extraction, and then purified by ion exchange chromatography to obtain an elastase. The resulting product was measured for its viability using a colorimetric method using Congo red elastin as a substrate.

【Materials】

1. Apparatus

(1) Tissue masher

(2) Electric mixer

(3) Buchner funnel

(4) Suction filter bottle

(5) Air pump

(6) Scissors

(7) Beaker

(8) Silk cloth (25cm × 25cm)

(9) Measuring cylinder

(10) Glass rod

(11) Test tube

(12) Pipette

(13) 721 spectrophotometer

(14) Milk thistle

(15) Vacuum dryer and vacuum pump

2. Reagents

(1) 0.1M pH 4.5 Acetate buffer

$NaAc \cdot 3H_2O$ 5.85g and 36% acetic acid solution 9.51ml (glacial acetic acid 3.42ml) were dissolved in water, diluted to 1000ml, and calibrated with pH meter.

(2) 1.0M pH 9.3 ammonium chloride buffer

26.8g of ammonium chloride was dissolved in 500ml of water and adjusted to pH 9.3 with concentrated chlorine water.

(3)pH 8.8 Boric acid buffer

Take 3.72g of boric acid and 13.43g of borax dissolved in water, dilute to 1000ml, and calibrate with pH meter.

(4)pH 6.0 Phosphate buffer

Take 6.071g of KH_2PO_4 and 0.215g of NaOH dissolved in 1000ml of water, and calibrate with pH meter.

(5)Acetone

(6)Congo red elastin

(7)Elastase pure product

(8)Amberlite CG_{50} resin

【Procedures】

1. Pretreatment and cell disruption

Take the frozen pancreas, cut off the fat, and cut into small pieces. Weigh 100g and add 50ml of acetic acid buffer (containing 0.05M $CaCl_2$). Stir with a tissue mincer and let stand to activate.

2. Extraction

200ml of a pH 4.5 0.1M acetate buffer was added, and the mixture was stirred at 25℃ (mechanical stirring) for 1.5 hours. The upper layer of fat and precipitate were removed by centrifugation (3000r/min, 15 minutes). Sift with a silk cloth and leave the filtrate.

3. Resin adsorption

100ml of distilled water and 40g (dry weight) of treated Amberlite CG_{50} resin were added to the filtrate, and the mixture was stirred and adsorbed at 20 to 25℃ for 2.5 hours. The upper liquid was decanted and the resin was washed with distilled water. Repeat washing 5 to 6 times.

4. Desorption

Add 50ml 1M pH 9.3 NH_4Cl solution to the resin and stir for 1 hour. Measure pH every 10 minutes during the elution process, the whole process must be guaranteed $5.2 < pH < 6.0$, otherwise it is adjusted with ammonia water. The eluate filtered through a Buchner funnel was pre-cooled for 15 minutes in a refrigerator or ice salt bath at pH 7.

5. Product collection

Three times the volume of cold acetone was added while stirring at -5℃ and stirring was continued for 2 minutes. Stand at low temperature for 20 minutes and centrifugate (3000r/min, 15 min). The precipitate was collected, and the precipitate was transferred to a centrifuge tube, washed by centrifugation with 10 times of cold acetone, and once with 5 times of diethyl ether, and centrifuged. P_2O_5 was placed in a vacuum desiccator, and the elastic enzyme powder was dried and weighed.

6. Vitality measurement

Vitality unit definition: The enzyme amount of hydrolyzed 1.0mg of Congo red elastin was defined as one unit of activity at pH 8.8, 37℃ for 20 minutes.

(1) Production of standard curve: Take 6 test tubes and operate according to the following table (Tab. 18-1).

Note:①Congo red elastin should be accurately weighed.

②The amount of enzyme in the standard solution is excessive to ensure complete hydrolysis.

Tab. 18 –1　Production of standard curve

Number	1	2	3	4	5	0
Congo red elastin(mg)	5	10	15	20	25	10
Elastase(ml)	5	5	5	5	5	0
pH 8.8 borate buffer(ml)	—	—	—	—	—	5
Hydrolysis at 37℃ for 60 minutes(intermittent stirring 30 times)						
Phosphate buffer(ml)	5	5	5	5	5	5
3000r/min × 10 minutes,take the supernatant after centrifugation						
A_{495nm}						

(2) Sample determination

Accurately weigh the sample about 5mg(can be less if the potency is high),put it in the milk thistle,add 5ml(first add a small amount)of pH 8.8 boron buffer to completely dissolve. Pipette 1ml into a large test tube,prepare 5 to 10 units of the test solution per ml with the above buffer solution, and take 3 test tubes to the following table(Tab. 18 –2).

Tab. 18 –2　Sample determination

Number	1	2	0
Congo red elastin(mg)	6	6	3
pH 8.8 borate buffer(ml)	4	4	5
Test enzyme solution(ml)	1	1	0
Hydrolysis at 37℃ for 20 minutes(intermittent stirring 20 times or more)			
Phosphate buffer(ml)	5	5	5
3000r/min × 10 minutes,take the supernatant after centrifugation			
A_{495nm}			

The average absorption value is taken from the standard curve to find the unit number,and then the dilution factor is used to convert the elastic enzyme specific activity,and the total yield is converted.

Attachment:

Resin treatment:Take Amberlite CG_{50} resin, rinse with water, add 5 times volume of 1mol/L HCl for 2 hours,wash with water until neutral. Add 5 volumes of 1mol/L NaOH to treat for 2 hours, wash to neutrality,and equilibrate overnight with pH 4.5 0.1mol/L acetate buffer.

【Questions】

1. What is the principle of elastase preparation?

2. What are the factors that affect the yield of elastase?

扫码“学一学”

实验十九　超氧化物歧化酶的分离纯化与活力检测

【目的要求】

1. 掌握　测定超氧化物歧化酶活性和比活力的方法。

2. 通过超氧化物歧化酶的分离纯化，了解有机溶剂沉淀蛋白质以及纤维素离子交换柱层析方法的原理。

【实验原理】

超氧化物歧化酶（Superoxide dismutase）简称SOD，它广泛存在于各类生物体内，按其所含金属离子的不同，可分为3种：铜锌超氧化物歧化酶（Cu·Zn－SOD）、锰超氧化物歧化酶（Mn－SOD）和铁超氧化物歧化酶（Fe－SOD）。SOD催化如下反应：

$$O_2 + O_2^- \cdot + 2H^+ \rightarrow H_2O_2 + O_2$$

在生物体内，它是一种重要的自由基清除剂，能治疗人类多种炎症、放射病及自身免疫性疾病，并具有抗衰老作用，对生物体有保护作用。

在血液里Cu·Zn－SOD与血红蛋白等共存于红细胞中，当红细胞破裂溶血后，用氯仿－乙醇处理溶血液，使血红蛋白沉淀，而Cu·Zn－SOD则留在水－乙醇均相溶液中。磷酸氢二钾极易溶于水，在乙醇中的溶解度甚低，将磷酸氢二钾加入水－乙醇均相溶液中时，溶液明显分层，上层是具有Cu·Zn－SOD活性的含水乙醇相，下层是溶解大部分磷酸氢二钾的水相（比重大）。用分液漏斗处理，收集上层具有SOD活性的含水乙醇相，再加入有机溶剂丙酮，使SOD沉淀。极性有机溶剂能使蛋白质脱去水化层，并降低介电常数从而增加带电质点间的相互作用，致使蛋白质颗粒凝集而沉淀。采用这种方法沉淀蛋白质时，要求在低温下操作，并且需要尽量缩短处理时间，避免蛋白质变性。

Cu·Zn－SOD的pI为4.95。将收集的SOD丙酮沉淀物溶于蒸馏水中，在pH 7.6的条件下，Cu·Zn－SOD带负电，过DE－32纤维素阴离子交换柱可得到进一步纯化。目前常用的活性测定方法为邻苯三酚自氧化法，该法所用试剂和仪器比较普通、测试方便、灵敏度高，是目前应用最广泛的测试方法，但对温度、pH、邻苯三酚浓度、SOD待测液存放时间等诸因素比较敏感，因此，测定时要严格控制这些因素。

本实验以新鲜猪血为原料，通过氯仿－乙醇除去杂蛋白，采用DE－32纤维素进行纯化，并用邻苯三酚法测定SOD的活力。

【实验材料】

1. 器材

（1）离心机		1台
（2）G_3漏斗		1个
（3）抽滤瓶		1个
（4）751型分光光度计		1台
（5）梯度混合器		1台
（6）玻璃柱	1.0cm×10cm	1根

（7）试管　　　　　　　　若干
（8）自动收集器　　　　　1 台
（9）紫外检测仪　　　　　1 台
（10）移液管　　　5ml、1ml　　若干
（11）量筒
（12）烧杯　　　　　　　　若干
（13）分液漏斗　　　　　　1 个

2. 材料

（1）新鲜猪血
（2）0.9% NaCl
（3）95% 乙醇
（4）氯仿
（5）$K_2HPO_4 \cdot 3H_2O$
（6）丙酮
（7）pH 7.6，2.5mmol/L 磷酸钾缓冲液
（8）pH 7.6，200mmol/L 磷酸钾缓冲液
（9）10mmol/L HCl
（10）6mmol/L 邻苯三酚（用 10mmol/L HCl 作溶剂配制），4℃下保存。
（11）2.5% 草酸钾
（12）DE－32 纤维素
（13）pH 8.2，100mmol/L Tris－二甲胂酸钠缓冲液（内含 2mmol/L 二乙基三氨基五乙酸）：以 200mmol/L Tris 二甲胂酸钠溶液（内含 4mmol/L 二乙基三氨基五乙酸）50ml 加 200mmol/L HCl 22.38ml，然后用重蒸水稀释至 100ml。

【实验方法】

1. 酶的制备

（1）分离血球　取新鲜猪血 500ml（加入抗凝剂 2.5% 草酸钾 50ml），3000r/min 离心 20 分钟除去血浆，收集红细胞约 250ml，加入等体积 0.9% NaCl 溶液，用玻璃棒搅起充分洗涤，3000r/min，离心 20 分钟，弃去上清液（如此反复 3 次），收集洗净的红细胞放入 800ml 烧杯中，加 250ml 重蒸水，将烧杯置于冰浴中搅拌溶血 40 分钟以上，得溶血液 500ml。

（2）向溶液缓慢加入在 4℃下预冷过的 95% 乙醇 125ml（0.25 倍体积），然后再缓慢加入在 4℃下预冷过的氯仿 75ml（0.15 倍体积），搅拌 15 分钟，室温下 3000r/min 离心 20 分钟，弃去沉淀（血红蛋白），收集上清液约 330ml（留样 2ml 测酶活和蛋白含量）。此即酶的粗提液。

（3）向酶粗提液加入 $K_2HPO_4 \cdot 3H_2O$（按 43g $K_2HPO_4 \cdot 3H_2O$/100ml 粗提液的比例），转移到分液漏斗，振摇后静置 15 分钟，见分层明显。收集上层乙醇－氯仿相（微混浊），室温下离心（3500r/min）25 分钟，弃去沉淀，得上清液约 150ml（留样 1.5ml 测酶活和蛋白含量）。

(4) 向上一步得到的上清液加入0.75倍体积在4℃下预冷过的丙酮，Cu·Zn－SOD便沉淀下来。室温下3500r/min离心20分钟，收集灰白色沉淀物。将此灰白色沉淀物溶于约5ml重蒸水中（呈悬浮状），在4℃下，对250ml pH 7.6，2.5mmol/L的磷酸钾缓冲液透析，每隔0.5小时以上，换透析外液1次，共换4～5次。透析内液如出现沉淀，需在室温下3500r/min离心25～30分钟，弃去沉淀，收集上清液约7ml（留样0.5ml）。

(5) DE－32纤维素柱层析

DE－32纤维素的处理：称量DE－32纤维素干品5～6g，用自来水浮选除去1～2分钟不下沉的细小颗粒，用G_3烧结漏斗抽干，滤饼放入烧杯中，加适量1mol/L NaOH溶液，搅匀后放置15分钟，用G_3烧结漏斗抽滤；水洗至中性，滤饼悬浮于1mol/L HCl溶液中，搅匀后放置10分钟后，用G_3烧结漏斗抽滤；水洗至中性，滤饼再悬浮于1mol/L NaOH溶液中，抽滤，水洗至中性；最后将滤饼悬浮于色谱柱平衡缓冲液中待用。

DE－32纤维素使用后的回收处理与上述步骤相同，只是不用HCl，所用NaOH浓度改为0.5mol/L。将上一步所得离心上清液过DE－32纤维素柱。柱体1.0cm×6cm，用pH 7.6，2.5mmol/L磷酸钾缓冲液作色谱柱平衡液，用pH 7.6，2.5～200mmol/L（100ml）的磷酸钾缓冲液进行梯度洗脱。流速30ml/h，每管收集3ml。

2. 酶蛋白浓度的测定 采用紫外吸收法，先测定不同已知浓度标准酪蛋白在280nm波长处的光吸收值。绘出标准曲线作定量的依据，再测定样品在280nm波长处的光吸收值，从标准曲线上查出待测样品的蛋白浓度。

3. 酶活性的测定 本实验采用邻苯三酚自氧化法。邻苯三酚自氧化的机理极为复杂，它在碱性条件下，能迅速自氧化，释放出O_2^-，生成带色的中间产物。反应开始后，反应液先变成黄棕色，几分钟后转绿，几小时后又转变成黄色，这是因为生成的中间物不断氧化的结果。这里测定的是邻苯三酚自氧化过程中的初始阶段，中间物的积累在滞留30～45秒后，与时间成线性关系，一般线性时间维持在4分钟的范围内。中间物在420nm波长处有强烈光吸收，当有SOD存在时，由于它能催化O_2^-与H^+结合生成O_2和H_2O_2，从而阻止了中间物的积累，因此，通过计算就可求出SOD的酶活性。

邻苯三酚自氧化速率受pH、浓度和温度的影响，其中pH影响尤甚，因此，测定时要求对pH严格掌握。

(1) 邻苯三酚自氧化速率的测定　在试管中按表19－1加入缓冲液和重蒸水，25℃下保温20分钟，然后加入25℃预热过的邻苯三酚（对照管用10mmol/L HCl代替邻苯三酚），迅速摇匀，立即倾入比色杯中，在420nm波长处测定A值，每隔30秒读数一次，要求自氧化速率控制在0.060A/min（可增减邻苯三酚的加入量，使速率正好是0.060A/min）。

(2) 酶活性的测定　酶活性的测定按表19－2加样，操作与测定邻苯三酚自氧化速率相同。根据酶活性情况可适当增减酶样品的加入量。

酶活性单位的定义：在1ml反应液中，每分钟抑制邻苯三酚自氧化速率达50%时的酶量定义为一个活性单位，即在420nm波长处测定时，0.030A/min为一个活性单位。若每分中抑制邻苯三酚自氧化速率在35%～65%范围，通常可按比例计算，若数值不在此范围时，应增减酶样品加入量。

表 19－1　邻苯三酚自氧化速率测定加样表

试　剂	对照管（ml）	样品管（ml）	最终浓度（mmol/L）
pH 8.2 100mmol/L Tris－二甲胂酸钠缓冲液（内含 2mmol/L 二乙基三氨基五乙酸）	4.5	4.5	50
重蒸水	4.2	4.2	
10mmol/L HCl	0.3	—	
6mmol/L 邻苯三酚	—	0.3	0.2
总体积	9	9	

表 19－2　酶活性测定加样表

试　剂	对照管（ml）	样品管（ml）	最终浓度（mmol/L）
pH 8.2 100mmol/L Tris－二甲胂酸钠缓冲液（内含 2mmol/L 二乙基三氨基五乙酸）	4.5	4.5	
酶溶液	—	0.1	
重蒸水	4.2	4.1	
6mmol/L 邻苯三酚	—	0.3	0.2
10mmol/L HCl	0.3	—	
总体积	9	9	

（3）活性和比活的计算公式

$$①每毫升酶液活性单位(U/ml) = \frac{\frac{0.06-酶样品管自氧化速率}{0.06}\times 100\%}{50\%} \times 反应液总体积 \times \frac{酶样品液稀释倍数}{酶样品液体积}$$

$$②总活性单位 = 每毫升酶液活性单位(U/ml) \times 酶原液总体积$$

$$③比活 = \frac{每毫升酶液活性单位(U/ml)}{每毫升蛋白浓度(mg/ml)} = \frac{总活性单位(U)}{总蛋白(mg)}$$

【思考题】

1. SOD 对人体有何生物学意义？
2. 有机溶剂沉淀 SOD 根据的原理是什么？

扫码“练一练”

EXPERIMENT 19　Purification and viability determination of superoxide dismutase

【Purpose】

1. To master the method of determining superoxide dismutase activity and specific activity.

2. To learn the principles of protein precipitation by organic solvents and cellulose ion exchange column chromatography.

【Principle】

SOD exists widely in all kinds of organisms. According to the different metal ions, SOD can be

divided into three types: Cu · Zn – SOD, Mn – SOD, Fe – SOD. SOD catalyzes the following reactions:

$$O_2 + O_2^- \cdot + 2H^+ \rightarrow H_2O_2 + O_2$$

In vivo, it is an important free radical scavenger, which can treat a variety of human inflammatory, radiation and autoimmune diseases. SOD also has anti – aging effect and protective effect on organisms.

In the blood, Cu · Zn – SOD and hemoglobin coexist in the red blood cells. When the red blood cells rupture and hemolysis, the hemolysis is treated with chloroform – ethanol to precipitate the hemoglobin, while Cu · Zn – SOD remains in the water – ethanol homogeneous solution. Potassium hydrogen phosphate is very soluble in water and very low in ethanol. When potassium hydrogen phosphate is added into the homogeneous solution of water – ethanol, the solution is obviously stratified. The upper layer is the aqueous ethanol phase with the activity of Cu · Zn – SOD, and the lower layer is the water phase which dissolves most of potassium hydrogen phosphate. The aqueous ethanol phase with SOD activity in the upper layer was collected by funnel treatment, and then the organic solvent acetone was added to precipitate SOD. Polar organic solvents can cause proteins to dehydrate and decrease dielectric constant, and increase the interaction between charged particles, resulting in agglutination and precipitation of protein particles. When using this method to precipitate protein, it is required to operate at low temperature, and the processing time should be shortened as far as possible to avoid protein denaturation.

The pI of Cu · Zn – SOD is 4.95. The collected SOD acetone precipitate was dissolved in distilled water. Under the condition of pH 7.6, Cu · Zn – SOD was negatively charged and further purified by DE – 32 cellulose anion exchange column. At present, pyrogallol autoxidation method is commonly used to determine the activity of pyrogallol. The reagents and instruments used in this method are common, convenient and sensitive. It is the most widely used test method at present. But it is sensitive to temperature, pH, pyrogallol concentration and storage time of SOD solution. Therefore, these factors should be strictly controlled in the determination.

In this experiment, fresh pig blood was used as raw material, impurity proteins were removed by chloroform – ethanol, purified by DE – 32 cellulose, and the activity of SOD was determined by pyrogallol method.

【Materials】

1. Apparatus

(1) Centrifuge

(2) G_3 Funnel

(3) Filtering flask

(4) 751 Spectrophotometer

(5) Gradient mixer

(6) Glass column 1.0cm × 10cm

(7) Tubes

(8) Automatic fraction collection

(9) Ultraviolet detector

(10) Pipette

(11) Measuring cylinder

(12) Beaker

(13) Separating funnel

2. Reagents

(1) Flesh blood

(2) 0. 9% NaCl

(3) 95% alcohol

(4) Chloroform

(5) $K_2HPO_4 \cdot 3H_2O$

(6) Acetone

(7) pH 7. 6, 2. 5mmol/L Potassium phosphate buffer

(8) pH 7. 6, 200mmol/L Potassium phosphate buffer

(9) 10mmol/L HCl

(10) 6mmol/L Pyrogallol keep at 4℃

(11) 2. 5% Potassium oxalate

(12) DE – 32 cellulose

(13) pH 8. 2, 100mmol/L Tris – Sodium dimethylarsenate buffer (including 2mmol/L Diethyl – triaminopentaacetic acid): adding 200mmol/L HCl 22. 38ml to 200mmol/L Tris – Sodium dimethylarsenate buffer (including 4mmol/L Diethyl – triaminopentaacetic acid), then using ddH_2O dilute to 100ml.

【Procedures】

1 Preparation of Enzyme

(1) Separation of the blood cells

500ml fresh pig blood (2. 5% potassium oxalate 50ml added with anticoagulant) was centrifuged for 30000r/min to remove plasma, and about 250ml red blood cells were collected. Add 0. 9% NaCl solution of equal volume, stir and wash thoroughly, centrifuge for 20 minutes 3000r/min, then discard the supernatant (repeated 3 times). Collect the red blood cells and put them into 800ml beaker. Add 250ml of re – steamed water. Put the beaker in an ice bath and stir the hemolysis for more than 40 minutes, and get 500ml of hemolysis.

(2) Slowly add 95% ethanol 125ml (0. 25 times volume) which is pre – cooled at 4℃ to the solution, then slowly add chloroform 75ml (0. 15 times volume) which is pre – cooled at 4℃, stir for 15 minutes, centrifuge for 30000r/min at room temperature, discard the precipitation (hemoglobin), and collect about 330ml of supernatant (keep 2ml of sample to measure enzyme activity and protein content). This is the crude extract of the enzyme.

(3) Add $K_2HPO_4 \cdot 3H_2O$ (43g $K_2HPO_4 \cdot 3H_2O$/100ml crude enzyme) to the solution, transfer it to the separating funnel and stand for 15 minutes after shaking. The stratification is obvious. Collection the upper ethanol – chloroform phase (micro – turbidity). Centrifugation at room tem-

perature(3500r/min) for 25 minutes, discarding precipitation and obtaining supernatant of about 150ml(keep 1.5ml of sample to measure enzyme activity and protein content).

(4) Adding 0.75 times volume of pre - cooled at 4℃ acetone to the supernatant obtained from the further step, the precipitates is the Cu · Zn - SOD. Centrifuge for 20 minutes at room temperature for 3500r/min to collect gray - white sediments. Dissolve the gray - white sediment in about 5ml of re - steamed water. The dialysis of 250ml KPO_3 buffer solution with pH 7.6 and 2.5mol/L at 4 C was carried out at intervals of more than 0.5 hours. The dialysis fluid was changed 1 time and 4 to 5 times altogether. If the precipitation occurs in dialysis fluid, it should be centrifuged for 25 to 30 minutes at room temperature for 3500r/min, discarding precipitation and collecting supernatant for about 7ml.

(5) DE - 52 Cellulose column chromatography

Pretreatment of DE - 32 cellulose: Weighing DE - 32 cellulose dried product 5 - 6g, using tap water flotation to remove the particles that do not sink for 1 - 2 minutes, drying with G_3 sintering funnel, putting filter cake into beaker, adding appropriate amount of 1mol/L NaOH solution, stirring and placing for 15 minutes, filtering with G_3 sintering funnel, washing with water to pH 7. The suspension was precipitated in 1mol/L HCl solution, stirred and placed for 10 minutes, then filtered by G_3 sintering funnel. Wash to neutral, re - suspend and precipitate in 1mol/L NaOH solution, drain, wash to neutral, and finally suspend in chromatographic equilibrium buffer.

The recovery of DE - 32 cellulose after use is the same as the above steps, the difference is the concentration of NaOH was changed to 0.5mol/L without HCl. The centrifugal supernatant obtained from the previous step was passed through DE - 32 cellulose column. The column was 1.0cm × 6cm, and the balance solution is 2.5mmol/L KPO_3 buffer with pH 7.6, The gradient elution was carried out with KPO_3 buffer with pH 7.6 and 2.5 - 200mmol/L(100ml). The flow rate was 30ml/h and 3ml per tube was collected.

2. Determination of enzyme protein concentration

Ultraviolet absorption method was used to determine the light absorption values of standard casein at 280nm wavelength at different known concentrations. The standard curve was drawn for quantitative basis, and then the light absorption value of the sample at 280nm wavelength was determined. The protein concentration of the sample was found from the standard curve.

3. Determination of Enzyme Activity

In this experiment, pyrogallol autoxidation was used. Under alkaline conditions, pyrogallol can rapidly self - oxidize, release O_2^- and form colored intermediates. After the reaction starts, the reaction solution turns yellow - brown, green after a few minutes, and yellow after a few hours, because the intermediate is oxidized continuously. The initial stage of pyrogallol autoxidation is determined here. The accumulation of intermediates is linear with time after retention of 30 to 45 seconds. The general linear time is within 4 minutes. The intermediate has strong light absorption at 420nm wavelength. When SOD exists, the activity of SOD can be calculated because it can catalyze the combination of O_2^- and H^+ to form O_2 and H_2O_2, thus preventing the accumulation of intermediate.

The autoxidation rate of pyrogallol is influenced by pH, concentration and temperature, especially pH. Therefore, the determination of pyrogallol requires strict control of pH.

(1) Determination of Autoxidation Rate of Pyrogallol

In the test tube, buffers and ddH_2O were added according to Tab. 19 – 1 below, and then pyrogallol preheated at 25℃ for 20 minutes (10mmol/L HCl instead of Pyrogallol in the control tube) was added. It was shaken rapidly and immediately poured into the colorimetric cup. A value was measured at 420nm wavelength and read every 30 seconds. The rate of autoxidation should be controlled at 0. 060 A/min.

(2) Determination of Enzyme Activity

The determination of enzymatic activity was added according to Tab. 19 – 2. The operation and determination of pyrogallol autoxidation rate were the same.

Definition of Enzyme Activity Unit: In 1ml reaction solution, the enzyme quantity is defined as an active unit when the autoxidation rate of pyrogallol is inhibited by 50% per minute. That is, 0. 030 A/min is an active unit at 420nm wavelength. If the autoxidation rate of pyrogallol is inhibited in the range of 35% – 65% per minute, it can be calculated in proportion. If the value is not in this range, the amount of enzymatic sample should be increased or decreased.

Tab. 19 – 1　Pyrogallol autoxidation rate determination sample addition table

reagent	Control tube/ml	Sample tube/ml	Final Concentrationmmol/L
pH 8. 2 100mmol/L Tris – Na – Dimethylarsenate Buffer (including 2mmol/L Diethyl triaminopentaacetic acid)	4. 5	4. 5	50
ddH_2O	4. 2	4. 2	
10mmol/L HCl	0. 3	—	
6mmol/L Pyrogallol	—	0. 3	0. 2
Total volume	9	9	

Tab. 19 – 2　Sample addition table for enzyme activity determination

reagent	Control tube/ml	Sample tube/ml	Final Concentrationmmol/L
pH 8. 2 100mmol/L Tris – Na – Dimethylarsenate Buffer (including 2mmol/L Diethyl triaminopentaacetic acid)	4. 5	4. 5	
Enzyme solution	—	0. 1	
ddH_2O	4. 2	4. 1	
6mmol/L Pyrogallol	—	0. 3	0. 2
10mmol/L HCl	0. 3	—	
Total volume	9	9	

(3) Formulas for calculating activity and specific activity

① $$\text{per ml activity unit} = \frac{\dfrac{0.06 - \text{the auto} - \text{xidatioryrate of sample}}{0.06} \times 100\%}{50\%} \times \text{total volume} \times \frac{\text{sample dilution multiple}}{\text{sample volume}}$$

② Total activity unit = activity unit per milliliter of enzyme solution (U/ml) × total volume of enzyme solution

③ Specific enzyme activity $= \dfrac{\text{per ml activity unit(U/ml)}}{\text{per ml protein amount(mg/ml)}} = \dfrac{\text{total activity unit}}{\text{total protein}}$

【Questions】

1. What is the biological significance of SOD to human body?
2. What is the principle of SOD precipitation by organic solvents?

扫码"学一学"

实验二十　银耳多糖的制备及分析

【实验目的】

1. 掌握　多糖类物质的一般鉴定方法。

2. 学习真菌多糖类的分离、纯化原理。

【实验原理】

银耳是我国传统的一种珍贵药用真菌，具有滋补强壮、扶正固本之功效。银耳中含有的多糖类物质则具有明显提高机体免疫功能、抗炎症和抗放射等作用。

多糖的纯化方法很多，但必须根据目的物的性质及条件选择合适的纯化方法。而且往往用一种方法不易得到理想的结果，因此必要时应考虑合用几种方法。①乙醇沉淀法：是制备黏多糖的最常用手段。乙醇的加入，改变了溶液的极性，导致糖溶解度下降。供乙醇沉淀的多糖溶液，其含多糖的浓度以1%～2%为佳。加完乙醇，搅拌数小时，以保证多糖完全沉淀。沉淀物可用无水乙醇、丙酮、乙醚脱水，真空干燥即可得疏松粉末状产品。②分级沉淀法：不同多糖在不同浓度的甲醇、乙醇或丙酮中的溶解度不同，因此可用不同浓度的有机溶剂分级沉淀分子大小不同的黏多糖。③季铵盐络合法：黏多糖与一些阳离子表面活性剂如十六烷基三甲基溴化铵（CTAB）和十六烷基氯化吡啶（CPC）等能形成季铵盐络合物。这些络合物在低离子强度的水溶液中不溶解，在离子强度大时，这种络合物可以解离、溶解、释放。

本实验采用固体法培养获得的银耳子实体，经沸水抽提、三氯甲烷－正丁醇法除蛋白质和乙醇沉淀分离可制得银耳多糖粗品，再用CTAB（溴化十六烷基三甲胺）络合法进一步精制可得银耳多糖精品。然后进行定性和定量测定及杂质含量测定。

【实验材料】

1. 器材

（1）布氏漏斗　1只
（2）500ml抽滤瓶　1只
（3）250ml分液漏斗　1只
（4）100ml量筒　2只
（5）10ml量筒　1只
（6）离心机　1台
（7）250ml烧杯　2只

（8）500ml 烧杯　　　　1 只
（9）1000ml 烧杯　　　1 只
（10）水浴锅　　　　　1 台
（11）透析袋
（12）滤纸
（13）层析缸
（14）搅拌器

2. 试剂

（1）银耳子实体：20g。
（2）2% CTAB：取 2g CTAB 溶于 100ml 蒸馏水中，摇匀备用。
（3）硅藻土
（4）活性炭
（5）2mol/L 氢氧化钠溶液，6.2mol/L 氯化钠溶液
（6）三氯甲烷－正丁醇溶液（4：1）
（7）95% 乙醇
（8）甲苯胺
（9）乙醚
（10）无水乙醇
（11）浓硫酸
（12）α－萘酚
（13）斐林试剂
A 液：将 34.5g 硫酸铜（$CuSO_4 \cdot 5H_2O$）溶于 500ml 水中；
B 液：将 125g 氢氧化钠和 137g 酒石酸钾钠溶于 500ml 水中；
临用时，将 A、B 两液等量混合。

【实验方法】

1. 提取　将 20g 银耳子实体和 800ml 水加入 1000ml 烧杯中，于沸水浴中加热搅拌 8 小时，离心去（3000r/min，25 分钟）残渣。上清液用硅藻土助滤，水洗，合并滤液后于 80℃水浴搅拌浓缩至糖浆状。然后加入 1/4 体积的氯仿—正丁醇溶液摇匀，离心（3000r/min，10 分钟）分层，再分液漏斗分出下层氯仿和中层变性蛋白，然后，重复去蛋白质操作两次。上清液用 2mol/L NaOH 调至 pH 7，加热回流，用 1% 活性炭脱色，抽滤，滤液扎袋，流水透析 48 小时。透析液离心（3000r/min，10 分钟），上清液于 80℃水浴浓缩至原体积的 1/3。然后加入 3 倍量 95% 乙醇，搅拌均匀后，离心（3000r/min，15 分钟），沉淀用无水乙醇洗涤 2 次，乙醚洗涤一次，真空干燥得银耳多糖粗品。

2. 纯化　取粗品 1g，溶于 100ml 水中，溶解后离心（3000r/min，10 分钟）除去不溶物，上清液加 2% CTAB 溶液至沉淀完全，摇匀，静置 4 小时，离心，沉淀用热水洗涤三次，加 100ml 2mol/L NaCl 溶液于 60℃解离 4 小时，离心（3000r/min，10 分钟），上清液扎袋流水透析 12 小时。透析液于 80℃水浴浓缩，加三倍量 95% 乙醇，搅拌均匀后，离心（3000r/min，10 分钟），沉淀再分别用无水乙醇、乙醚洗涤，真空干燥，得精品银耳多糖。

3. 理化性质分析　将银耳多糖精品分别加入水、乙醇、丙酮、乙酸乙酯和正丁醇中，

观察其溶解性。另在浓硫酸存在下观察银耳多糖与 α－萘酚的作用，于界面处观察颜色变化。

4. 含量测定 多糖在浓硫酸中水解后，进一步脱水生成糖醛类衍生物，与蒽酮作用形成有色化合力，进行比色测定。另外，以 Folin 酚法测定银耳多糖样品中蛋白质含量，以紫外分光光度法测定样品中核酸的含量。

5. 银耳多糖纸层析 以正丙醇－浓氨水－水（40∶60∶5）为展开剂，分别将银耳多糖粗品和精品溶于水中，使浓度为0.5%，点样于层析滤纸上，展层后吹干，以0.5%甲苯按乙醇溶液染色，95%乙醇漂洗。

6. 结果记录 ①粗多糖和精多糖的含量；②记录并分析层析结果。

扫码“练一练”

【思考题】

1. 多糖类物质按其来源和组分可分别分为几种？不同材料来源的多糖其提取方法是否相同？
2. CTAB 为什么能与多糖类物质发生沉淀反应？
3. 以热水提取多糖是否会破坏多糖的结构？

EXPERIMENT 20 Preparation and analysis of tremella polysaccharide

【Purpose】

1. To master the method for the general identification of polysaccharides.
2. To learn the principle of separation, purification of fungus polysaccharides.

【Principle】

Tremella fuciformis is a kind of traditional Chinese treasured medical fungus. It has a great effect on strengthening with tonics and supporting the healthy energy. The polysaccharide of *Tremella fuciformis* has an obvious effect on immune function, antiinflammatory and antiradiation.

There are many methods for polysaccharide purification, which must correspond to theproperties and purification conditions of target substances. And it usually fails to produce ideal results merely using any single method, and a combination of different methods are therefore required. ①Ethanol precipitation. Ethanol precipitation is frequently used in preparing mucopolysaccharide. The addition of alcohol changes the polarity of the solution, leading to the decrease of solubility of mucopolysaccharide. The best concentration of the mucopolysaccharide solution by alcohol precipitation is 1% – 2%. Then beat for a couple of hours after the addition of alcohol in order for the complete precipitation. Absolute alcohol, acetone, ether are used for the precipitation dehydration, and puff powder production is obtained after vacuum drying. ②Fractional precipitation. Solubilities of different mucopolysaccharides vary in different concentrations of methol, alcohol and acetone. solutions, thus mucopolysaccharides with different molecular weights could be separated by fractional precipitation in organic solutions of different concentrations. ③Quaternary ammonium salt coordination. Mucopolysaccharides could bind to cation surfactant as such CTAB and CPC to produce quaternary ammonium slat

complexes. These complexes are not soluble in low - ion - strength water solution. As ion strength increases, the complexes would deassociate, and become dissolved and released.

Boiling solid cultivated tremella in boiling water and chloroform - n - butanol method is used to remove the protein and ethanol is used to separate the precipitate. Use CTAB (Cetyl trimethyl Ammonium Bromide) complex method to get purified tremellan. Several other tests are required to determine the quality and purity.

【Materials】

1. Apparatus

(1) Buchner funnel

(2) Filter flask

(3) Separating funnel

(4) Measuring cylinder (100ml)

(5) Measuring cylinder (10ml)

(6) Centrifugal machine

(7) Beakers

(8) Water bath

(9) dialysis bag

(10) Filter papers

(11) Developing tank

(12) Mixer

2. Reagents

(1) 20g of Tremella material

(2) 2% CTAB: Dissolve 2g of CTAB in 100ml of distilled water and well shake it up.

(3) Diatomaceous earth

(4) Activated carbon

(5) 2mol/L NaOH solution, 2mol/L NaCl solution

(6) Trichloromethane - n - butanol solution (4 : 1)

(7) 95% ethanol, Toluidine, Aether

(8) Absolute alcohol, concentrated oil of vitriol

(9) Concentrated sulfuric acid, a - naphthol;

(10) Fehling's reagent.

Solution A: Dissolve 34.5g of copper sulfate (with 5 molecules of crystallized water) into 500ml water.

Solution B: Dissolve 125g of NaOH and 137g of potassium tartrate in 500ml water.

Mix equal share of solution A and B when use.

【Procedures】

1. Extraction

Dispose 20g of tremella and 800ml of water in 1000ml beaker, stir while bathing the beaker in boiling water for 8 hours. Remove the residue by centrifugalization (3000r/min, 25 minutes). Filter the supernatant with diatomaceous earth. Wash and collect the filtrate, bath the filtrate in the 80℃ water, keep stirring until the filtrate is enriched into syrup. Add 1/4 volume of chloroform – n – butanol solution. Shake it up and centrifugalize (3000r/min, 10 minutes) the mixture, remove the lower layer of chloroform and the middle layer of denatured protein with the separate funnel, repeat the deprotein operation twice. Adjust the pH of supernatant layer to 7 with 2mol/L NaOH solution, then heat and reflux it, decolor it with 1% activated carbon. Suction the liquid and fill the filtrate into the dialysising bag, dialyse with running water for 48 hours. Centrifugalize the dialysate (3000r/min, 10 minutes), and enrich the supernatant to 1/3 of its volume in the 80℃ of water. Add triple amount of 95% alcohol, mix it up and centrifugalize the mixture (3000r/min, 15 minutes). Wash the precipitate twice with absolute alcohol, once with ether; dry the precipitate by vacuum to obtain crude tremella.

2. Purification

Dissolve 1g of crude tremella in 100ml of water, remove the precipitate by centrifugalization (3000r/min, 10 minutes). Add 2% CATB solution to the supernatant to precipitate solid substances completely. Let the mixture stay for 4 hours. Centrifugalize the liquid and wash the precipitate with hot water 3 times. Add 100ml of 2mol/L NaCl solution and dissociate the mixture at 60℃ for 4 hours. Centrifugalize (3000r/min, 10 minutes) it. Dialyse the supernatant in dialysising bag with running water for 12 hours. Enrich the dialysate by 80℃ bath. Add triple amount of 95% alcohol, shake it up and centrifugalize (3000r/min, 10 minutes) the mixture. Wash the precipitate separately with absolute alcohol and ether. Dry it vacuumly to obtain purified tremella.

3. Determination of the physical & chemical property

Dissolve the tremella sample and observe its solubility separately in water, alcohol, acetone, ethyl acetate and n – butanol. The effect of Tremella polysaccharides and α – naphthol was observed in the presence of concentrated sulfuric acid, and color change was observed at the interface.

4. Quantitive determination

After hydrolyzed in concentrated sulfuric acid, polysaccharide is further dehydrated into aldose derivation that can form colored compound with anthrone. See test 42 for details of color comparison test. Folin phenol method is used to determine the protein concentration while UV spectophotometry is used for determination of the concentration of ribonucleic acid.

5. Polysaccharide paper chromatography

The Developing agent contain with porpanol – concentrated ammonia water – water (40 : 60 : 5). The crude and purified tremella polysaccharide are respectively dissolved in water, with the concentration of 0.5%. The samples are put on the chromatography filter paper, dried after layer development, dyed with 0.5% toluene in ethanol solution rinsed with 95% ethanol.

6. Record the Result

①Work out the content of crude tremella and purified trmella.

②Record and analyze the chromatography results.

【Questions】

1. How many types of polysaccharides are there according to the classification of source and

component? Do they have the same method of extraction?

2. Explain the precipitating reaction of CATB with polysaccharides.

3. Will the structures of polysaccharides be ruined if extracted with hot water?

实验二十一　氯化血红素的制备及含量测定

扫码“学一学”

【实验目的】

1. 掌握　氯化血红素制备的原理。

2. 了解　血红素的药用价值。

【实验原理】

血红素（Heme）是高等动物血液、肌肉中的红色色素，由原卟啉与Fe^{2+}结合而成，它与珠蛋白结合成血红蛋白。在体内的主要生理功能是载氧，帮助呼出CO_2，还是cty p450、cty c、过氧化酶的辅基。血红素不溶于水，溶于酸性丙酮及碱性水中，在溶液中易形成聚合物，临床上常用作补铁剂和抗贫血药及食品中色素添加剂，另外可用于制备原卟啉来治疗癌症。

氯化血红素（Hemin）是天然血红素的体外纯化形式。实验室常用酸性丙酮分离提取法制备氯化血红素。首先使血球在酸性丙酮中溶血，抽提后再经浓缩、洗涤、结晶得到氯化血红素。工业上制取氯化血红素常用冰乙酸结晶法。血球用丙酮溶血后，制取血红蛋白，再用冰乙酸提取。在NaCl存在下，氯化血红素沉淀析出。

卟啉环系化合物在400nm处有强烈吸收，称Soret带，其最大吸收波长（λ_{max}）对各种卟啉化合物是特征的，但溶剂对λ_{max}也有影响，采用0.25% Na_2CO_3作溶剂。在600nm处有特征峰吸收，光吸收值与氯化血红素浓度的关系符合朗比定律。

【实验材料】

1. 器材

（1）烧杯	1000ml	1只
	500ml	2只
	250ml	2只
（2）抽滤瓶	500ml	1只
（3）布氏漏斗	8cm	1只
（4）三颈瓶	500ml	1只
（5）电动搅拌机		1台
（6）球形冷凝管	30cm	1只
（7）温度计	200℃	1支
（8）离心机		1台
（9）分液漏斗	500ml	1只
（10）小试管		20支

2. 试剂

（1）新鲜猪血 500ml

（2）0.8% 柠檬酸三钠 20ml

（3）丙酮

（4）冰乙酸

（5）NaCl（固体）

（6）KCl（固体）

（7）浓 HCl

（8）20% 氯化锶

（9）0.25% Na_2CO_3

【实验方法】

1. 酸性丙酮抽提 0.8% 柠檬酸三钠抗凝猪血 200ml，3000r/min 离心 15 分钟，倾去上层血浆，制得血球，加 2～3 倍的蒸馏水，充分溶胀后，沸水浴 20～30 分钟，纱布过滤，滤渣加入含 3% 盐酸的丙酮溶液 200ml，振摇抽提 30 分钟，抽滤，将滤液用旋转蒸发仪浓缩至原体积的 1/4～1/3，加入 20% 氯化锶至终浓度 2%，静置 15 分钟，3000r/min 离心 10 分钟，沉淀用水、95% 乙醇、乙醚各洗涤一次，真空干燥后得氯化血红素粗品，称重，计算收率。

2. 冷乙酸结晶法 0.8% 柠檬酸三钠抗凝猪血 500ml，3000r/min 离心 15 分钟，倾去上层血浆，下层红细胞，加丙酮 200ml 搅拌，过滤，得红色血红蛋白。

取 500ml 带温度计、冷凝器、搅拌插口的三颈烧瓶，加入 300ml 冰乙酸，加热升温，再加入 16g NaCl、8g KCl，在搅拌下加入 100g 血红蛋白，在 105℃继续搅拌 10 分钟，冷却，静置过夜，离心收集沉淀的氯化血红素结晶，用冰乙酸和 0.1mol/L 醋酸洗涤，再用水洗至中性，过滤，干燥后得氯化血红素粗品，称重，计算收率。

3. 含量测定 取标准氯化血红素，用 0.25% Na_2CO_3 配置成浓度 0.08mg/ml 备用。取制备所得氯化血红素，用 0.25% Na_2CO_3 配置成 0.1mg/ml 备用。按下表稀释，在 600nm 处测定 OD 值，以 0.25% Na_2CO_3 溶剂作空白，根据所得数据计算氯化血红素含量。

	标准氯化血红素										制备血红素			
	0	1	2	3	4	5	6	7	8	9	10	11	12	13
Hemin 溶液（ml）	0	0.4	0.8	1.2	1.6	2.0	2.4	2.8	3.2	3.6	4.0	1.0	2.0	3.0
0.25% Na_2CO_3（ml）	4.0	3.6	3.2	2.8	2.4	2.0	1.6	1.2	0.8	0.4	0	3.0	2.0	1.0
Hemin 含量（mg/ml）	0													
OD_{600nm}														

扫码“练一练”

【思考题】

1. 血红素临床应用有哪些？

2. 影响氯化血红素收率的因素有哪些？

EXPERIMENT 21　Preparation and quantitative analysis of Hemin

【Purpose】

1. To master the principles and methods of the preparation and determination of hemin.
2. To know the medicinal values of heme.

【Principle】

Heme is the red pigment in the blood and muscle of higher animals. It is formed by protoporphyrin and Fe^{2+}, which combines with globin to form hemoglobin. The main physiological function in the body is to carry oxygen and help exhale CO_2, as well as the coradicals of cty p450, cty c and peroxidase. Heme is insoluble in water and soluble in acidic acetone and alkaline water. It is easy to form polymers in solution. It is commonly used as iron supplement, anemia drug and pigment additive in food.

Hemin is the purified form of natural heme *in vitro*. In laboratory, hemin is prepared by acid acetone separation and extraction method. Firstly, hemolysis of blood cells in acidic acetone was conducted, and then hemin was obtained by concentration, washing and crystallization after extraction. The production of hemin in industry is usually carried out by crystallization of glacial acetic acid. After hemolysis with acetone, hemin is extracted by glacial acetic acid from hemoglobin and then precipitated out in the presence of NaCl.

The porphyrin ring has strong absorbance at λ400nm, called the Soret band. The maximum absorbance wavelength (λ_{max}) is characteristic for most porphyrin compounds, but the solvent is λ_{max} dependent. The solution is 0. 25% Na_2CO_3. There is characteristic absorption peak at λ600nm, and the relationship between the light absorption value and the concentration of hemin conforms to Lambert – Beer.

【Materials】

1. Apparatus

(1) Beaker
(2) Suction flask
(3) Buchner funnel
(4) Three – necked bottle
(5) Electric mixer
(6) Spherical condensing tube
(7) Thermometer
(8) Centrifuge
(9) Separation funnel
(10) Test tubes

2. Reagents

(1) Fresh pig blood

(2) 0.8% trisodium citrate

(3) Acetone

(4) Glacial acetic acid

(5) NaCl (solid)

(6) KCl (solid)

(7) Concentrated HCl

(8) 20% strontium chloride

(9) 0.25% Na_2CO_3

【Procedures】

1. Acid acetone extraction

200ml of fresh pig blood (containing 0.8% trisodium citrate) was centrifuged at 3000r/min 15 minutes, the deposit was obtained and then added 2 – 3 times of distilled water. After fully swelling and then boiling water bath 20 – 30 minutes, the gauze filter was used to collect filter residue and then added 200ml of acetone solution containing 3% hydrochloric acid. After shaking out 30 minutes and then suction filtration, the filtrate was condensed by rotary evaporation apparatus to 1/4 – 1/3 of the original volume. The sample was added 20% strontium chloride to final concentration of 2% and resting for 15 minutes, and then was centrifuged at 3000r/min 10 minutes. The precipitation was washed with water, 95% ethanol, and ethyl ether for each time. After vacuum drying, the crude hemin was obtained, weighed and the yield was calculated.

2. Cold acetic acid crystallization

500ml of fresh pig blood (containing 0.8% trisodium citrate) was centrifuged at 3000r/min 15 minutes, and then the upper plasma was discarded. The lower red blood cells were added 200ml acetone, stirred and filtered to obtain red hemoglobin. Take 500ml three – necked flask with thermometer, condenser and stirring socket, was added 300ml ice acetic acid and heating, and then added 16g NaCl, 8g KCl, 100g hemoglobin under stirring. After continue stirring for 10 minutes in 105℃, the sample was cooled down and resting overnight. Hemin crystallization was collected by centrifuge and then washed with ice acetic acid and 0.1mol/L acetic acid. The sample was washed with water to neutral and then filtered. The crude hemin was obtained, weighed and the yield was calculated.

3. Content determination

Prepare 0.08mg/ml hemin solution by adding standard hemin with 0.25% Na_2CO_3 solution. Prepare 0.1mg/ml hemin solution by adding prepared hemin from pig blood with 0.25% Na_2CO_3 solution and diluted the sample as the following table. The OD value was measured at 600nm with 0.25% Na_2CO_3 solvent as blank. The hemin content was calculated according to the obtained data.

	Standard hemin										Prepared hemin			
	0	1	2	3	4	5	6	7	8	9	10	11	12	13
Heminsolution(ml)	0	0.4	0.8	1.2	1.6	2.0	2.4	2.8	3.2	3.6	4.0	1.0	2.0	3.0
0.25% Na_2CO_3(ml)	4.0	3.6	3.2	2.8	2.4	2.0	1.6	1.2	0.8	0.4	0	3.0	2.0	1.0
Hemin concentration(mg/ml)	0													
OD_{600nm}														

【Questions】

1. What does heme clinical application have?
2. What are the factors affecting the yield of hemin?

实验二十二　发酵法制备辅酶 Q_{10}及其检测

扫码“学一学”

【实验目的】

1. 掌握　微生物发酵法生产辅酶 Q_{10}的基本原理及其注意事项。

2. 了解　辅酶 Q_{10}的基本性质。

【实验原理】

辅酶 Q 又称为泛醌，是一类在生物界广泛存在的辅酶，其主要结构是苯醌环，带有不同长度单位的异戊二烯，在哺乳动物中以 10 个单位的辅酶 Q_{10}最为常见，化学结构式如图 22－1。辅酶 Q_{10}存在于线粒体内膜，是呼吸链中的重要递氢体，帮助脱氢酶参与呼吸链的电子传递作用，辅酶 Q_{10}同时具有自由基清除剂的作用，帮助保护细胞膜和膜上蛋白抵御过氧化物。

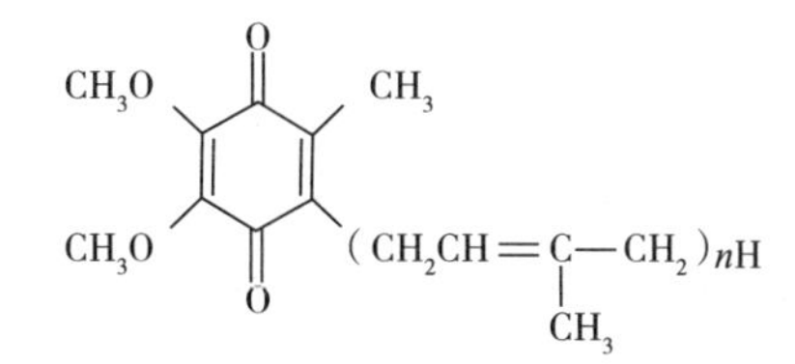

图 22－1　辅酶 Q_{10}

人工合成的辅酶 Q_{10}是一种良好的生化药物，可以治疗心血管疾病、肌肉萎缩、牙周炎以及早期充血性心力衰竭等。辅酶 Q_{10}为黄色或淡橙黄色、无臭无味结晶性粉末，易溶于氯仿、苯、四氯化碳，溶于丙酮、乙醚、石油醚；微溶于乙醇，不溶于水和甲醇。遇光易分解成微红色物质，对温度和湿度较稳定，熔点 49℃。

辅酶 Q_{10}的制备方法有三种：化学合成法、微生物发酵法和动植物组织提取法。目前国内大多采用动植物组织提取法。国外多用微生物发酵法，尤其是日本早在 1977 年就实现了微生物发酵法工业生产辅酶 Q_{10}。化学合成法合成条件苛刻，步骤繁多且化学合成的辅酶 Q_{10}的异戊二烯单体大多为顺式结构，生物活性不好，副产物多，提纯成本高。动植物组织提取法主要是从提取细胞色素 C 后的猪心残渣提取。动物组织中辅酶 Q_{10}含量低，每千克新鲜猪心的收率仅为75mg，并受原材料和来源限制，规模化生产受到一定制约。相比之下，利用微生物发酵生产辅酶 Q_{10} 有以下几个优点：①发酵产物为天然品，生物活性好，易被人体吸收；②没有原材料的制约，适合工业化生产。

【实验材料】

1. 器材

（1）摇床

（2）离心机

（3）恒温水浴锅

（4）高压灭菌锅

（5）HPLC

（6）分析天平

（7）旋转蒸发仪

2. 试剂

（1）菌种

（2）LB 培养基

（3）丙酮

（4）无水乙醇

（5）甲醇

（6）石油醚

（7）乙醚

【实验方法】

1. 发酵条件

（1）种子液培养　从辅酶 Q_{10} 产生菌甘油管中取400μl 至 50ml LB 液体培养基中，置于30℃水浴摇床 170r/min 振摇培养 12 小时。

（2）发酵液培养　用无菌移液管吸取种子液按 2% 接种量（2ml）转接至无菌发酵液培养基中，置于 30℃水浴摇床 170r/min 振摇培养 22 小时。

发酵结束后将发酵培养液 8000r/min 离心 10 分钟，弃上清收集菌体，置冰箱备用。

2. 提取　取菌体 2g 加入 100ml 丙酮，搅拌 10 分钟后，8000r/min 离心 3 分钟，离心取上清。减压蒸馏至干，用无水乙醇洗下，备用。

3. 浓缩　合并 5 份提取液，以薄膜蒸发器减压浓缩。水浴温度不超过 50℃，直至蒸干，趁热以 0.5ml 石油醚溶解。

4. 硅胶柱层析纯化　将 1.2cm×20cm 层析柱下端以少量玻璃棉塞住，用漏斗装入 11g 硅胶与 20ml 石油醚的混合液，装柱完毕后，再以 10ml 石油醚过柱，以稳定柱床，用滴管小心将浓缩液上柱，用乙醚：石油醚（2∶8）洗脱，小试管收集洗脱液，每管 2ml，对各管进行薄层鉴定。

5. 薄层鉴别　（1）薄层色谱板的制备　称取薄层层析用硅胶 2g 至乳钵中，加入 0.5% CMC－Na 5ml 共研磨片刻，迅速用角匙均匀平铺于洗净烘干的玻片上（5.5cm×12cm 两片），1 小时后轻轻移入烘箱，于 105℃活化 1 小时。

（2）点样　用毛细管分别吸取上述各管收集液在层析板端 1.5cm 处点样，并以辅酶 Q_{10}标准液为对照，点样量 10～15μl，吹干。

（3）展开　将薄层板竖直放入盛有展开剂的展开瓶中，盖上瓶盖。展开剂选用乙醚：石油醚 =2：8，点样处必须在液面以上；当展开剂上升至离薄层板上缘 2cm 处取出吹干。

（4）显色　置于碘缸中碘蒸汽显色，用铅笔画出各点的位置，并计算辅酶 Q_{10}的 R_f值。

6. 结晶　合并薄层鉴定较纯的管，以薄膜蒸发器减压浓缩。水浴温度不超过 50℃，直至蒸干，趁热以数滴无水乙醇将油状物移至小试管，普通冰箱静置过夜结晶，称重。

7. 含量测定（高效液相色谱法）

（1）HPLC 色谱条件　流动相：甲醇：乙醇 =3：7；检测波长：275nm；流速：1ml/min；柱温：35℃；色谱柱：反向 C_{18}柱；进样量：10μl。

（2）标准曲线的绘制　辅酶 Q_{10}的氧化型较还原型稳定，故多以氧化型存在，其氧化型在 275nm 处有特征吸收峰，在一定范围内峰面积与辅酶 Q_{10}含量成线性关系。

标准曲线的绘制

试管号	1	2	3	4	5	6
辅酶 Q_{10}标准液（100μg/ml）	0	1	2	3	4	5
无水乙醇	5	4	3	2	1	0
OD_{275}						

分别吸取上述 6 种标准溶液 10μl 进样，HPLC 定量，以峰面积（y）对 CoQ_{10}的浓度（x，μg/ml）作曲线图。作出标准曲线或回归方程。

（3）吸取提取液 10μl 进样，HPLC 定量，根据峰面积计算进样辅酶 Q_{10}溶液的浓度（μg/ml）及本次发酵液中辅酶 Q_{10}总产量（μg），单位菌体辅酶 Q_{10}含量（μg/g 菌重）。

【思考题】

1. 采用微生物法生产辅酶 Q_{10}的优点有哪些？
2. 影响辅酶 Q_{10}提取效率的因素有哪些？提取过程中有何注意点？

扫码“练一练”

EXPERIMENT 22　Preparation and analysis of coenzyme Q_{10} by fermentation

【Purpose】

1. To master the principle of preparation coenzyme of Q_{10} by fermentation.
2. To learn about the properties of coenzyme Q_{10}.

【Principle】

Coenzyme Q(CoQ), also named ubiquinone, ubidecarenone, or coenzyme Q_{10}, is a well – known ubiquitous coenzyme found in the biological system which is synthesized by the conjugation of benzoquinone ring with isoprenoid chain of variable length (Fig. 22 – 1). Coenzyme Q_{10} is found at the inner mitochondrial membrane, which transfer electron from dehydrogenases bounded to membrane to complexⅢ of ETC and reduce them. Coenzyme Q_{10} also act as free radical scavenger which help to protect membrane protein and phospholipids from lipid peroxidation by regeneration of tocopherol.

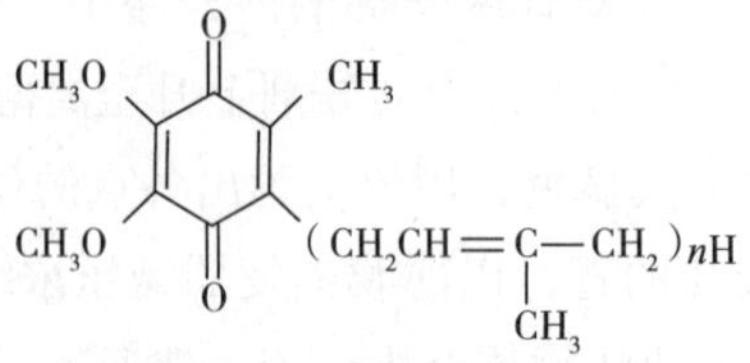

Fig. 22 – 1 Coenzyme Q_{10}

Quinone species like coenzyme Q_{10} can be artificially synthesized which helps to treat various human diseases like cardiomyopathy, muscular dystrophy, periodontal disease, and in the early stages of congestive heart failure. Coenzyme Q_{10} is yellow or light orange yellow, odorless and tasteless crystalline powder, soluble in chloroform, benzene, carbon tetrachloride, acetone, ether, petroleum ether, slightly soluble in ethanol, insoluble in water and methanol. When exposed to light, it is easy to decompose into reddish substance, which is stable to temperature and humidity with melting point of 49℃.

There are three methods to produce CoQ_{10}: chemical synthetic, extraction from animal and plant tissues and microbial fermentation. At present, the extraction from animal and plant tissues are still common in China. However, microbial fermentation is widely used abroad. The industrial production of coenzyme Q_{10} by microbial fermentation was popular in Japan as early as 1977. The isoprene monomers of coenzyme Q_{10} synthesized by chemical synthesis method are mostly cis – structure, with harsh synthetic conditions, numerous steps and high purification cost. The method of extracting animal and plant tissues is mainly from pig heart residue after extracting cytochrome C. Coenzyme Q_{10} content in animal tissues is low, the yield of fresh pig heart is only 75mg per kilogram, and limited by raw materials and sources, large – scale production is restricted. In contrast, the production of coenzyme Q_{10} by microbial fermentation has the following advantages: ①Fermentation products are natural products with good biological activity and easy to be absorbed by human body; ②No restrictions on raw materials, suitable for industrial production.

【Materials】

1. Apparatus

(1) Shaker

(2) Centrifuge

(3) Constant temperature water bath

(4) High temperature sterilizing oven

(5) HPLC

(6) Analytical balance

(7) Rotary evaporator

2. Reagents

(1) E. coil

(2) LB liquid medium.

(3) Acetone

(4) Ethanol

(5) Methanol

(6) petroleum

(7) ether

【Procedures】

1. Fermentation conditions:

1) Take 400μl broth from glycerol pipe to 50ml LB liquid medium, 30℃ for 12 hours (170r/min).

2) Take 2ml Inocula, access to 200ml liquor culture solution in 500ml triangular flask, culture at 30℃ for another 22 hours (170r/min).

3) Centrifuged 8000r/min 10 minutes in 4℃ for collection broth.

2. extraction

Weight 2g of bacteria, then added with 100ml acetone, stirred for 10 minutes, centrifuged for 3 minutes at 8000r/min, discard the precipitate. Vacuum distillation to dry, then wash with alcohol, reserve.

3. Concentration

Five extracts were combined to concentrate by vacuum evaporator. The water bath temperature shall not exceed 50℃ until it is dried, and dissolved in 0.5ml petroleum ether.

4. Purification by silica gel column chromatography

Use the glass wool to plug the 1.2cm × 20cm column, then fill the mixture of 11g silica gel and 20ml petroleum by a funnel. After the column was filled, 10ml petroleum ether was used to cross the column to stabilize the column bed. Carefully put the concentration on the column by dropper. Use ether : petroleum ether (2 : 8) as eluent. The eluent was collected in a small test tube, and each tube was fed with 2ml petroleum ether. TLC identification was performed.

5. TLC

(1) Preparation of thin-layer chromatographic plate

Weigh 2g silica gel for thin-layer chromatography into a latex bowl, add 0.5% CMC-Na 5ml to grind for a moment, quickly spread it evenly on the washed and dried slide (5.5cm × 12cm two pieces), then move it into oven gently after 1 hour, and activate it at 105℃ for 1 hours.

(2) Spotting

The capillary was used to absorb the samples collected from the above tubes at 1.5cm of the end of the chromatographic plate, and the standard solution of coenzyme Q_{10} was used as the control. The sample volume was 10 ~ 15μl, then dried it.

(3) Spread out

Place the sheet vertically in the expanding bottle with the expander and cover the bottle. The expander is ether : petroleum ether = 2 : 8, the point must be above the liquid level; when the expander rises to 2cm away from the edge of the thin layer, it is removed and dried.

(4) Colour rendering

Place iodine vapor in iodine cylinder to colour. Draw the position of each point with pencil, and calculate the R_f value of coenzyme Q_{10}.

7. Crystal

Thin - layer identification of the more pure tube, to thin - film evaporator vacuum concentration. The temperature of the water bath is no more than 50℃ until it is steamed and dried. The oil is moved to a small test tube with a few drops of absolute ethanol while it is hot. The ordinary refrigerator is placed overnight to crystallize and weigh.

8. Content determination (HPLC)

(1) HPLC chromatographic conditions mobile phase: methanol : ethanol = 3 : 7, detection wavelength: 275nm, flow rate: 1ml/min, column temperature: 35℃ column: reverse C_{18}, injection volume: 10ml.

(2) Drawing Standard Curve Coenzyme Q_{10} is more stable in oxidative form than in reductive form, so it mostly exists in oxidative form. The characteristic absorption peaks were observed at 275nm, and the peak area was linearly related to the content of coenzyme Q_{10} in a certain range.

Tube number	1	2	3	4	5	6
coenzyme Q_{10} Standard solution (100μg/ml)	0	1	2	3	4	5
Absolute ethanol	5	4	3	2	1	0
OD_{275}						

The concentration of CoQ_{10} (x, μg/ml) was plotted with peak area (y) by HPLC. Make standard curve or regression equation.

(3) The concentration (g/ml) of coenzyme Q_{10} solution, the total production (g) of coenzyme Q_{10} in the fermentation broth and the content of coenzyme Q_{10} per unit cell (g/g bacterial weight) were calculated according to the peak area.

【Questions】

1. What are the advantages of microbial production of coenzyme Q_{10}?

2. What factors affect the extraction efficiency of coenzyme Q_{10}? What are the points of attention in the extraction process?

扫码"学一学"

实验二十三　青霉素钾盐的制备及含量测定

【实验目的】

1. 掌握　溶媒萃取法分离原理；青霉素碘量法含量测定的原理和方法。

2. 了解　产黄青霉培养发酵生产青霉素的基本方法；溶媒萃取法纯化青霉素的操作方法。

3. 学习青霉素钾盐的制备方法。

【实验原理】

最早发现的产青霉素菌种是点青霉菌（*Penicillium notatum*），生产能力很低，表面培养只有几个单位，远不能满足工业生产的需求。后发现适合于深层培养的产黄青霉菌（*P. chrysogenum*），经一系列诱变、杂交、育种，目前工业上青霉素的发酵单位已超过每毫升6万单位。

青霉菌能利用的碳源有葡萄糖、蔗糖、淀粉、天然油脂等。乳糖能被产生菌缓慢利用而维持青霉素分泌的有利条件，故为最佳碳源，但价格高。目前生产上用的碳源多为工业用葡萄糖。在氮源上，玉米浆含多种氨基酸及β-苯乙胺等，后者为苄青霉素生物合成提供侧链的前体，因而是青霉素发酵常用的一种氮源。

青霉素（本身是一种有机弱酸，$pK_a=2.75$）在pH 2左右呈游离酸状态，易溶于醇类、酮类、醚类和酯类，但在水中溶解度很小。另一方面，在pH 7左右，青霉素能与碱金属或碱土金属及有机胺类结合成盐，其在有机溶剂和水中的溶解性跟青霉素游离酸有很大的不同：青霉素的金属盐十分易溶于水和易溶于甲醇，几乎不溶于乙醚、氯仿或醋酸戊酯。从滤液中提取青霉素就是利用青霉素以游离酸或成盐状态存在时，在水及溶媒中的溶解度不同这一性质。在一定温度下达到平衡时，青霉素在两相间浓度的关系服从分配定律，青霉素在酸性条件下以游离酸的状态转入溶媒相，在碱性条件下以盐的状态反萃取（reverse extraction）到水相，经过第二次转入溶媒相后，掺入醋酸钾，预期获得青霉素钾盐结晶，如下图所示：

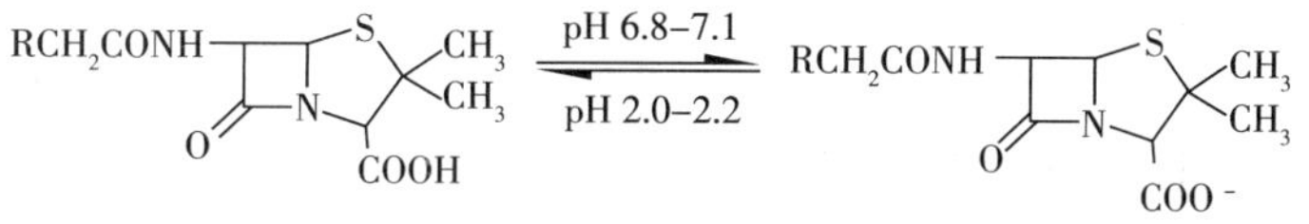

RCH_2CONH–(β-内酰胺噻唑烷环, S, CH_3, CH_3, O, N, COOH) $\underset{pH\ 2.0-2.2}{\overset{pH\ 6.8-7.1}{\rightleftharpoons}}$ RCH_2CONH–(β-内酰胺噻唑烷环, S, CH_3, CH_3, O, N, COO^-)

青霉素在水溶液中，当pH > 7，β-内酰胺环水解而生成青霉噻唑酸。青霉噻唑酸能和4分子的碘起作用，而青霉素不吸收碘。碘量法测定青霉素含量就是根据这一原理。

在pH 4.5时，青霉噻唑酸与碘按上式进行定量反应，一个分子的青霉素要消耗4分子的碘，过量的碘用硫代硫酸钠标准溶液回滴定，即可计算出青霉素的效价。测定样品的同时作空白平行实验，以消除存在的产物和其他耗碘物质的影响。

【实验材料】

1. 器材

（1）分液漏斗	60ml	1只
（2）筛板漏斗		1只
（3）烧杯	50ml	2只
（4）抽滤瓶	250ml	1只
（5）量杯	10ml	1只
	50ml	1只
（6）铁架台		1个
（7）铁圈		1只
（8）吸量管	1ml	4支

	5ml	2 支
（9）碘量瓶	250ml	1 只
（10）碱式滴定管		1 根
（11）试管	15ml	1 支
（12）三角烧瓶	250ml	1 个
	500ml	1 个
（13）玻璃棒		1 支

（14）滤纸

（15）恒温箱

（16）恒温振荡培养箱

（17）恒温水浴锅

（18）真空泵

2. 试剂

（1）1mol/L H_2SO_4：取 5.56ml 硫酸缓慢的加到 80ml 水中，边加边连续搅拌，用水定容到 100ml。

（2）10% KI

（3）0.5% 淀粉指示剂：取可溶性淀粉 5g，用适量蒸馏水搅匀，然后倒入煮沸的蒸馏水中，搅拌，煮沸 2 分钟，加入蒸馏水使总体积达 1000ml，冷却备用。

（4）0.01mol/L $Na_2S_2O_3$标准溶液

配制：取 26g $Na_2S_2O_3 \cdot 5H_2O$ 和 0.2g Na_2CO_3 溶于新沸过放冷的蒸馏水中，稀释 1000ml，转入棕色瓶中，放置 7～10 天。使用前过滤，标定，用新沸过放冷的蒸馏水稀释 10 倍，即为 0.01mol/L $Na_2S_2O_3$标准溶液。

标定：将基准 KIO_3研细，在 105℃干燥至恒重，称取约 0.8g，精密称定，加水溶解，转移到 1000ml 容量瓶，定容至刻度。精密吸取 10ml 置 250ml 碘量瓶中，加 10% KI 10ml，1mol/L $H_2SO_4$10ml 和 20ml 水，密塞摇匀，于暗处放置 3～10 分钟（使反应进行完全），用 $Na_2S_2O_3$溶液滴定至淡黄色，加淀粉指示剂 1ml，继续滴定至蓝色消失，并将结果用空白试验校正。

（5）36% KI 溶液

（6）0.01mol/L 碘液：取 13g 碘放入一盛有 KI 溶液（36%）100ml 的研钵中研细后使完全溶解，然后转移到烧杯中，加 3 滴盐酸，加蒸馏水适量，使完全溶解，过滤，用蒸馏水稀释至 1000ml，使用时稀释 10 倍，即为 0.01mol/L 碘液。碘液须盛于棕色玻璃瓶中密塞。

（7）pH 4.5 醋酸缓冲液：称取无水醋酸钠 21.6g，加入醋酸 19.6ml，稀释至 2000ml。

（8）1mol/L NaOH

（9）1mol/L HCl

（10）2% Na_2CO_3溶液

（11）10% 硫酸

（12）50% 醋酸钾乙醇溶液（W/V）：称取 50g 醋酸钾，以乙醇溶解并定容到 100ml。

（13）pH 试纸：广泛试纸；pH 0.5～4；pH 6～9。

3. 培养基

（1）斜面培养基：取去皮土豆 20g，切成小块，加入 100ml 水煮沸 20 分钟，4 层纱布过滤，滤液加入2g葡萄糖，2g琼脂粉，补水到100ml，加热使琼脂粉溶解，分装入15ml试管，加塞，于121℃灭菌20分钟，在琼脂凝固前取出斜躺冷却成斜面。

（2）种子培养基：取1g葡萄糖，3.8g玉米浆，装入250ml三角烧瓶，加50ml水，搅拌溶解后调pH 5.8，加入0.25g碳酸钙，用8层纱布塞口，于121℃灭菌20分钟。

（3）发酵培养基：取13g乳糖，4g玉米浆，0.3g磷酸二氢钾，0.4g硫酸铵，0.15g硫酸亚铁，装入500ml三角烧瓶，加100ml水，搅拌溶解后调pH 5.8，加入1g碳酸钙，用8层纱布塞口，于121℃灭菌20分钟。

【实验方法】

1. 产黄青霉菌的培养发酵　将青霉素产生菌产黄青霉菌接种于斜面培养基，25℃培养7天。从培养好的斜面培养基表面挖取$2cm^2$琼脂块，接入50ml种子培养基中，25℃，220r/min条件下培养44小时，然后取10ml接种于100ml发酵培养基中，25℃，220r/min条件下培养7～8天。

2. 溶媒萃取法纯化青霉素　发酵液过滤，取30ml滤液，用10% H_2SO_4调pH 2.0～2.2（此步骤应尽可能迅速），然后倒入分液漏斗中，加入1/2总体积量的（15ml）醋酸丁酯，振摇10分钟，静置10分钟，分出水相，再加入1/5总体积量的（6ml）醋酸丁酯于水相中，操作同前，弃去水相。合并两次丁酯相，并加入体积分别为10ml、5ml、2ml的2% $NaHCO_3$进行分批反萃取，每次振摇10分钟，静置10分钟，弃去丁酯层，合并三次水相。然后用10% H_2SO_4调水相pH 2.0～2.2，再分两次加入总体积量1/2、1/5的醋酸丁酯，振摇，静置分层后弃去水相，于酯相中加入少量无水硫酸钠，搅拌片刻，过滤。最后滤液中加入50%醋酸钾乙醇溶液1ml，30℃水浴搅拌10分钟，可析出青霉素钾盐，抽滤得湿晶体，自然风干后称重。

3. 样品制备　精密称取制备青霉素钾样品35mg，置50ml容量瓶中，用蒸馏水溶解，定容至50ml。

4. 样品滴定　准确吸取上述配好的样品5ml，放入250ml碘量瓶中，加入1mol/L NaOH 1ml，于室温放置20分钟，然后依次加入1mol/L HCl 1ml，pH 4.5醋酸缓冲液5ml，0.01mol/L碘标准溶液20ml，于暗处放置20分钟，加0.5%淀粉指示剂1ml，用0.01mol/L $Na_2S_2O_3$标准溶液滴定至蓝色消失，记录其毫升数$V_{样}$。

5. 空白滴定　取青霉素钾样品液5ml，放入250ml碘量瓶中，依次加入pH 4.5醋酸缓冲液5ml、0.01mol/L碘标准溶液20ml，于暗处放置20分钟，加0.5%淀粉指示剂1ml，用0.01mol/L $Na_2S_2O_3$标准溶液滴定至蓝色消失，记录其毫升数$V_{空}$。

6. 计算结果

$$\text{青霉素 G 钾盐效价(U/ml)} = \frac{(V_{空} - V_{样}) \times M \times 372.5 \times 1598}{8 \times 5}$$

式中，$V_{空}$——空白消耗$Na_2S_2O_3$标准溶液的毫升数；$V_{样}$——样品消耗$Na_2S_2O_3$标准溶液的毫升数；M——$Na_2S_2O_3$标准溶液的摩尔浓度；372.5——青霉素G钾盐的分子量；1598——1毫克青霉素G钾盐的理论效价；8——1个青霉素分子消耗8个碘原子（4分子碘）；5——青霉素溶液取样量（ml）。

实验结果要求计算所得青霉素钾盐的效价及纯度，计算产量。

【思考题】

1. 加入少量无水硫酸钠丁酯相中的目的是什么？

2. 青霉素G游离酸及青霉素G钾（钠）盐在水中及有机溶媒中的溶解特性如何？在本实验中是如何应用这些特性进行分离提取的？

3. 除碘量法外，青霉素的含量测定还可以用什么方法？比较各方法的优缺点。

EXPERIMENT 23 Preparation and analysis of penicillin potassium salt

【Purpose】

1. To master the isolation principle of solvent extraction and the principles and methods of penicillin iodometric titer determination.

2. To understand the basic method of producing penicillin by fermentation of Penicillium chrysogenum, and the operation of purifying penicillin by solvent extraction.

3. To learn how to prepare penicillin potassium salt.

【Principle】

The earliest discovered penicillin - producing strain was *Penicillium notatum*, which had a very low production capacity, only a few units from surface culture. This is far from meeting the requirements of industrial production. After finding a *P. chrysogenum* suitable for deep culture, after a series of mutagenesis, hybridization and breeding, the fermentation unit of penicillin in the industry has exceeded 60000 units per ml.

The carbon sources that can be utilized by *Penicillium* are glucose, sucrose, starch, natural oils, etc. Lactose can be slowly used by the bacteria to maintain the favorable conditions for penicillin secretion, so it is the best carbon source, but the price is high. At present, most of the carbon sources used in production are industrial glucose. For the nitrogen source, corn syrup contains a variety of amino acids and β - phenethylamine, the latter provides a precursor for the side chain of benzylpenicillin biosynthesis, and thus is a nitrogen source commonly used for penicillin fermentation.

Penicillin (an organic weak acid itself, p*K*a = 2.75) is free acid at pH 2, soluble in alcohols, ketones, ethers and esters, but with little solubility in water. On the other hand, at about pH 7, penicillin can form a salt with alkali metal or alkaline earth metal and organic amine. Its solubility in organic solvent and water is very different from penicillin free acid. The metal salt of penicillin is very soluble in water and soluble in methanol, almost insoluble in ether, chloroform or amyl acetate. Extraction of penicillin from the filtrate is a property in which the solubility in water and solvent is different when penicillin is present in the form of free acid or salt. When equilibrium is reached at a certain temperature, the relationship between the concentrations of penicillin in the two phases obeys the law of distribution. Penicillin is transferred to the solvent phase in the state of free acid under acidic conditions and is reversely extracted to the aqueous phase in the state of salt under al-

kaline conditions. After transferring to the solvent phase, with potassium acetate incorporated, the penicillin potassium salt crystals are expected to obtain, as shown in the following figure.

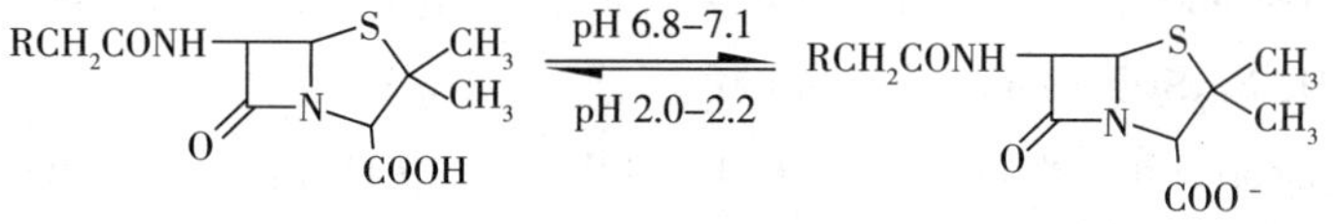

RCH_2CONH — S — CH_3, CH_3 — N — O — COOH $\underset{\text{pH 2.0–2.2}}{\overset{\text{pH 6.8–7.1}}{\rightleftharpoons}}$ RCH_2CONH — S — CH_3, CH_3 — N — O — COO^-

When penicillin is in aqueous solution, pH > 7, the β – lactam ring is hydrolyzed and penicillin thiazolyl is formed. Penicillin thiazole can act with 4 molecules of iodine, while penicillin does not absorb iodine. Theiodometric method for determining penicillin content is based on this principle.

At pH 4. 5, penicillin and iodine are quantitatively reacted according to the above formula. One molecule of penicillin consumes 4 molecules of iodine, and excess iodine is titrated with sodium thiosulfate standard solution to calculate the potency of penicillin. A blank parallel experiment should be performed while the samples are measured to eliminate the effects of existing products and other iodine – depleting substances.

【Materials】

1. Apparatus

(1) Separatory funnel

(2) Sieve plate funnel

(3) Beaker

(4) Filter bottle

(5) Measuring cup

(6) Iron stand

(7) Iron ring

(8) Pipette

(9) Iodine measuring bottle

(10) Basic burette

(11) Tubes

(12) Triangular flask

(13) Glass rod

(14) Filter paper

(15) Thermostat

(16) Constant temperature shaking incubator

(17) Constant temperature water bath

(18) Vacuum pump

2. Reagents

(1) 1mol/L H_2SO_4

Add 5. 56ml of sulfuric acid slowly to 80ml of water, continuously stir, adjust the volume to 100ml with water.

(2) 10% KI

(3) 0.5% starch indicator

Weigh 5g of soluble starch, mix well with distilled water, then pour into boiling distilled water, stir, boil for 2 minutes, add distilled water to make the total volume up to 1000ml, cool and set aside.

(4) 0.01mol/L $Na_2S_2O_3$ standard solution

Preparation: dissolve 26g $Na_2S_2O_3 \cdot 5H_2O$ and 0.2g Na_2CO_3 in fresh boiling and cooling down distilled water, dilute 1000ml, transfer to brown bottle, and place for 7 – 10 days. Filter and calibrate before use. Diluted 10 times with fresh boiling and cooling down distilled water.

Calibration: grind the reference KIO_3, dry to constant weight at 105℃, weigh 0.8g, accurately, dissolve KIO_3 in water, transfer to a 1000ml volumetric flask, and make up to the mark. Accurately pipette 10ml into 250ml iodine volumetric flask, add 10% KI 10ml, 1mol/L H_2SO_4 10ml and 20ml water, shake it tightly, place it in the dark for 3 ~ 10 minutes (to make the reaction complete), titrate with $Na_2S_2O_3$ solution to light Yellow, add 1ml of starch indicator, continue titration until blue disappears, and correct the results with a blank test.

(5) 36% KI solution

(6) 0.01mol/L iodine solution

Take 13g iodine into a mortar containing 100ml of KI solution (36%), then completely dissolve it, then transfer it to a beaker, add 3 drops of hydrochloric acid, add distilled water, Dissolve completely, filter, dilute to 1000ml with distilled water, and dilute 10 times when used, that is, 0.01mol/L iodine solution. The iodine solution should be stored in a brown glass bottle.

(7) pH 4.5 Acetic acid buffer

Weigh 21.6g of anhydrous sodium acetate, add 19.6ml of acetic acid, and dilute to 2000ml.

(8) 1mol/L NaOH

(9) 1mol/L HCl

(10) 2% Na_2CO_3 solution

(11) 10% sulfuric acid

(12) 50% potassium acetate ethanol solution (*W/V*)

Dissolve 50g potassium acetate in ethanol and dilute it to 100ml.

(13) pH test strip

Extensive pH test strip; pH 0.5 to 4; pH 6 to 9.

3. Medium

(1) Inclined medium

Take 20g peeled potatoes, cut into small pieces, add 100ml of water and boil for 20 minutes, filter with 4 – layer gauze, add 2g glucose to the filtrate, 2g agar powder, make up water to 100ml, heat to dissolve the agar powder, and dispense The tube was placed in a 15ml tube, stoppered, and sterilized at 121℃ for 20 minutes. Take the tubes out and slant to form a slope before solidification.

(2) Seed medium

Take 1g glucose, 3.8g corn syrup, add 250ml erlenmeyer flask, add 50ml water, stir to dissolve and adjust pH 5.8, add 0.25g calcium carbonate, plug with 8 layers of gauze, sterilize at 121℃ for 20 minutes.

(3) Fermentation medium

Take 13g lactose, 4g corn syrup, 0.3g potassium dihydrogen phosphate, 0.4g ammonium sulfate, 0. 15g ferrous sulfate, into a 500ml triangular flask, add 100ml of water, stir to dissolve and adjust pH 5. 8. Add 1g calcium carbonate, cover the triangular flask with 8 – layer gauze and sterilize at 121℃ for 20 minutes.

【Procedures】

1. Culture and fermentation of Penicillium chrysogenum

Inoculatepenicillin – producing fungi into the slant medium and culture at 25℃ for 7 days. Excavate $2cm^2$ agar block from the surface of the cultured slanted medium, and then add into 50ml seed culture medium, culture at 25℃, 220r/min for 44 hours, then inoculate 10ml into 100ml fermentation medium, 25℃, 220r/min. Culture for 7 to 8 days.

2. Purification of penicillin by solvent extraction

Filter the fermentation broth, take 30ml the filtrate, and adjust the pH to 2. 0 – 2. 2 with 10% H_2SO_4 (this step should be as fast as possible), then pour the solution into a separatory funnel. Add a total volume of 1/2 (15ml) of butyl acetate. Shake for 10 minutes, let stand for 10 minutes, separate the aqueous phase, and then add 1/5 of the total volume of (6ml) butyl acetate in the aqueous phase, operate the same as before, discard the aqueous phase. Combine the butyl ester phase twice. Add 10ml, 5ml, 2ml of 2% $NaHCO_3$ for batch back extraction, shake for 10 minutes each time, stand for 10 minutes, discard the butyl ester layer, and combine three aqueous phases. Then, adjuste pH of the aqueous phase to 2. 0 – 2. 2 with 10% H_2SO_4, then add 1/2 and 1/5 of butyl acetate in two portions, shake, stand layering, and discard the aqueous phase in the ester phase. Add a small amount of anhydrous sodium sulfate, stir for a while, and filter. Then, add 1ml 50% potassium acetate ethanol solution to the filtrate, and stir the mixture in a water bath at 30℃ for 10 minutes to precipitate a penicillin potassium salt, and obtain the wet crystal by suction filtration. Finally, air – dry and weigh.

3. Sample Preparation

Weigh 35mg penicillin potassium sample, place in a 50ml volumetric flask, dissolve in distilled water, and make up to 50ml.

4. Sample titration

Accurately absorb 5ml of the above prepared sample, put it into 250ml iodine measuring flask, add 1ml/L NaOH 1ml, stand for 20 minutes at room temperature. Then add 1ml/L HCl 1ml, pH 4. 5 acetate buffer 5ml, 0. 01mol/L 20ml of iodine standard solution. Place it in the dark for 20 minutes. Add 0. 5ml 0. 5% starch indicator, titrate with a 0. 01mol/L $Na_2S_2O_3$ standard solution until the blue color disappear, and record the number of milliliters of V_{sample}.

5. Blank titration

Take 5ml penicillin potassium sample solution, put it into 250ml iodine volumetric flask, add 5ml pH 4. 5 acetate buffer, 20ml 0. 01mol/L iodine standard solution. Place it in the dark for 20min. Add 0. 5ml 0. 5% starch indicator, titrate with a 0. 01mol/L $Na_2S_2O_3$ standard solution until the blue color disappear, and record the number of milliliters of V_{empty}.

6. Calculation results

Calculation formula:

$$\text{Potency of Penicillin G potassium salt(U/ml)} = \frac{(V_{empty} - V_{sample}) \times M \times 372.5 \times 1598}{8 \times 5}$$

V_{empty}——the number of milliliters of blank consumed $Na_2S_2O_3$ standard solution.

V_{sample}——the number of milliliters of sample consumption $Na_2S_2O_3$ standard solution.

M——molar concentration of $Na_2S_2O_3$ standard solution.

372. 5——molecular weight of penicillin G potassium salt.

1598——theoretical titer of 1mg penicillin G potassium salt.

8——1 penicillin molecule consumes 8 iodine atoms(4molecules of iodine).

5——Sampling amount of penicillin solution(ml).

The experimental results require calculation of the potency and purity of the penicillin potassium salt and calculation of the yield.

【Questions】

1. What is the purpose of adding a small amount of anhydrous butyl sulfate phase?

2. What is the solubility characteristic of penicillin G free acid and penicillin G potassium(sodium)salt in water and organic solvents? How do you apply these features for isolation and extraction in this experiment?

3. What method can be used to determine thecontent of penicillin in addition to the iodometric method? Compare the advantages and disadvantages of each method.

第二节　生物技术类药物的制备

实验二十四　重组水蛭素Ⅲ的制备与分析

扫码“学一学”

【实验目的】

1. 掌握　基因工程菌发酵培养条件、分泌表达目的蛋白及表达蛋白分析鉴定的方法。

2. 了解　以基因工程菌种 *E. coli*/rHV3 为材料进行工程菌发酵培养、外分泌表达小分子多肽类药物重组水蛭素Ⅲ及表达蛋白制备和抗凝血酶活力测定、SDS－PAGE 电泳分析鉴定的工作原理。

【实验原理】

水蛭素（Hirudin）是一类由医用水蛭唾液腺中分泌出的低分子量的酸性单链多肽。1904 年，Markwardt 首先将其分离纯化，并于 1970 年确定它是迄今发现的最强的凝血酶特异性抑制剂，即使在较低的浓度也可充分地阻止血液的凝固。药理学及临床研究表明，水蛭素能有效地预防静动脉血栓的形成及弥散性血管凝结，不引起过敏、免疫反应，不会造成循环功能障碍。可用于治疗不稳定性心绞痛（USA）、急性心肌梗死（AMI）、血管成形术、术后血栓形成、血液透析、体外循环、弥散性血管内凝血（DIC）等，是很好的抗凝抗栓药物。

水蛭素的分子质量约为7000Da，含有65～66 个氨基酸残基，水蛭素肽链的二级和三级结构对其抗凝活性起决定性作用，其 N 端的 3 个二硫键则是决定分子二级和三级结构及其稳定性的关键，将二硫键氧化或分子发生蛋白降解，则失去抗凝活性。若 C 端羧基被酯化或失去 C 端氨基酸，也会失去与凝血酶结合的能力。

水蛭素干燥状态下稳定，室温下水中可稳定存在 6 个月，80℃下加热 15 分钟不被破坏。pH 值升高则稳定性下降，在 0. 1mol/L 盐酸溶液或 0. 1mol/L 氢氧化钠溶液中可稳定 15 分钟。水蛭素不被胰蛋白酶水解，对 α－糜蛋白酶也有一定的耐受性，但番木瓜蛋白酶、胃蛋白酶和枯草杆菌蛋白酶 A 则可使之失去活性。

由于天然水蛭素来源非常困难，每条水蛭只含 20μg 的水蛭素，不能满足临床应用。为此，国外多家公司及研究单位从 80 年代末就开始研究采用基因工程技术生产重组水蛭素。目前，在国外已上市的水蛭素产品有 Hoechs 公司的重组水蛭素（Lepirudin，商品名 Refludon），Novartis 公司的重组水蛭素产品（Desirudin）。1997 年，本院合成了重组水蛭素Ⅲ基因并构建了大肠杆菌外分泌表达体系，分泌型重组水蛭素Ⅲ基因工程菌的表达产物重组水蛭素Ⅲ（recombine hirudinⅢ，简写为 rHV3）相对分子质量为 7010. 8Da。

此工程菌为外分泌表达体系，其基因表达产物 rHV3 易于纯化。但该表达系统会产生一些杂质蛋白和色素，影响下游的分离、纯化。提取纯化工艺采用了大孔吸附树脂疏水层析和离子交换层析方法，较好地解决了杂质蛋白和色素的分离。大孔吸附树脂作为 rHV3 的第一个提取步骤，高分子量杂蛋白被基本去除，而离子交换则是去色素的关键。吸附及离子交换时采用合适的洗脱条件是提高活力收率的重要因素。该纯化方法工艺简单、rHV3 收率和纯度均较高，适用于 rHV3 的大规模制备。

本实验以此工程菌为材料，先进行工程菌发酵培养，再将发酵液离心，从工程菌发酵液上清中分离纯化 rHV3。发酵上清液经过大孔吸附树脂 HP20 处理后，用 DEAE－52 阴离子纤维素交换层析，得到较高纯度表达产物 rHV3。rHV3 进行抗凝血酶活力分析和 15% SDS－PAGE 电泳分析。rHV3 生物活性的测定参照 Markwardt 的凝血酶滴定方法进行。比活力测定方法为：精确称取纯化产物 10mg，溶解于 1ml H_2O 中，采用 Lowry 法测定蛋白质含量，并测定抗凝血酶活力，rHV3 比活力＝抗凝活力（ATU）/蛋白含量。

【实验材料】

1. 器材

（1）玻璃层析柱　1. 6cm×20cm 或/和 2. 6cm×40cm　1 套
（2）恒流泵　1 台
（3）自动部分收集器　1 台
（4）记录仪　1 台
（5）恒温水浴锅　1 台
（6）高速台式离心机　1 台
（7）冷冻离心机　1 台
（8）旋涡混合器　1 台
（9）752 紫外分光光度计　1 台
（10）稳压电泳仪　1 台
（11）摇床　1 台
（12）垂直板式电泳槽　1 套

(13) 电炉 1只
(14) 秒表 1只
(15) 高压灭菌锅 1台
(16) 电磁搅拌器 1台
(17) 酸度计 1台
(18) 发酵罐 1台
(19) 微量移液器(20μl, 200μl, 1000μl) 各1支
(20) 标准净化工作台 1台
(21) 真空泵和真空干燥器 1套
(22) 电子天平(精确到10mg) 1台
(23) 照相器材 1套
(24) 基因工程菌菌种:*E. coli*/pHV3为本实验室保存
(25) 大孔吸附树脂HP20 若干
(26) DEAE-52阴离子纤维素 若干
(27) Eppendorf离心管架 1只
(28) 96孔酶标板 1只
(29) 脱色摇床 1台
(30) 灭菌的移液器枪头 若干
(31) 灭菌的Eppendorf管(1.5ml微量离心管) 若干
(32) 试剂瓶(5000ml, 1000ml, 500ml, 250ml, 100ml) 若干
(33) 量筒(2000ml, 1000ml, 100ml, 10ml) 各1只
(34) 烧杯(2000ml, 1000ml, 500ml, 250ml, 50ml) 各3只
(35) 布氏漏斗(8cm)和吸滤瓶(1000ml) 各1只
(36) 滤纸 若干
(37) 玻棒 2根
(38) 试管(5ml, 150支; 15ml, 16支)
(39) 三角瓶(250ml) 2只
(40) 大塑料桶 1只
(41) 石英比色杯 1对

2. 试剂

重组水蛭素Ⅲ制备和抗凝血酶活力:

(1) 培养基

1) 种子培养基(LB液体培养基):1%胰蛋白胨(Tryptone), 0.5%酵母提取物(Yeast Extract), 1% NaCl, 加蒸馏水配制, pH 7.0, 分装, 高压蒸汽(1.03×10^5Pa)灭菌20分钟。接种前在无菌条件下加入抗生素, 氨苄青霉素的终浓度为100μg/ml。

2) 发酵培养基:1%蛋白胨, 0.5%酵母提取物, 4%谷氨酸钠, 10%麦芽汁。pH 6.5, 摇瓶装液量为12%。

(2) 氨苄青霉素(Amp)

(3) 0.5%牛血纤维蛋白原:0.05mol/L Tris-HCl缓冲液(pH 7.4)配制。

(4) 凝血酶溶液:500 NIH单位加5ml蒸馏水。

（5）低分子量标准蛋白（17.5～94kDa）

（6）20mmol/L 乙酸

（7）盐酸

（8）异丙醇

（9）20mmol/L 哌嗪

（10）氯化钠

（11）三（羟甲基）胺基甲烷（Tris）

（12）乙醇

（13）氢氧化钠

（14）谷氨酸钠

（15）15% SDS－PAGE 电泳分析：同实验九。

【实验方法】

1. 工程菌发酵培养

（1）从平板上挑取菌种接种于新鲜的含氨苄青霉素（终浓度为60μg/ml）的 LB 液体培养基中，37℃，摇床转速 220r/min，培养 18 小时。

（2）摇瓶培养　再按 3% 接种量转接于发酵培养基中，37℃，250r/min，30 小时，离心收集上清。

（3）补料－分批式发酵　30L 的发酵罐装入 15L 发酵培养基，初始 pH 为 6.5，121℃下灭菌 20 分钟，待温度降至 37℃ 时加氨苄青霉素至最终浓度为 60μg/ml，接入种子液 1000ml，37℃培养，调节搅拌速度及通气量，控制溶氧在 10%～20%，通过补充酸、碱物质控制 pH 在 6.0～7.5，同时通过流加葡萄糖或蛋白胨调节培养基中营养成分，延长活力表达时间。培养 9～16 小时。

2. 重组水蛭素Ⅲ分离纯化制备

（1）发酵上清液制备　下罐后发酵液经 12000r/min 高速离心除去菌体，收获发酵上清液。采用 Lowry 法和 Markwardt 的凝血酶滴定法分别检测目的蛋白重组水蛭素Ⅲ含量和活性及进行 15% SDS－PAGE 电泳分析。计算收率。

（2）大孔吸附层析　大孔吸附树脂 HP20 处理装柱，用 20mmol/L 乙酸平衡。调节发酵上清液 pH 至 5.0～6.0，于抽气泵中脱去气泡，以 1ml/min 的流速将样品上柱。依次用 20mmol/L 乙酸洗涤和 50mmol/L pH 8.5 Tris－HCl 洗涤，再用含 10%～40% 异丙醇的 20mmol/L 乙酸梯度洗脱，分部收集器收集，1.5ml/min，每管收集 6ml，测定每管的 A_{280} 值和重组水蛭素Ⅲ活性，绘制洗脱曲线。收集显示活性组分，检测重组水蛭素Ⅲ含量和活性及进行 15% SDS－PAGE 电泳分析。计算收率。

大孔吸附柱处理方法：首先用蒸馏水漂洗，去除漂浮的细小颗粒，再用 95% 乙醇反复浸泡使其充分溶胀和初步除杂。然后用 95% 乙醇在布氏漏斗中淋洗至乙醇在 254nm 的紫外吸收值小于 0.03 为止。再用蒸馏水洗尽乙醇。最后分别用 5% 盐酸和 5% 氢氧化钠浸泡 3 小时，均分别洗至 pH 为中性。

（3）DEAE－52 阴离子交换柱层析　DEAE－52 阴离子纤维素柱经 20mmol/L 哌嗪平衡。大孔吸附层析产物液样品经 10mmol/L 哌嗪调节至 pH 6.0 后上样。用平衡缓冲液洗涤至基线平稳。再用含 0.1～0.4mol/L NaCl 的 20mmol/L 哌嗪（pH 6.0）溶液梯度洗脱，分

部收集器收集，1.5ml/min，每管收集6ml，测定每管的A_{280}值和重组水蛭素Ⅲ活性，绘制洗脱曲线。收集显示活性组分，冷冻干燥后即得高纯度的重组水蛭素Ⅲ冻干粉。检测重组水蛭素Ⅲ含量和活性及进行15% SDS－PAGE电泳分析。计算收率。

3. 重组水蛭素Ⅲ生物活性测定 于酶标板小孔中加0.5%牛血［0.05mol/L Tris－HCl缓冲液（pH 7.4）配制］200μl，再加入重组水蛭素Ⅲ溶液10～100μl，充分混匀。用微量进样器吸取标准的凝血酶溶液（100 NIH单位），时间间隔为1分钟，若在1分钟内纤维蛋白原发生凝固，即说明已达滴定终点。由凝血酶的消耗量换算出重组水蛭素Ⅲ的单位数。由于水蛭素与凝血酶是1∶1结合，故每消耗一个凝血酶单位（NIH）相当于一个抗凝血酶单位（ATU）。

4. 重组水蛭素Ⅲ比活力测定方法 精称纯化产物10mg，溶解于1ml H_2O中，采用Lowry法测定蛋白质含量，并测定抗凝血酶活力，经以下公式换算即得单位质量的rHV3比活力。

rHV3比活力＝抗凝活力（ATU）/蛋白含量。

5. 15%SDS－PAGE电泳分析样品纯度 电泳方法同实验九。

（1）按常规制备电泳用聚丙烯酰胺凝胶，浓度15%。

（2）将发酵液上清、解吸液和离子交换洗脱液样品处理后，进行15% SDS－PAGE电泳析。

（3）电泳2～3小时后，取出凝胶，用0.15%考马斯亮蓝－R250进行染色。过夜后倾出染液，不断脱色，起初每20分钟换液1次，1小时后每小时换液1次，直到区带明显，画出电泳区带图谱。

6. 结果处理

（1）用箭头图表示重组水蛭素Ⅲ制备的操作流程。

（2）绘制洗脱曲线。

（3）重组水蛭素Ⅲ的纯化过程分析

重组水蛭素Ⅲ的纯化过程分析

纯化步骤	总蛋白（mg）	总活力（ATU）	比活（ATU/mg）	纯化倍数	收率（%）
发酵上清液					
大孔吸附层析					
DEAE－52纤维素柱层析					

（4）电泳图谱分析。

扫码“练一练”

【思考题】

1. 影响重组水蛭素Ⅲ纯度和收率的因素有哪些？如何加以控制？
2. 试说明电泳图谱，并用电泳图谱判断所制得的重组水蛭素Ⅲ的纯度。
3. 如何在原核内成功表达小分子多肽类药物而不被宿主内蛋白酶水解？

Section 2 Preparation of biotechnological drugs

EXPERIMENT 24 Preparation and analysis of Recombinant Hirudin Ⅲ

【Purpose】

1. To master the fermentation conditions of genetic engineering bacteria and the method of secretory expression of target protein and expression protein analysis and identification.

2. To know the principle of genetic engineering strain *E. coli*/pHV3 as the material for engineering fermentation fermentation, exocrine expression of small molecule peptide drugs, recombinant hirudin Ⅲ, preparation of expressed protein, determination of antithrombin activity and identification by SDS - PAGE.

【Principle】

Hirudin is a low - molecular - weight acidic single - chain polypeptide secreted from the salivary glands of the medical leeches. In 1904, Markwardt first isolated and purified it, and in 1970 it was determined to be the strongest to date. Thrombin - specific inhibitors adequately block blood coagulation even at lower concentrations. Pharmacology and clinical studies have shown that hirudin can effectively prevent the formation of arterial thrombosis and disseminated vascular coagulation, does not cause allergies, immune reactions, and does not cause circulatory dysfunction. Can be used to treat unstable angina (USA), acute myocardial infarction (AMI), angioplasty, postoperative thrombosis, hemodialysis, extracorporeal circulation, disseminated intravascular coagulation (DIC), etc., is a good anticoagulant Tethered drug.

The molecular weight of hirudin is about 7000Da, containing 65 - 66 amino acid residues. The secondary and tertiary structure of the hirudin peptide chain plays a decisive role in its anticoagulant activity, and the three disulfide bonds at the N - terminus are determined to be the key to the secondary and tertiary structure of the molecule and its stability. Oxidation of the disulfide bond, or the degradation of the molecularly produced protein, can make the anticoagulant activity lost. If the C - terminal carboxyl group is esterified, or the C - terminal amino acid is lost, the ability to bind to thrombin is also lost.

The hirudin is stable in the dry state, and can be stably stored in water for 6 months at room temperature, and is not destroyed when heated at 80℃ for 15 minutes. When the pH is increased, the stability is lowered, and it can be stabilized for 15 minutes in 0.1mol/L hydrochloric acid solution or 0.1mol/L sodium hydroxide solution. Hirudin is not hydrolyzed by trypsin and is also tolerant to α - chymotrypsin, but papain, pepsin and subtilisin A can inactivate it.

Since the source of natural hirudin is very difficult, each leech contains only 20μg of hirudin, which is not suitable for clinical applications. To this end, many foreign companies and research units have been studying the use of genetic engineering technology to produce recombinant hirudin

since the late 1980s. At present, the hirudin products that have been marketed abroad include Hoechs' recombinant hirudin(Lepirudin, trade name Refludon) and Novartis's recombinant hydrated product(Desirudin). In 1997, the recombinant synthon Ⅲ gene was synthesized and the E. coli exocrine expression system was constructed. The recombinant product of the secreted recombinant hirudin Ⅲ genetic engineering product, recombinant hirudin Ⅲ (abbreviated as rHV3), has a relative molecular mass of 7010.8Da.

This engineered bacterium is an exocrine expression system, and its gene expression product rHV3 is easy to be purified. However, the expression system produces some impurity proteins and pigments that affect downstream separation and purification. The extraction and purification process use the macroporous adsorption resin hydrophobic chromatography and ion exchange chromatography to solve the separation of impurity proteins and pigmentsbetter. Macroporous adsorption resin as the first extraction step of rHV3, high molecular weight hybrid proteins are basically removed, and ion exchange is the key to depigmentation. The use of suitable elution conditions for adsorption and ion exchange is an important factor in increasing viability yield. The purification method has simple process and high yield and purity of rHV3, and is suitable for large-scale preparation of rHV3.

In this experiment, the engineering bacteria was used as the material, and the engineering bacteria were first fermented and cultured, and then the fermentation liquid was centrifuged to separate and purify rHV3 from the supernatant of the engineering fermentation broth. The fermentation supernatant was treated with macroporous adsorption resin HP20, and then subjected to DEAE-52 anion cellulose exchange chromatography to obtain a higher purity expression product rHV3. rHV3 was analyzed for antithrombin activity and 15% SDS-PAGE electrophoresis. The determination of rHV3 biological activity was carried out in accordance with Markwardt's thrombin titration method. The specific activity was determined by accurately weighing 10mg of the purified product, dissolving it in 1ml of H_2O, determining the protein content by the Lowry method, and measuring the antithrombin activity, rHV3 specific activity = anticoagulant activity(ATU)/protein content.

【Materials】

1. Apparatus

(1) Glass chromatography column: 1.6cm × 20cm or/and 2.6cm × 40cm

(2) Constant current pump

(3) Automatic part collector

(4) Recorder

(5) Constant temperature water bath

(6) High-speed desktop centrifuge

(7) Refrigerated centrifuge

(8) Vortex mixer

(9) 752 ultraviolet spectrophotometer

(10) Regulator electrophoresis system

(11) Shaker

(12) Vertical plate electrophoresis tank

(13) Electric stove

(14) Timer

(15) Autoclave

(16) Electromagnetic stirrer

(17) Acidity meter

(18) Fermentation tank

(19) Micropipette(20μl,200μl,1000μl)

(20) Standard purification workbench

(21) Vacuum pump and vacuumdryert

(22) Electronic balance(accurate to 10mg)

(23) Photographic equipment

(24) Genetically engineered bacteria: *E. coli*/pHV3 is preserved in this laboratory

(25) Macroporous adsorption resin HP20

(26) DEAE－52 anionic cellulose

(27) Eppendorf centrifuge tube holder

(28) 96－well ELISA plate

(29) Decolorization shaker

(30) Sterilized pipette tips

(31) Sterilized Eppendorf tubes(1.5ml microcentrifuge tubes)

(32) Reagent bottle(100ml,250ml,500ml,1000ml,5000ml)

(33) Measuring cylinder(10ml,100ml,1000ml,2000ml)

(34) Beaker(50ml,250ml,500ml,1000ml,2000ml)

(35) Buchner funnel(8cm) and suction filter bottle(1000ml)

(36) Filter paper

(37) Glass rod

(38) Test tubes(5ml,15ml)

(39) Triangle bottle(250ml)

(40) Large plastic bucket

(41) Quartz cuvette

2. Reagents

(1) Medium

Seed medium(LB liquid medium): 1% Tryptone, 0.5% Yeast Extract, 1% NaCl, prepared with distilled water, pH 7.0, divided, autoclaved for 20 minutes (1.03×10^5 Pa). Antibiotics were added under sterile conditions prior to inoculation, and the final concentration of ampicillin was 100μg/ml.

Fermentation medium: 1% peptone, 0.5% yeast extract, 4% sodium glutamate, 10% wort. pH 6.5, the shake flask volume is 12%.

(2) Ampicillin(Amp)

(3) 0.5% bovine fibrinogen: use 0.05mol/L Tris－HCl buffer(pH 7.4) to prepare

(4) Thrombin solution: 500 NIH unit plus 5ml of distilled water

(5) Low molecular weight standard protein(17.5～94 kDa)

(6) 20mmol/L acetic acid

(7) Hydrochloric acid

(8) Isopropanol
(9) 20mmol/L piperazine
(10) Sodium chloride
(11) Tris (hydroxymethyl) aminomethane (Tris)
(12) Ethanol
(13) Sodium hydroxide
(14) Sodium glutamate
(15) 15% SDS - DAGE analysis

【Procedures】

1. Engineering bacteria fermentation culture

(1) The strain was picked from the plate and inoculated into fresh LB liquid medium containing ampicillin (final concentration of 60μg/ml) at 37℃, shaking speed 220r/min, and cultured for 18 hours.

(2) Shake flask culture

Further transfer to the fermentation medium according to 3% inoculum, and the supernatant was collected by centrifugation at 37℃, 250r/min, 30 hours.

(3) Feed - batch fermentation

30L fermenter was charged with 15L fermentation medium, initial pH was 6.5, sterilization was carried out at 121℃ for 20 minutes, and ampicillin was added to the final concentration of 60μg/ml when the temperature was lowered to 37℃. Add 1000ml of seed liquid, incubate at 37℃, adjust the stirring speed and aeration, control the dissolved oxygen at 10% - 20%, control the pH at 6.0 - 7.5 by adding acid and alkali substances, and adjust the medium by adding glucose or peptone. Medium nutrients, prolonging the expression time of vitality. Culture for 9 to 16 hours.

2. Separation and purification of recombinant hirudin Ⅲ

(1) Preparation of fermentation supernatant

The fermentation broth was centrifuged at 12000r/min to remove the cells, and the fermentation supernatant was harvested. The content and activity of recombinant hirudin Ⅲ of the target protein were detected by Lowry method and Markwardt's thrombin titration method respectively and analyzed by 15% SDS - PAGE electrophoresis. Calculate the yield.

(2) Macroporous adsorption chromatography

The macroporous adsorption resin HP20 was packed in a column and equilibrated with 20mmol/L acetic acid. The pH of the fermentation supernatant was adjusted to 5.0 - 6.0, air bubbles were removed from the air pump, and the sample was applied to the column at a flow rate of 1ml/min. It was washed with 20mmol/L acetic acid and 50mmol/L pH 8.5 Tris - HCl, and then eluted with a gradient of 20mmol/L acetic acid containing 10% ~ 40% isopropanol. Collected by a fraction collector, 1.5ml/min, 6ml was collected per tube, and the A_{280} value and recombinant hirudin Ⅲ activity of each tube were measured, and an elution curve was drawn. The active components were collected and assayed, and the content and activity of recombinant hirudin Ⅲ were measured and analyzed by 15% SDS - PAGE electrophoresis. Calculate the yield.

Large pore adsorption column treatment method: firstly, rinse with distilled water to remove floating fine particles, and then repeatedly soak with 95% ethanol to fully swell and preliminary impurity removal. It was then rinsed with 95% ethanol in a Buchner funnel until the UV absorbance at 254nm was less than 0. 03. Wash the ethanol with distilled water. Finally, they were immersed in 5% hydrochloric acid and 5% sodium hydroxide for 3 hours, respectively, and washed to pH neutral.

(3) DEAE – 52 anion exchange column chromatography

The DEAE – 52 anionic cellulose column was equilibrated with 20mmol/L piperazine. The macroporous adsorption chromatography product sample was adjusted to pH 6. 0 after 10mmol/L piperazine and loaded. Wash with equilibration buffer until the baseline is stable. It was further eluted with a gradient of 20mmol/L piperazine (pH 6. 0) containing 0. 1 ~ 0. 4mol/L NaCl, collected by a fractional collector, 1. 5ml/min, 6ml per tube, and the A_{280} value of each tube was determined. Recombinant hirudin Ⅲ activity was plotted and the elution profile was plotted. The active ingredient was collected and lyophilized to obtain a high – purity recombinant hydrated dry hydrated hydrated lyophile powder. The content and activity of recombinant hirudin Ⅲ were measured and analyzed by 15% SDS – PAGE electrophoresis. Calculate the yield.

3. Recombinant hirudin Ⅲ biological activity assay

Add 50% of bovine blood [0. 05mol/L Tris – HCl buffer (pH 7. 4)] to the wells of the microplate, and add 10 to 100μl of the recombinant hirudin Ⅲ solution, and mix well. A standard thrombin solution (100 NIH units) was pipetted using a microinjector at a time interval of 1 minutes. If fibrinogen solidifies within 1min, the titration end point has been reached. The number of units of recombinant hirudin Ⅲ was converted from the consumption of thrombin. Since hirudin is 1 : 1 bound to thrombin, each thrombin unit (NIH) is equivalent to an antithrombin unit (ATU).

4. Method for determining specific activity of recombinant hirudin Ⅲ

10mg of the purified product was dissolved in 1ml of H_2O, the protein content was determined by the Lowry method, and the antithrombin activity was measured, and the specific mass per unit mass of rHV3 was obtained by the following formula.

rHV3 specific activity = anticoagulant activity (ATU)/protein content.

5. Analysis of sample purity by 15% SDS – PAGE electrophoresis

The electrophoresis method was the same as experiment 9.

(1) A polyacrylamide gel for electrophoresis was prepared as usual at a concentration of 15%.

(2) The fermentation supernatant supernatant, desorption solution and ion exchange eluate sample were treated and subjected to 15% SDS – PAGE electrophoresis.

(3) After electrophoresis for 2 to 3 hours, the gel was taken out and stained with 0. 15% Coomassie Brilliant – R250. After overnight, the dyeing solution was poured out, and the color was decolorized continuously. At the beginning, the liquid was changed once every 20 minutes, and the liquid was changed once every hour after 1 hour until the zone was obvious, and the electrophoresis zone map was drawn.

6. Result processing

(1) Indicate the operation flow of the preparation of recombinant hirudin Ⅲ by an arrow diagram.

(2) Draw an elution curve.

(3) Analysis of purification process of recombinant hirudin Ⅲ.

Analysis of Purification Process of Recombinant Hirudin Ⅲ

Procedures	Total protein (mg)	Total vitality (ATU)	Specific activity (ATU/mg)	Purification factor	Yield (%)
Fermentation supernatant					
Macroporous adsorption chromatography					
DEAE – 52 cellulose column chromatography					

(4) Electrophoresis analysis

【Questions】

1. What are the factors affecting the purity and yield of recombinant hirudin Ⅲ? How to control?

2. Explain the electrophoresis pattern, and judge the purity of the prepared recombinant hirudin Ⅲ by electrophoresis pattern.

3. How to successfully express small molecule peptide drugs in the pronucleus without being hydrolyzed by host proteases?

扫码"学一学"

实验二十五　重组门冬酰胺酶Ⅱ的制备与分析

【实验目的】

1. 掌握　以基因工程菌种 *E. coli*/pANS2 为材料进行工程菌发酵培养、分泌表达酶蛋白重组门冬酰胺酶Ⅱ及其制备和酶活力测定、SDS – PAGE 电泳分析鉴定的工作原理和技术方法。

2. 了解　基因工程菌发酵培养、分泌表达酶蛋白及表达酶蛋白纯化、分析鉴定的基本原理。

【实验原理】

L – 门冬酰胺酶（L – asparaginase，EC3. 5. 1. 1）广泛存在于动物的组织、细菌、植物和部分啮齿类动物的血清中，但在人体的各种组织器官中的分布未见报道。L – 门冬酰胺酶活性形式为一同源四聚体，每一亚基由 330 个氨基酸组成，分子质量为 34564Da，能专一地水解 L – 门冬酰胺生成 L – 门冬氨酸和氨。由于某些肿瘤细胞缺乏 L – 门冬酰胺酶合成酶，细胞的存活需要外源 L – 门冬酰胺的补充，如果外源门冬酰胺被降解，则由于蛋白质合成过程中氨基酸的缺乏，导致肿瘤细胞的死亡。因此，L – 门冬酰胺酶是一种重要的抗肿瘤药物，多年来一直是小儿急性成淋巴细胞性白血病（Acute Lumphoblastic Leukaemia，ALL）和淋巴肉瘤（Lymphosarcoma）临床化疗中重要的配伍药物之一。

用微生物，特别是大肠埃希菌来生产 L – 门冬酰胺酶已有广泛深入的研究。大肠埃希菌能产生 2 种天冬酰胺酶，即 L－门冬酰胺酶Ⅰ和 L－门冬酰胺酶Ⅱ，分别由其染色体上基因 ansA 和 ansB 编码。只有 L－门冬酰胺酶Ⅱ才有抗肿瘤活性。美、德、日均有 L－门冬酰胺酶Ⅱ商品出售。我国天津生化厂曾于 1974 年投产，后因菌种退化，产量降低而停产。国内目前用药全靠进口。1995 年，本院构建了高效分泌表达 L－门冬酰胺酶Ⅱ的基因工程菌，重组 L－门冬酰胺酶Ⅱ（rL－ASPⅡ）在大肠埃希菌细胞中合成后分泌到细胞周质中。表达的 rL－ASPⅡ相对分子质量为 138356Da，pI 为 4.85，紫外最大吸收值在 278nm 处。

本实验以此工程菌为材料采用蔗糖溶液渗透振扰提取法和酶解法联用提取周质中的 rL－ASPⅡ，再用硫酸铵分级沉淀、DEAE－52 阴离子纤维素交换层析等步骤提取和纯化，得到较高纯度表达产物 rL－ASPⅡ。并对 rL－ASPⅡ进行酶活力分析和 12% SDS－PAGE 电泳分析。

rL－ASPⅡ酶活性的测定参照 Peterson 的方法修订。取 1ml 0.04mol/L L－门冬酰胺、1ml 0.1mol/L pH 8.4 硼酸－硼酸钠缓冲液，0.2ml 细胞悬液或酶液于 37℃保温 15 分钟，速加 1ml 50% 三氯乙酸终止反应，4000r/min 离心沉淀细胞。取上清液 1ml，加奈氏试剂 2ml 和双蒸水 7ml，放置 15 分钟后，于 500nm 波长比色。在上述规定条件下，将每分钟催化 L－门冬酰胺水解释放 1μg 分子氨所需的酶量定义为 1 个酶活力单位（$OD_{500}=0.07$ 时为 1U）。

扫码“看一看”

比活力测定方法为：精称纯化产物 10mg，溶解于 1ml H_2O 中，采用 Lowry 法测定蛋白质含量，并测定酶活力，经以下公式换算即得单位质量的 rL－ASPⅡ比活力。rL－ASPⅡ比活力＝酶活力（U）/蛋白含量。

【实验材料】

1. 器材

（1）基因工程菌菌种：*E. coli*/pANS2 为本实验室保存。

（2）其他同前实验二十四器材。

2. 试剂　重组 L－门冬酰胺酶Ⅱ制备和酶活力测定：

（1）培养基

1）种子培养基（LB 液体培养基）

2）发酵培养基（玉米浆培养基）：玉米浆 50g/L，牛肉浸膏 35g/L，谷氨酸钠 10g/L，pH 7.0，高压灭菌。接种前在无菌条件下加入抗生素氨苄青霉素。

（2）氨苄青霉素（Amp）

（3）破壁液：45% 蔗糖，10mmol/L EDTA，200mg/L 溶菌酶，pH 7.5。

（4）奈氏试剂：称碘化钾 5g 加入 5ml 蒸馏水中，边搅拌边滴加 25% 氯化汞饱和液至稍有红色沉淀出现。再加 40ml 50% 氢氧化钠溶液，最后用蒸馏水稀释至 100ml，混匀入棕色试剂瓶中置暗处保存。

（5）1mol/L 氯化锰（$MnCl_2$）

（6）5mol/L 磷酸缓冲液（pH 6.4）

（7）pH 8.4 硼酸缓冲液：0.858g 硼砂（MW＝381.37），0.680g 硼酸（MW＝61.83），用蒸馏水稀释至 100ml。

（8）溶菌酶

（9）蔗糖

（10）氯化钠

（11）三（羟甲基）氨基甲烷（Tris）

（12）乙醇

（13）氢氧化钠

（14）硫酸铵

（15）谷氨酸钠

（16）四甲基乙二胺（EDTA）

（17）盐酸

（18）玉米浆

（19）低分子量标准蛋白（14.4～94kDa）

（20）12% SDS－PAGE电泳分析：同实验九。

【实验方法】

1. 工程菌发酵培养

（1）从平板上挑取菌种接种于新鲜的含氨苄青霉素（终浓度为100μg/ml）的LB液体培养基中，37℃，摇床转速220r/min，培养至OD_{600}为0.48～0.6。

（2）摇瓶培养　再按2%接种量转接于发酵培养基（三角瓶20%装液量）中，37℃，250r/min，20小时，12000r/min离心收集细胞。

（3）批式发酵培养　20L的发酵罐装入14L发酵培养基，初始pH为7.0，121℃下灭菌20分钟，待温度降至37℃时加氨苄青霉素至最终浓度为100μg/ml，接入种子液700ml，37℃培养12～14小时，罐压力0.05MPa。

2. 重组L－门冬酰胺酶Ⅱ分离纯化制备

（1）蔗糖溶液渗透振扰法和酶解法联用提取酶　将发酵收获的菌体细胞悬浮于5倍体积的破壁液中，在30℃温和振荡70分钟后搅拌下倾入大量水中，4℃，边搅拌边加入1mol/L $MnCl_2$（7.5%，V/V），沉淀核酸和菌体碎片，8000r/min，离心30分钟，收集上清液即得酶提取液。采用Lowry法和Peterson修订法分别检测目的蛋白重组L－门冬酰胺酶Ⅱ含量和酶活性及进行12% SDS－PAGE电泳分析。计算收率。

（2）硫酸铵分级沉淀　在酶提取液中加入硫酸铵至55%饱和度，调pH到7.0，冰水浴磁力搅拌60分钟，4℃静置过夜，12000r/min离心10分钟除去沉淀。再在上清液中加入硫酸铵至90%饱和度，调pH至等电点附近（约pH 5.0），搅拌60分钟，4℃静置过夜，12000r/min离心10分钟，收集沉淀。采用Lowry法和Peterson修订法分别检测目的蛋白重组L－门冬酰胺酶Ⅱ含量和酶活性及进行12% SDS－PAGE电泳分析。计算收率。

（3）透析脱盐　上述沉淀物用5mol/L磷酸缓冲液（pH 6.4）溶解，透析脱盐。

（4）DEAE－52阴离子交换柱色谱　DEAE－52纤维素柱经5mmol/L磷酸缓冲液（pH 6.4）平衡后，上样品脱盐酶液，用相同缓冲液洗涤至基线平稳，改用50mmol/L磷酸缓冲液（pH 6.4）洗脱，分部收集器收集，1.5ml/min，每管收集6ml，测定每管的A_{280}值和酶活性。绘制洗脱曲线。收集显示酶活性组分，冷冻干燥后即得高纯度的L－门冬酰胺酶冻干粉。采用Lowry法和Peterson修订法分别检测目的蛋白重组L－门冬酰胺酶Ⅱ含量和酶活性及进行12% SDS－PAGE电泳分析。计算收率。

3. 重组L－门冬酰胺酶Ⅱ酶活性测定　吸取1ml菌液入EP管，12000r/min离心5分钟，收集菌体弃去上清，加入蒸馏水1ml洗涤，混悬，12000r/min离心5分钟，弃上清。重复上述操作2～3次，加入pH 8.4硼酸缓冲液1ml混悬。

准备一组（2支）蒸馏水准备管：取10ml的洁净试管2只，每管加3.5ml蒸馏水。另取2支10ml的洁净试管，分别加入0.04mol/L天冬酰胺底物，0.1mol/L pH 8.4硼酸缓冲液各1ml，37℃水浴5分钟，其中一只加入细胞悬液20μl，另一支为对照管。反应15分钟后分别加入50%三氯乙酸1ml终止反应。再从终止反应管中各支取500μl进蒸馏水准备管，每管分别加入与25% NaOH以1∶1配好的奈氏液1ml显色，500nm处测OD值。

计算：　　酶活力(U)＝[(OD×1000)/(0.07×15x)]×(3＋x/1000)

x为取样体积（本实验为20μl）

4. 比活力测定方法　精确称取纯化产物10mg，溶解于1ml H_2O中，采用Lowry法测定蛋白质含量，并测定酶活力，经以下公式换算即得单位质量的rL－ASPⅡ比活力。

rL－ASPⅡ比活力＝酶活力（U）/蛋白含量。

5. 12%SDS－PAGE电泳分析样品纯度　电泳方法同实验九。

（1）按常规制备电泳用聚丙烯酰胺凝胶，浓度12%。

（2）将发酵收获的细胞、酶提取液、硫酸铵分级沉淀和离子交换洗脱液样品处理后，进行电泳分析。

（3）电泳2～3小时后，取出凝胶，用0.15%考马斯亮蓝－R250进行染色。过夜后倾出染液，不断脱色，起初每20分钟换液1次，1小时后每小时换液1次，直到区带明显，画出电泳区带图谱。

6. 结果处理

（1）用箭头图表示重组L－门冬酰胺酶Ⅱ制备的操作流程。

（2）绘制洗脱曲线。

（3）重组L－门冬酰胺酶Ⅱ的纯化过程分析

重组L－门冬酰胺酶Ⅱ的纯化过程分析

纯化步骤	总蛋白（mg）	总活力（U）	比活（U/mg）	纯化倍数	收率（%）
粗提酶液					
硫酸铵沉淀					
DEAE－52 纤维素柱层析					

（4）电泳图谱分析。

【思考题】

1. 试说明电泳图谱，并用电泳图谱判断所制得的重组L－门冬酰胺酶Ⅱ的纯度。
2. 影响重组L－门冬酰胺酶Ⅱ纯度和收率的因素有哪些？如何加以控制？
3. 硫酸铵分级沉淀的原理？

扫码“练一练”

EXPERIMENT 25 Preparation and analysis of recombinant Asparaginase Ⅱ

【Purpose】

1. To master fermentation culture of *E. coli*/pANS2, secretory expression of recombinant asparaginase Ⅱ, principle and technical method of preparation, enzyme activity assay and SDS - PAGE electrophoresis analysis of enzyme proteins.

2. To know the basic principles of fermentation culture of genetically engineered bacteria, secretory expression, purification, analysis and identification of enzyme proteins.

【Principle】

L - asparaginase (EC3. 5. 1. 1) widely exists in animals' tissue, bacteria, plants, and serum of some rodents, but the distribution in various tissues and organs of the body has not been reported. The active form of L - asparaginase is a homologous tetramer, each subunit consists of 330 amino acids with a molecular weight of 34564Da, which can hydrolyze L - asparaginase to L - aspartic acid and ammonia. Due to the lack of L - asparaginase synthetase in some tumor cells, the survival of the cells requires the supplement of exogenous L - asparaginase. If exogenous asparaginase is degraded, the lack of amino acids in the protein synthesis process will lead to the death of tumor cells. Therefore, L - asparaginase is an important anti - tumor drug and has been one of the important combination drugs for clinical chemotherapy of Acute Lumphoblastic Leukaemia (ALL) and Lymphosarcoma for many years.

The production of L - asparaginase by microorganisms, especially *E. coli*, has been studied extensively. *E. coli* can produce two kinds of asparagine enzymes, named L - asparaginase Ⅰ and L - asparaginase Ⅱ, respectively by the chromosome gene ansA and ansB coding. Only L - asparaginase Ⅱ has antitumor activity. America, Germany and Japan have L - asparaginase Ⅱ goods for sale. Tianjin biochemical factory was put into production in 1974, but stopped production due to the degradation of bacteria and the decrease of yield. At present, domestic drugs rely on imports. In 1995, we built a genetic engineering bacteria that has more efficient secretory expression of L - asparaginase Ⅱ. Recombinant L - asparaginase Ⅱ (rL - ASP Ⅱ) secreted into periplasm after synthesis in *E. coli* cells. The relative molecular mass of rL - ASP Ⅱ is 138356Da, pI is 4. 85, Maximum UV absorption value is at 278nm.

We use the engineering bacterium as material and combine sucrose solution infiltration vibration extraction and enzymatic hydrolysis to extract the rL - ASP Ⅱ of periplasm, and then get high purity express product rL - ASP Ⅱ through ammonium sulfate precipitation, DEAE - 52 cellulose anion exchange chromatography. The rL - ASP Ⅱ enzyme activity analysis and 12% SDS - PAGE electrophoresis analysis both are carried out.

Theenzyme activity determination of rL - ASP Ⅱ refers to Peterson's method. 1ml 0. 04mol/L L - asparagine, 1ml 0. 1mol/L pH 8. 4 boracyl - sodium borate buffer, and 0. 2ml cell suspension or enzyme solution were mixed and then incubated at 37℃ for 15 minutes. 1ml 50% trichloroacetic

acid was quickly added to terminate the reaction, and then 4000r/min centrifugation was used to precipitate the cells. 1ml supernatant solution, 2ml nessler's reagent and 7ml double - steamed water were mixed and placed for 15 minutes, and then measured the absorbance at the wavelength of 500nm. Under the above conditions, the amount of enzyme needed to release 1μg molecular ammonia by L - asparagine hydrolysis per minute was defined as 1unit of enzyme activity (1U when OD_{500} = 0.07).

【Materials】

1. Apparatus

(1) Genetically engineered strains: *e. coli*/pANS2 are preserved in our laboratory.

(2) Other equipment is the same as devices in experiment 24.

2. Reagents

Preparation of Recombinant L - asparaginase Ⅱ and determination of enzyme activity.

(1) Culture medium

1) Seed medium (LB liquid medium)

2) Fermentation medium (corn pulp medium): corn pulp 50g/L, beef extract 35g/L, sodium glutamate 10g/L, pH 7.0, autoclave sterilization. The antibiotic ampicillin was added under aseptic conditions before inoculation.

(2) Ampicillin (Amp)

(3) Wall breaking solution: 45% sucrose, 10mmol/L EDTA, 200mg/L lysozyme, pH 7.5.

(4) Nessler's reagent: potassium iodide 5g was added to 5ml distilled water, and 25% mercuric chloride saturated solution was added while stirring until slightly red precipitation appeared. Add 40ml 50% sodium hydroxide solution, dilute it to 100ml with distilled water, mix it well, put it in brown reagent bottle and store it in dark place.

(5) 1mol/L manganese chloride ($MnCl_2$)

(6) 5mol/L phosphate buffer (pH 6.4)

(7) Boric acid buffer solution: 0.858g borax (M_W = 381.37), 0.680g boric acid (M_W = 61.83), diluted to 100ml with distilled water.

(8) Lysozyme

(9) Sucrose

(10) Sodium chloride

(11) Tris

(12) Ethanol

(13) Sodium hydroxide

(14) Ammonium sulfate

(15) Sodium glutamate

(16) Tetramethylenediamine (EDTA)

(17) Hydrochloric acid

(18) Corn pulp

(19) Low molecular weight protein marker (14.4 ~ 94kDa)

(20)12% SDS－PAGE analysis: same as experiment 9.

【Procedures】

1. Fermentation culture of engineering bacteria

(1) Single colony was selected from the plate and inoculated in LB liquid medium containing ampicillin (final concentration: 100μg/ml), 37℃, rotary speed is 220r/min, and cultured to OD_{600} for 0.48～0.6.

(2) Shaker culture

The cells were transferred to the fermentation medium at the inoculation rate of 2% (20% of the tripod bottle), and cultured at 37℃ 250r/min for 20 hours, and then centrifuged at 12000r/min to collect the cells.

(3) Batch fermentation culture

20L fermentation tank was loaded into 14L fermentation medium with initial pH of 7.0 and sterilized at 121℃ for 20 minutes. When the temperature dropped to 37℃, ampicillin was added to the final concentration of 100μg/ml. The seed solution was put into 700ml and cultured at 37℃ for 12－14 hours with tank pressure of 0.05MPa.

2. Isolation and purification of recombinantL－asparaginase Ⅱ

(1) Combine sucrose solution infiltration vibration extraction and enzymatic hydrolysis to extract enzyme.

The harvested bacterial cells were suspended in the wall breaking liquid with a volume of 5 times, stirred and poured into a large amount of water after a moderate oscillation at 30℃ for 70 minutes. At 4℃, 1mol/L $MnCl_2$ (7.5%, *V/V*) was added while stirring, nucleic acid and bacterial fragments were precipitated, and the supernatant was centrifuged at 8000r/min for 30 minutes to collect the enzyme extract. Recombinant L－asparaginase Ⅱ content and enzyme activity was detected by Lowry method and Peterson revision method and 12% SDS－PAGE electrophoresis analysis. Calculate the yield.

(2) Ammonium sulfate precipitation

Ammonium sulfate was added to the enzyme extract to 55% saturation, and the pH was adjusted to 7.0. After magnetic stirring in ice water bath for 60 minutes, the enzyme was stand overnight at 4℃ and centrifuged for 10min at 12000r/min to remove the precipitation. Then add ammonium sulfate to the supernatant to 90% saturation, adjust pH to near the isoelectric point (about pH 5.0), stir for 60 minutes, let it sit overnight at 4℃, centrifuge at 12000r/min for 10 minutes, and collect precipitation. Lowry method and Peterson revision method and 12% SDS－PAGE electrophoresis analysis. Calculate the yield.

(3) Dialysis desalination

The precipitate was dissolved with 5mol/L phosphate buffer (pH 6.4) and desalted by dialysis.

(4) DEAE－52 anion exchange column chromatography

After the DEAE－52 cellulose column was balanced with 5mmol/L phosphate buffer (pH 6.4), the desalting enzyme solution of the above sample was washed with the same buffer until the baseline was stable, and then eluted with 50mmol/L phosphate buffer (pH 6.4), which was

collected by partial collectors, 1.5ml/min, and 6ml per tube. The A_{280} value and enzyme activity of each tube were measured. Draw an elution curve. The lyophilized recombinant L – asparaginase Ⅱ powder with high purity was obtained after freeze – drying. Recombinant L – asparaginase Ⅱ content and enzyme activity was detected by Lowry method and Peterson revision method and 12% SDS – PAGE electrophoresis analysis. Calculate the yield.

3. Enzyme activity determination of recombinant L – asparaginase Ⅱ

1ml of bacterial liquid was centrifuged at 12000r/min for 5 minutes, discarded the supernatant, added distilled water 1ml for washing, suspended, and then centrifuged at 12000r/min for 5 minutes, and discarded the supernatant. Repeat the above operation for 2 – 3 times and added 1ml pH 8.4 boric acid buffer for suspension.

Prepared a set of two distilled water preparation tubes: took two 10ml clean test tubes and added 3.5ml distilled water per tube. Another two 10ml clean test tubes were added with 1ml 0.04mol/L asparagine substrate and 1ml 0.1mol/L pH 8.4 boric acid buffer, 37℃ water bath for 5min, one of which was added with 20μl cell suspension, and the other one was used for control tube. After reaction for 15 minutes, 1ml 50% trichloroacetic acid was added to terminate the reaction respectively. After that, 500μl was extracted from each termination reaction tube and put into distilled water preparation tube. In each tube, 1ml nessler's reagent which mixed with 25% NaOH according to 1 : 1 ratio was added to cause chromogenic reaction, and OD value was measured at 500nm.

Calculation: enzyme activity(U) = $[(OD \times 1000)/(0.07 \times 15x)] \times (3 + x/1000)$

x is sample volume(20μl in this experiment)

4. Specific activity determination

Accurately weigh purified products 10mg, dissolved in 1ml H_2O, Lowry method was used for determination of protein content, and the determination of enzymatic activity. Specific activity of rL – ASP Ⅱ for unit mass was calculated by the following formula.

Specific activity of rL – ASP Ⅱ = enzyme activity(U)/protein content

5. 12% SDS – PAGE analysis of sample purity

The electrophoresis method is the same as experiment 9.

(1) Prepare polyacrylamide gel for electrophoresis at a concentration of 12%.

(2) After processing the cells harvested by fermentation, enzyme extract, ammonium sulfate fractional precipitation and ion exchange eluent samples, electrophoresis analysis was conducted.

(3) After 2 – 3 hours of electrophoresis, the gel was removed and stained with 0.15% coomassie brilliant blue – R250. After overnight, the dye was poured out and decolorized continuously. At the beginning, the liquid was changed once every 20 minutes, and once every hour after 1 hour, until the band was obvious, and the electrophoresis band map was drawn.

6. The result processing

(1) Using the arrow diagram illustrate recombinant L – asparaginase Ⅱ preparation process.

(2) Draw the elution curve.

(3) Recombinant L – asparaginase Ⅱ purification process analysis.

Recombinant L－asparaginase Ⅱ purification process analysis

purification process	protein contet (mg)	enzyme activity (U)	specific activity (U/mg)	purification fold	recovery rate (%)
enzyme extract					
ammonium sulfate precipitation					
DEAE－52 cellulose column chromatography					

(4) Electrophoresis analysis.

【Questions】

1. Explain electrophoresis map and judge the purity of recombinant L－asparaginase Ⅱ according to electrophoresis map.

2. What are the factors of the recombinant L－asparaginase Ⅱ purity and yield? How to control it?

3. What is the principle of ammonium sulfate fractional precipitation?

扫码"学一学"

实验二十六　重组人干扰素α的制备及含量测定

【实验目的】

1. 掌握　干扰素的制备方法

2. 了解　干扰素在生物制药领域的应用

【实验原理】

干扰素（IFN）是一种广谱抗病毒剂。干扰素系统是动物细胞普遍存在的一种防御系统。正常情况下，细胞并不自发产生干扰素，组织和血清中也不含干扰素。当病毒、细菌等病原微生物、细菌脂多糖、多核苷酸、有丝分裂原等作用于细胞膜后，细胞开始合成干扰素，并分泌到细胞外。干扰素具有广泛的抗病毒和免疫调节作用，主要是通过细胞表面受体作用使细胞产生抗病毒蛋白，从而抑制病毒的复制；同时还可增强自然杀伤细胞、巨噬细胞和T淋巴细胞的活力，从而起到免疫调节作用，并增强抗病毒能力。干扰素的受体分布广泛，几乎所有的有核细胞表面都存在，因此干扰素的作用范围也非常广泛。干扰素是由多种细胞产生的一组可溶性分泌型糖蛋白。根据产生细胞、受体和活性等因素，将干扰素分为Ⅰ型和Ⅱ型。Ⅰ型干扰素又称抗病毒干扰素，分为α－型、β－型和ω型；Ⅱ型干扰素又称免疫干扰素，只有一种γ－型。按生产途径，将用人血细胞生产的干扰素称为天然干扰素，用基因重组技术生产的称为重组型干扰素。

干扰素的生产方法主要有两种：生物来源提取法和基因重组技术法，前者是早期的生产方法，这种方法得到的干扰素具有天然的分子结构和生物活性，但是成本太高，难以大规模生产。目前基因工程提取法是生产干扰素的主要方法，具有无污染、安全性高、纯度高、生物活性高、成本低等优点，从而进入大规模的产业化生产阶段，并广泛用于临床。IFNα、IFNβ和IFNγ都有基因工程产品。干扰素的表达系统有原核生物表达系统和真核生物表达系统，前者主要是大肠埃希菌，后者有酵母菌、哺乳动物细胞等。

干扰素 α 可采用双抗体夹心 ELISA 法进行检测。首先将抗人 IFNα 抗体包被于酶标板上，再将标准品或实验制备的样品中的人 IFNα 会与包被抗体结合，没有结合的游离成分被洗去，然后依次加入生物素标记的抗人 IFNα 抗体和辣根过氧化物酶标记的亲和素。抗人 IFNα 抗体与结合在包被抗体上的人 IFNα 结合、生物素与亲和素特异性结合，最终形成免疫复合物，没有结合的游离成分被洗去。加入显色底物（TMB），TMB 在辣根过氧化物酶的催化下呈现蓝色，加入终止液后变为黄色。用酶标仪在 450nm 波长处测 OD 值，IFNα 浓度与 OD_{450} 值成正比关系，因此通过绘制标准曲线计算出样品中人干扰素 α 的浓度。

【实验材料】

1. 器材

（1）培养箱　　1 台
（2）摇床　　1 台
（3）层析　　1 套
（4）稳压电泳仪　　1 台
（5）电泳槽　　1 个
（6）高压灭菌锅　　1 个
（7）超净工作台　　1 台

2. 试剂

（1）菌株：BL21/pBV220/IFN
（2）LB 液体培养基
（3）1% 胰蛋白胨
（4）0.5% 酵母提取物
（5）1% NaCl（pH 7.0）
（6）氨苄青霉素
（7）镍柱亲和层析试剂盒
（8）人干扰素 -α 酶联免疫分析试剂盒

【实验方法】

1. 工程菌诱导表达　挑取保存在甘油管中的工程菌 BL21/pBV220/IFN，划线于含氨苄青霉素的 LB 平板上（氨苄青霉素终浓度为 100μg/ml），37℃过夜培养后，挑取单菌落接种于 5ml LB 培养液中，37℃，200r/min 活化过夜。

取已经活化的菌液 5ml，加入 100ml 含有氨苄青霉素的 LB 液体培养基中，30℃，220r/min 培养 5.5 小时，然后立刻升温到 42℃诱导培养 3 小时，SDS - PAGE 检测干扰素的表达量。

2. 重组干扰素 -α 的分离纯化　取 50ml 诱导表达液，4℃ 5000r/min 离心 5 分钟，收集细胞，加入 8ml LB 培养基重悬细胞。在冰上操作，超声破碎细胞，总时间位 30 ~ 45 分钟。4℃ 12000r/min 离心裂解液 15 分钟，收集上清液过 Ni - NTA 亲和层析介质。将含多聚 his 标签的澄清样品上样至柱中，流速控制为 0.5 ~ 1ml/min，收集流出液以待后续分析。以

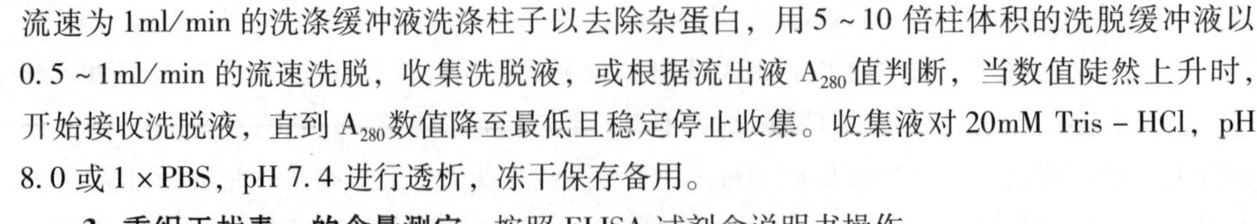

流速为1ml/min的洗涤缓冲液洗涤柱子以去除杂蛋白，用5～10倍柱体积的洗脱缓冲液以0.5～1ml/min的流速洗脱，收集洗脱液，或根据流出液A_{280}值判断，当数值陡然上升时，开始接收洗脱液，直到A_{280}数值降至最低且稳定停止收集。收集液对20mM Tris－HCl，pH 8.0或1×PBS，pH 7.4进行透析，冻干保存备用。

3. 重组干扰素α的含量测定 按照ELISA试剂盒说明书操作。

【思考题】

扫码“练一练”

1. 工程菌表达的干扰素和天然来源的干扰素有哪些不同?
2. 干扰素的检测方法有哪些?

EXPERIMENT 26 Preparation and content determination of recombinant human Interferon α

【Purpose】

1. To master the preparation method of interferon.
2. To know the application of interferon in biopharmaceuticals.

【Principle】

Interferon(IFN) is a broad－spectrum antiviral agent. The interferon system is a defensive system that is ubiquitous in animal cells. Under normal conditions, cells do not spontaneously produce interferon, and interferon is not present in tissues and serum. When pathogenic microorganisms such as viruses and bacteria, bacterial lipopolysaccharides, polynucleotides, and mitogens act on the cell membrane, the cells begin to synthesize interferon and secrete it to the outside of the cell. Interferon has a wide range of antiviral and immunomodulatory effects, mainly through the action of cell surface receptors to produce antiviral proteins, thereby inhibiting viral replication; and also enhancing the viability of natural killer cells, macrophages and T lymphocytes, thereby playing an immunomodulatory role and enhancing antiviral ability. Interferon receptors are widely distributed and almost all nucleated cell surfaces are present, so interferon has a wide range of effects. Interferon is a group of soluble secreted glycoproteins produced by a variety of cells. Interferon is classified into type Ⅰ and type Ⅱ based on factors such as production of cells, receptors, and activity. Type I interferon, also known as antiviral interferon, is classified into α－type, β－form and ω－type; type Ⅱ interferon is also known as immunointerferon, and only one γ－type. According to the production route, the interferon produced by human blood cells is called natural interferon, and the interferon produced by genetic recombination technology is called recombinant interferon.

There are two main methods for producing interferon: biological source extraction method and gene recombination technology method. The former is an early production method. The interferon obtained by this method has natural molecular structure and biological activity, but the cost is too high and it is difficult to produce in large scale. At present, the genetic engineering extraction method is the main method for producing interferon, and has the advantages of no pollution, high safety, high purity,

high biological activity, low cost, etc. , and thus enters a large – scale industrial production stage and is widely used in clinical practice. IFNα, IFNβ and IFNγ all have genetically engineered products. The expression system of interferon has a prokaryotic expression system and a eukaryotic expression system. The former is mainly Escherichia coli, and the latter is yeast, mammalian cells, etc.

IFNα can be detected by double antibody ELISA. First, anti – human IFNα antibodies are coated on the plate, then human IFNα in the standard substance or the sample prepared by the experiment is combined with the coated antibodies, and the unbound free components are washed away, then biotin – labeled anti – human IFNα antibodies and horseradh – peroxidase – labeled avidin are added successively. The anti – human IFNα antibody binds to the human IFNα bound to the coated antibody, biotin binds specifically to avidin, and eventually forms an immune complex, with unbound free components washed away. When the chromogenic substrate (TMB) is added, the TMB turns blue under the catalysis of horseradish peroxidase and turns yellow when the termination solution is added. OD value is measured at the wavelength of 450nm with the microplate reader, and IFNα concentration is in direct proportion to OD_{450} value. Therefore, the concentration of human interferon α in the sample is calculated by drawing the standard curve.

【Materials】

1. Apparatus

(1) Incubator

(2) Shaker

(3) Chromatography column

(4) Regulator

(5) Electrophoresis

(6) Electrophoresis tank

(7) Autoclave

(8) Clean bench

2. Reagents

(1) Strain: BL21/pBV220/IFN

(2) LB liquid medium

(3) 1% tryptone

(4) 0.5% yeast extract

(5) 1% NaCl (pH 7.0)

(6) Ampicillin

(7) Nickel column affinity chromatography kit

(8) Human interferon – α enzyme – linked immunoassay kit

【Procedures】

1. Engineering bacteria induced expression

The engineering bacteria BL21/pBV220/IFN stored in the glycerol tube were picked and streaked on the LB plate containing ampicillin (the final concentration of ampicillin was 100μg/

ml). After incubating at 37℃ overnight, single colonies were picked and inoculated in 5ml of LB medium, activated at 37℃, 200r/min overnight. 5ml of the activated bacterial solution was taken, added to 100ml of LB liquid medium containing ampicillin, cultured at 30℃, 220r/min for 5.5 hours, and then immediately warmed to 42℃ for 3 hours, and the expression level of interferon was detected by SDS - PAGE.

2. Isolation and purification of recombinant interferon α

50ml of the induced expression solution was taken, centrifuged at 5000r/min for 5 minutes at 4℃, the cells were collected, and the cells were resuspended by adding 8ml of LB medium. Operate on ice, ultrasonically disrupt the cells for a total time of 30 - 45 minutes. The lysate was centrifuged at 12000r/min for 15 minutes at 4℃, and the supernatant was collected through a Ni - NTA affinity chromatography medium. A clarified sample containing the poly - his tag was loaded into the column at a flow rate of 0.5 - 1ml/min and the effluent was collected for subsequent analysis. Wash the column with a wash buffer at a flow rate of 1ml/min to remove the heteroprotein, elute with 5 - 10 column volumes of elution buffer at a flow rate of 0.5 - 1ml/min, collect the eluent. Or judge according to the effluent A_{280}, when the value rises abruptly, begin to receive the eluent until the A_{280} value is minimized and the collection stops steadily. The collected solution was dialyzed against 20mM Tris - HCl, pH 8.0 or 1 × PBS pH 7.4, and stored by lyophilization.

3. Content determination of recombinant interferon α

Follow the ELISA kit instructions.

【Questions】

1. What are the differences between interferon expressed by engineered bacteria and interferon from natural sources?

2. What are the detection methods of interferon?

扫码"学一学"

实验二十七　人促红细胞生成素的制备及活性测定

【实验目的】

1. 掌握　人促红细胞生成素的制备方法及其活性测定方法。

2. 了解　人促红细胞生成素的性质及其应用。

【实验原理】

人促红细胞生成素（erythropoietin，EPO）是调控人红细胞生成的主要调节因子。EPO是由肾脏和肝脏分泌的一种糖蛋白激素类内源性生理物质，其主要作用于骨髓造血细胞，可促进红祖细胞增殖、分化直至最终成熟生成红细胞。肾功能受到损害时，如当人体发生急性、慢性肾衰竭或肾切除情况下，EPO的产生受阻，将出现贫血症状。因此EPO对肾性贫血患者的病症有很好的疗效。

在基因重组技术诞生之前，EPO主要从贫血患者的尿和绵羊血中提取，产率非常低，

且极不稳定，理化和生物学性质难以测定，亦限制了大规模应用。1985 年，Lin 等首先从人类基因库中分离 EPO 基因，测定其核苷酸序列，并在哺乳动物细胞中获得表达。1989 年 6 月，美国 FDA 正式批准安进公司通过 DNA 重组技术研制的重组人促红细胞生成素（rhEPO）上市。人 EPO 基因克隆和表达的成功，使得 rhEPO 的大批量制备成为现实，并于临床上广泛试用，经过大量的临床试验证明，rhEPO 不仅表现出对肾源性贫血有较好的疗效，而且也表现出对非肾源性贫血有良好的治疗前景。

EPO 基因以单拷贝形式位于人类的 7 号染色体的 q11 - q22 区域，其存在于一个长度为 5.4kb 的 HindⅢ - BamHⅠ限制性酶切片段中。EPO 为单链多肽，由 166 个氨基酸残基组成，含两对二硫键（Cys7 - 161 和 Cys29 - 33）以及四个糖基化位点，其中 3 个 N 糖基化位点（Asn24，Asn38，Asn83）和 1 个 O 糖基化位点（Ser126），通过这四个糖基化位点与四个富含唾液酸的糖链相连。EPO 的分子量为 30 ~ 34kDa，其中多肽仅为 18kDa，糖基化在 EPO 的生物学活性中发挥重要作用。

IgG 类免疫球蛋白是人体血液中最丰富的蛋白，其半衰期可达 21 天。已有研究表明，将 IgG 的 Fc 片段与其他蛋白融合，可显著增加该蛋白在体内的生物活性和半衰期。EPO 在体内的半衰期仅为 4 ~ 11.2 小时，对病人而言这就需要一周内接受 2 ~ 3 次的注射。将人选择性裂解活性最弱的人 IgG 的 Fc 片段连接到 EPO 的 C 末端，构建 rhEPO - Fc 融合蛋白，大大提高 EPO 在体内的半衰期，同时保持与天然 EPO 相似的生物活性，并实现在哺乳动物细胞 CHO 中的高效表达。

目前，EPO 的活性测定有体内法和体外法，体外法又包括集落形成法、放射性免疫法、酶联免疫吸附法。体内法是用红细胞增多症大鼠为材料，通过测定红细胞中的 Fe 计算 rhEPO 的活性，这是检测 rhEPO 的标准方法，但此法烦琐、耗时太长。目前对于 EPO 体外活性检测最常用的方法是酶联法（ELISA），具有方便、准确、经济等优点。

【实验材料】

1. 器材

（1）量筒	1000ml	1 个
（2）烧杯	1000ml	1 个
（3）移液枪	10μl、200μl、1000ml	各 1 个
（4）玻璃棒		1 支
（5）酶标仪		1 台
（6）冷冻离心机		1 台
（7）PCR 扩增仪		1 台
（8）水平电泳仪		1 台
（9）凝胶成像仪		1 台
（10）层析柱	1.5cm × 25cm	1 支
（11）生物安全柜		1 个
（12）分析天平		1 个
（13）pH 计		1 个
（14）恒流泵		1 个
（15）自动集液器		1 台

（16）蛋白在线检测系统

2. 试剂和材料

（1）大肠埃希菌 JM109

（2）T4 噬菌体 DNA 聚合酶、DNA 连接酶

（3）限制性内切酶 HindⅢ、BamHⅠ、NheⅠ和 EcoRⅠ

（4）RT-PCR 试剂盒

（5）琼脂糖凝胶电泳试剂盒

（6）琼脂糖凝胶 DNA 回收试剂盒

（7）质粒：pUC18、pcDNA3.1。

（8）质粒提取试剂盒

（9）磷酸钙法试剂

（10）中国仓鼠卵巢细胞（CHO）

（11）重组人促红细胞生成素的无血清培养基

（12）细胞培养所需耗材（细胞瓶、孔板等）

（13）PB、NaCl、Na_2HPO_4

（14）醋酸缓冲液

（15）Mabselect SuRe 亲和层析柱、Capto adhere 复合型阴离子交换层析柱

（16）RT-PCR 试剂

（17）鼠抗 EPO IgG

（18）兔抗 EPO IgG

（19）羊抗兔 IgG-HRP

【实验方法】

1. EPO 基因及人 IgG2 Fc 片段基因的获得 采用 RT-PCR 技术从正常人血液提取的总 RNA 中扩增 EPO 的 cDNA 片段。根据 EPO 基因序列设计的引物为：

5′引物：5′CTC CAA GCT TGC TAG CAT GGG GGT GCA CGA ATG TCC TG 3′

3′引物：5′GCG GAT CCT CTG TCC CCT GTC CTG CAG GCC TC 3′

5′引物中引入 HindⅢ和 NheⅠ酶切位点，3′引物中引入 BamHⅠ酶切位点。

RT-PCR 反应条件如下：

将 RNA 2μl（1μg/μl）、Oligo dT（18）1μl、0.1% ddH_2O 9μl 混合后 65℃水浴 5 分钟，再冰浴 5 分钟后加入 5×Reaction buffer 4μl、RNase inhibitor（20U/μl）1μl、dNTPs（10mmol/L）2μl、M-MLV（200U/μl）1μl。混匀，42℃ 60 分钟，70℃ 5 分钟，冰浴 5 分钟，-20℃保存备用。

2×Taq Master Mix+loading dye 25μl，5′引物（20μmol·L-1）1μl，3′引物（20μmol/L）1μl，cDNA 2μl，ddH_2O 21μl。PCR 反应：94℃变性 30 秒；60℃退火 30 秒；68℃延伸 45 秒，30 个循环。68℃再延伸 12 分钟。

反应完成后，1% 琼脂糖凝胶电泳检测 RT-PCR 产物，并测定序列（琼脂糖凝胶电泳同实验八）。

采用相同的方法从人淋巴细胞总 RNA 中扩增 IgG2 Fc 片段的 cDNA 片段。

根据 IgG2 Fc 片段基因序列设计的引物为：

5′引物：5′CGG GAT CCG AGC GCA AAT GTT GTG TCG AGT GC 3′

3′引物：5′GGA ATT CAT TTA CCC GGA GAC AGG GAG AGG 3′

5′引物中引入 BamH Ⅰ酶切位点，3′引物中引入 EcoR Ⅰ酶切位点。

2. 构建重组质粒 在扩增 EPO cDNA 的5′引物中引入了 HindⅢ酶切位点，3′引物中引入了 BamH Ⅰ酶切位点；在扩增 IgG2 Fc 片段 cDNA 的5′引物中引入 BamH Ⅰ酶切位点，3′引物中引入 EcoR Ⅰ酶切位点。因此，可以直接将 EPO 的 cDNA 片段直接插入克隆载体 pUC18 的多克隆位点 HindⅢ/BamH Ⅰ中，IgG2 Fc 片段插入到 pUC18 的多克隆位点 BamH Ⅰ/EcoR Ⅰ中，即可得 EPO－Fc 融合片段。测序鉴定后，再用 Nhe Ⅰ和 EcoR Ⅰ回切插入表达载体 pcDNA3.1 的多克隆位点 Nhe Ⅰ/EcoR Ⅰ中，即可得真核表达质粒 pEPO－Fc。

（1）重组质粒 pUC18－EPO 的构建 将扩增的 EPO cDNA 片段插入 pUC18 的 HindⅢ/BamH Ⅰ位点，将连接产物 pUC18－EPO 转化大肠埃希菌 JM109，随机挑选 12 个克隆用 HindⅢ/BamH Ⅰ进行酶切筛选，电泳检测是否回切出目的片段，并测序鉴定。

（2）重组质粒 pUC18－EPO－Fc 的构建 以克隆成功的 pUC18－EPO 为基础，将扩增的 IgG2 Fc cDNA 片段插入 pUC18 的 BamH Ⅰ/EcoR Ⅰ位点，将连接产物 pUC18－EPO－Fc 转化大肠埃希菌 JM109，随机挑选 12 个克隆用 BamH Ⅰ/EcoR Ⅰ进行酶切筛选，电泳检测是否回切出目的片段，并测序鉴定。

（3）真核表达质粒 pEPO－Fc 的构建 将 EPO－Fc 片段从克隆成功的 pUC18－EPO－Fc 克隆中用 Nhe Ⅰ/EcoR Ⅰ酶切下来，插入到 pcDNA3.1 的多克隆位点 Nhe Ⅰ/EcoR Ⅰ中，得到真核表达质粒 pEPO－Fc。将连接产物转化大肠埃希菌 JM109，随机挑选 12 个克隆用 Nhe Ⅰ/EcoR Ⅰ进行酶切筛选，电泳检测是否回切出目的片段，并测序鉴定。

3. CHO－EPO－Fc 稳定表达细胞株的构建、筛选和鉴定 取 10μg 大量制备的真核表达质粒 pEPO－Fc，采用磷酸钙法转染 CHO 细胞。两天后，按 1：5 的比例传代，加入 0.4mg/ml 的 G418 进行筛选，10 天可见克隆形成。随机消化边缘明显分开、细胞状态良好的单克隆接种于 24 孔板中（第一轮筛选）。取三天后的培养上清用 ELISA 检测融合蛋白表达情况，从中选出表单阳性的克隆接种于 24 孔板（第二轮筛选），培养 4 天后取上清用 ELISA 检测融合蛋白表达情况，从中选出表达较高的克隆进一步用有限稀释法进行筛选。

分别接种于 96 孔板（5 细胞/孔/200μl），待细胞长满后（大约 13 天后），取培养上清用 ELISA 检测融合蛋白表达情况（第三轮筛选）。挑选表达较高的克隆接种于 24 孔板，各平行接 2 孔，待细胞长满后，其中一孔用于保种，另外一孔接种 6 孔板。待 6 孔板长满后，提取基因组 DNA 和总 RNA，用 PCR 鉴定整合入基因组的插入片段的存在情况，用 RT－PCR 鉴定插入片段的转录情况。

取正确的克隆接种于 96 孔板（1 细胞/孔/200μl），待细胞长满后（大约 15 天后），取培养上清用 ELISA 检测融合蛋白表达情况（第四轮筛选）。将其中表达最高的克隆分别扩增、保种。

4. rhEPO－Fc 融合蛋白的表达和纯化 大规模培养转化后的 CHO 细胞，在培养液中分泌表达 rhEPO－Fc 融合蛋白。将含有 rhEPO－Fc 融合蛋白的无血清培养液经深层过滤后，采用 Mabselect SuRe 亲和层析进行吸附纯化，然后用 0.5mol/L Na_2HPO_4调整蛋白溶液的 pH 至中性，得到粗纯化的 rhEPO－Fc 融合蛋白液。具体步骤如下。

（1）平衡 用 20mmol/L PB＋150mmol/L NaCl（pH 7.3）缓冲液平衡亲和柱至基线平

稳，流速2.5cm/min。

（2）上样　上样前紫外吸收值调为零，深层过滤后样品直接上样。

（3）洗涤1　用20mmol/L PB + 150mmol/L NaCl（pH 7.3）缓冲液洗涤亲和柱至峰降到基线，流速2.5cm/min。

（4）洗涤2　用20mmol/L PB（pH 7.3）缓冲液洗涤亲和柱至峰降到基线，流速2.5cm/min。

（5）洗脱　以20mmol/L 醋酸缓冲液（pH 4.0）进行洗脱，流速2.0cm/min，洗脱下来的样品用0.5mol/L Na_2HPO_4调节pH至7.0，过滤后保存。

（6）再生　20mmol/L 醋酸缓冲液（pH 3.0）。

将得到的粗纯化rhEPO - Fc融合蛋白液采用Capto adhere复合型阴离子交换层析进一步纯化，得到高纯度rhEPO - Fc融合蛋白液。具体步骤如下。

（1）平衡　用20mmol/L PB + 500mmol/L NaCl（pH 7.0）缓冲液预平衡至基线后，再用20mmol/L PB（pH 7.0）平衡亲和柱至基线平稳，流速2.5cm/min。

（2）上样　粗纯化的融合蛋白液，流速2.5cm/min上样。

（3）洗涤　用20mmol/L PB（pH 7.0）缓冲液洗涤亲和柱至峰降到基线，流速2.5cm/min。

（4）洗脱　以20mmol/L PB + 300mmol/L NaCl（pH 7.0）缓冲液进行洗脱，流速2.5cm/min。

（5）再生　20mmol/L PB + 500mmol/L NaCl（pH 7.0），流速2.5cm/min。

5. rhEPO - Fc融合蛋白的活性测定

（1）将1μg/ml的鼠抗EPO溶液加入酶标板中，每孔200μl，轻震摇匀，用膜封口，放置4℃过夜，或37℃放置2小时后，用含0.05% Tween - 20的磷酸盐缓冲液（pH 6.8）洗板三次，拍干。

（2）每孔加入250μl含有0.5%牛血清白蛋白的碳酸盐缓冲液（pH 9.6）进行封闭，37℃放置2小时后，用含0.05% Tween - 20的磷酸盐缓冲液（pH 6.8）洗板三次，拍干。

（3）精确吸取200μl稀释成合适浓度的标准品、样品溶液加入酶标板中，轻震摇匀，用膜封口，37℃放置2小时后，用0.05% Tween - 20的磷酸盐缓冲液（pH 6.8）洗板三次，拍干。

（4）每孔加入1μg/ml的兔抗EPO 200μl，37℃放置2小时后，用含0.05% Tween - 20的磷酸盐缓冲液（pH 6.8）洗板三次，拍干。

（5）每孔加入1μg/ml的酶标羊抗兔溶液200μl，37℃放置2小时后，用含0.05% Tween - 20的磷酸盐缓冲液（pH 6.8）洗板五次，拍干。

（6）每孔加入新鲜配制的显色液200μl，放置于暗处，显色10分钟。每孔加入1M硫酸溶液100μl终止显色反应。酶标仪读数，记录数据，计算结果。

扫码“练一练”

【思考题】

1. 为什么选择哺乳动物生产体系？

2. 细胞转染和细菌转化有何异同点？

EXPERIMENT 27　Preparation and activity assay of human erythropoietin

【Purpose】

1. To master the preparation method of EPO, andy the method for measuring the activity of EPO.

2. To know the nature of EPO and its applications.

【Principle】

Human erythropoietin(EPO) is a major regulator of human erythropoiesis. EPO is a glycoprotein hormone endogenous physiological substance secreted by the kidney and liver. It mainly acts on bone marrow hematopoietic cells and can promote the proliferation and differentiation of red progenitor cells until the final maturation. When kidney function is impaired, such as when acute, chronic renal failure or nephrectomy occurs, the production of EPO is impeded and anemia symptoms will occur. Therefore, EPO has a good effect on the symptoms of patients with renal anemia.

Before the birth of genetic recombination technology, EPO was mainly extracted from the urine and sheep blood of anemia patients. The yield was very low and extremely unstable. The physical and chemical properties were difficult to measure and limited the large – scale application. In 1985, Lin et al first isolated the EPO gene from the human gene pool, determined its nucleotide sequence, and obtained expression in mammalian cells. In June 1989, the US FDA officially approved the launch of recombinant human erythropoietin(rhEPO) developed by Amgen through DNA recombination technology. The successful cloning and expression of human EPO gene makes the large – scale preparation of rhEPO a reality and widely used drug in clinical trials. After a large number of clinical trials, rhEPO not only shows better curative effect on nephrogenic anemia, but also shows good therapeutic prospects for non – renal anemia.

The EPO gene is located in the q11 – q22 region of human chromosome 7 in a single copy. It is present in a 5.4 kb HindⅢ – BamHⅠ restriction fragment and consists of 193 amino acids encoding a mature protein with a molecular weight of 18kDa. EPO is a single – chain polypeptide consisting of 166 amino acid residues with, two pairs of disulfide bonds(Cys7 – 161 and Cys29 – 33) and four glycosylation sites, of which three N – glycosylation sites(Asn24, Asn38, Asn83) and an O – glycosylation site(Ser126) are linked to four sialic acid – rich sugar chains via these four glycosylation sites. The molecular weight of EPO is 30 – 34kDa, and the polypeptide is only 18 kDa. Glycosylation plays an important role in the biological activity of EPO.

IgG – like immunoglobulins are the most abundant proteins in human blood and have a half – life of up to 21 days. Studies have shown that merging the Fc fragment of IgG with other proteins can significantly increase the biological activity and half – life of the protein *in vivo*. The half – life of EPO in the body is only 4 to 11.2 hours, which requires 2 – 3 injections per week for the patient. The Fc fragment of human IgG with the weakest selective cleavage activity was ligated to the C – terminus of EPO to construct a rhEPO – Fc fusion protein, which greatly improved the half – life

of EPO *in vivo* while maintaining the biological activity similar to that of natural EPO.

At present, the activity of EPO is determined by *in vivo* method and in vitro method, and *in vitro* methods include colony formation method, radioactive immunoassay, and enzyme - linked immunosorbent assay. The *in vivo* method uses the polycythemia rat as a material to calculate the activity of rhEPO by measuring Fe in red blood cells. This is a standard method for detecting rhEPO, but this method is cumbersome and time consuming. At present, the most commonly used method for detecting in vitro activity of EPO is enzyme - linked method (ELISA), which has the advantages of convenience, accuracy and economy.

【Materials】

1. Apparatus

(1) Measuring cylinder

(2) Beaker

(3) Pipette

(4) Glass rods

(5) Microplate reader

(6) Refrigerated centrifuges

(7) PCR amplification instrument

(8) Horizontal electrophoresis apparatus

(9) Gel imager

(10) Column: 1.5cm × 25cm

(11) Biological safety cabinets

(12) Analytical balances

(13) pH meter

(14) Constant current pump

(15) Automatic liquid collector

(16) Protein online detection system.

2. Reagents and materials

(1) *Escherichia coli* JM109

(2) T4 phage DNA polymerase, DNA ligase

(3) Restriction enzymes HindⅢ, BamHⅠ, NheⅠ and EcoRⅠ

(4) RT - PCR kit

(5) Agarose gel electrophoresis kit

(6) Agarose gel DNA recovery kit

(7) Plasmid: pUC18, pcDNA3.1

(8) Plasmid extraction kit

(9) Calcium phosphate reagent

(10) Chinese hamster ovary cells (CHO)

(11) Serum - free medium for recombinant human erythropoietin

(12) Consumables for cell culture (cell bottles, well plates, etc.)

(13) PB, NaCl, Na_2HPO_4

(14) Acetate buffer

(15) Mabselect SuRe affinity chromatography column, Capto adhere complex anion exchange chromatography column

(16) RT - PCR reagents

(17) Mouse anti - EPO IgG

(18) Rabbit anti - EPO IgG

(19) Sheep anti - rabbit IgG - HRP

【Procedures】

1. Acquisition of EPO gene and human IgG2 Fc fragment gene

Amplify a cDNA fragment ofEPO from total RNA which was extracted from normal human blood by RT - PCR. Primers are designed according to the EPO gene sequence:

5′ Primer: 5′ CTC CAA GCT TGC TAG CAT GGG GGT GCA CGA ATG TCC TG 3′

3′ Primer: 5′ GCG GAT CCT CTG TCC CCT GTC CTG CAG GCC TC 3′

The HindⅢ and NheⅠ restriction sites are introduced into the 5′ primer, and the BamHⅠ restriction site is introduced into the 3′ primer.

The RT - PCR reaction conditions are as follows:

2μl RNA (1μg/μl), Oligo dT(18) 1μl, 0.1% ddH_2O 9μl is mixed and incubated at 65℃ for 5 minutes, then ice bath for 5 minutes, then 5 × Reaction buffer 4μl, RNase inhibitor (20U/μl) 1μl, dNTPs (10mmol/L) 2μl, M - MLV (200U/μl) 1μl. Mix well, store at 42℃ for 60 minutes, 70℃ for 5 minutes, ice bath for 5 minutes, and store at -20℃.

2 × Taq Master Mix + loading dye 25μl, 5′ primer (20μmol/L) 1μl, 3′ primer (20μmol/L) 1μl, cDNA 2μl, ddH_2O 21μl. PCR reaction: denaturation at 94℃ for 30 seconds; annealing at 60℃ for 30 seconds; extension at 68℃ for 45 seconds, 30 cycles. Extend for another 12 minutes at 68℃.

After the reaction is completed, the RT - PCR product is detected by 1% agarose gel electrophoresis, and the sequence is determined. (Agarose gel electrophoresis is the same as experiment 8)

Amplify the cDNA fragment of the IgG2 Fc fragment from the total RNA of human lymphocytes by the same method.

The primers designed according to the IgG2 Fc fragment gene sequence are:

5′ Primer: 5′ CGG GAT CCG AGC GCA AAT GTT GTG TCG AGT GC 3′

3′ Primer: 5′ GGA ATT CAT TTA CCC GGA GAC AGG GAG AGG 3′

A BamHⅠ restriction site is introduced into the 5′ primer, and an EcoRⅠ restriction site is introduced into the 3′ primer.

2. Construction of recombinant plasmid

A HindⅢ restriction site is introduced into the 5′ primer of the amplified EPO cDNA, and a BamHⅠ restriction site is introduced into the 3′ primer; a BamHⅠ restriction site is introduced into the 5′ primer of the amplified IgG2 Fc fragment cDNA. The EcoRⅠ restriction site is introduced into the 3′ primer. Therefore, the EPO cDNA fragment can be directly inserted into the multiple cloning site HindⅢ/BamHⅠ of the cloning vector pUC18, and the IgG2 Fc fragment can be inserted

into the multiple cloning site BamH Ⅰ/EcoR Ⅰ of pUC18 to obtain an EPO – Fc fusion fragment. After sequencing, the eukaryotic expression plasmid pEPO – Fc is obtained by re – cutting Nhe Ⅰ and EcoR Ⅰ into the multiple cloning site Nhe Ⅰ/EcoR Ⅰ of the expression vector pcDNA3. 1.

(1) Construction of recombinant plasmid pUC18 – EPO

The amplified EPO cDNA fragment is inserted into the Hind Ⅲ/BamH Ⅰ site of pUC18, and the ligation product pUC18 – EPO is transformed into *Escherichia coli* JM109. 12 clones are randomly selected and screened by Hind Ⅲ/BamH Ⅰ for screening. Fragments are identified by sequencing.

(2) Construction of recombinant plasmid pUC18 – EPO – Fc

Based on the successfully cloned pUC18 – EPO, the amplified IgG2 Fc cDNA fragment is inserted into the BamH Ⅰ/EcoR Ⅰ site of pUC18, the ligation product pUC18 – EPO – Fc is transformed into *E. coli* JM109, and 12 clones were randomly selected for BamH Ⅰ/EcoR Ⅰ performed enzymatic cleavage screening, and electrophoresis is to be performed to detect whether the target fragment was cut back and identified by sequencing.

(3) Construction of eukaryotic expression plasmid pEPO – Fc

Digest The EPO – Fc fragment with the Nhe Ⅰ/EcoR Ⅰ clone from the cloned pUC18 – EPO – Fc clone and insert into the multiple cloning site Nhe Ⅰ/EcoR Ⅰ of pcDNA3. 1 to obtain the eukaryotic expression plasmid pEPO – Fc. The ligation product is then transformed into *Escherichia coli* JM109, and randomly select 12 clones and screen by Nhe Ⅰ/EcoR Ⅰ for electrophoresis to detect whether the target fragment is cut back and identify by sequencing.

3. Construction, screening and identification of CHO – EPO – Fc stable expression cell lines

Take 10μg of the highly prepared eukaryotic expression plasmid pEPO – Fc and transfect into CHO cells by the calcium phosphate method. After two days, passage the cells at a ratio of 1∶5, and add 0. 4mg/ml of G418 for screening, and observe colony formation for 10 days. Monoclones with clearly digested edges and well – preserved cells then are to be seeded in 24 – well plates (first round of screening). Use the culture supernatant after three days to detect the expression of the fusion protein by ELISA, and select the positive clones and inoculate into 24 – well plates (second round screening). After 4 days of culture, take the supernatant for detection of fusion protein expression by ELISA. Further screen the clones with higher expression by limiting dilution.

Theninoculate into 96 – well plates (5 cells/well/200μl), and after the cells are over (about 13 days later), take the culture supernatant for detection of fusion protein expression by ELISA (third round screening). The clones with higher expression is selected and inoculated into 24 – well plates, each of which is connected in parallel with 2 wells. After the cells are overgrown, use one of the wells for seed conservation, and the other one is inoculated with 6 – well plates. After the 6 – well plate is over, extract the genomic DNA and total RNA, and identify the presence of the inserted insert into the genome by PCR, and the transcription of the inserted fragment is identified by RT – PCR.

Inoculate the correct clones inoculated intothe 96 – well plates (1 cell/well/200μl), and after the cells are over (about 15 days later), take the culture supernatant for detection of fusion protein expression by ELISA (fourth round of screening). Amplify the clones with the highest expression and maintain separately.

4. Expression and purification of rhEPO – Fc fusion protein

The transformed CHO cells were cultured in a large scale, and the rhEPO – Fc fusion protein will be secreted and expressed in the culture solution. Subject the serum – free medium containing rhEPO – Fc fusion protein to deep filtration, and then adsorb and purify by Mabselect SuRe affinity chromatography, and then adjust the pH of the protein solution to neutral with 0.5mol/L Na_2HPO_4 to obtain crude purified rhEPO – Fc fusion protein solution. Specific steps are as follows.

(1) Equilibrium

Equilibrate the affinity column with 20mmol/L PB + 150mmol/L NaCl (pH 7.3) buffer to a stable baseline with a flow rate of 2.5cm/min.

(2) Loading

The UV absorption value is adjusted to zero before loading, and the sample is directly loaded after deep filtration.

(3) Wash 1

Wash the affinity column with 20mmol/L PB + 150mmol/L NaCl (pH 7.3) buffer to the peak drop to the baseline, the flow rate of 2.5cm/min.

(4) Wash 2

Wash the affinity column with 20mmol/L PB (pH 7.3) buffer to the peak drop to the baseline, the flow rate of 2.5cm/min.

(5) Elution

Elution is to be carried out with 20mmol/L acetate buffer (pH 4.0) at a flow rate of 2.0cm/min, and the eluted sample is adjusted to pH 7.0 with 0.5mol/L Na_2HPO_4, and stored after filtration.

(6) Regeneration

20mmol/L acetate buffer (pH 3.0).

The obtained crude purified rhEPO – Fc fusion protein solution is further purified by Capto adhere complex anion exchange chromatography to obtain a high purity rhEPO – Fc fusion protein solution. Specific steps are as follows.

(1) Equilibration

After pre – equilibration to baseline with 20mmol/L PB + 500mmol/L NaCl (pH 7.0) buffer, equilibrate the affinity column with 20mmol/L PB (pH 7.0) to a baseline stable flow rate of 2.5cm/min.

(2) Loading

Crude purified fusion protein solution, loading at a flow rate of 2.5cm/min.

(3) Washing

the affinity column was washed with 20mmol/L PB (pH 7.0) buffer until the peak dropped to the baseline, the flow rate was 2.5cm/min.

(4) Elution

Elution with 20mmol/L PB + 300mmol/L NaCl (pH 7.0) buffer at a flow rate of 2.5cm/min.

(5) Regeneration

20mmol/L PB + 500mmol/L NaCl (pH 7.0), flow rate of 2.5cm/min.

5. rhEPO – Fc fusion protein activity assay

(1) Add 1μg/ml of mouse anti – EPO solution to the plate, 200μl per well, shake gently, seal with membranc, placc at 4℃ overnight, or place at 37℃ for 2 hours, with 0.05% Tween – 20 Wash the plate three times with phosphate buffer (pH 6.8) and pat dry.

(2) 250μl of 0.5% bovine serum albumin in carbonate buffer (pH 9.6) was added to each well for blocking, and after standing at 37℃ for 2 hours, phosphate buffer (pH 6.8) containing 0.05% Tween – 20 was used. Wash the plate three times and pat dry.

(3) Accurately absorb 200μl of the standard and sample solution diluted to the appropriate concentration, add the sample solution to the microplate, shake gently, seal with a membrane, and place at 37℃ for 2 hours, then use 0.05% Tween – 20 phosphate buffer (pH 6.8) Wash the plate three times and pat dry.

(4) Add 1μg/ml of rabbit anti – EPO 200μl per well, and place at 37℃ for 2 hours, then wash the plate three times with phosphate buffer (pH 6.8) containing 0.05% Tween – 20, and pat dry.

(5) 200μl of 1μg/ml enzyme – labeled goat anti – rabbit solution was added to each well, and after standing at 37℃ for 2 hours, the plate was washed five times with phosphate buffer (pH 6.8) containing 0.05% Tween – 20, and patted dry.

(6) Add 200μl of freshly prepared coloring solution to each well and place in a dark place for 10 minutes. The color reaction was terminated by adding 100μl of a 1M sulfuric acid solution to each well. Read the microplate reader, record the data, and calculate the result.

【Questions】

1. What is the reason of choosing a mammalian production system?
2. What are the similarities and differences between cell transfection and bacterial transformation?

扫码“学一学”

实验二十八　抗 TNFα 抗体的制备与分析

【实验目的】

1. 掌握　proteinA 亲和纯化法制备抗体的原理和步骤。

2. 学习抗体鉴定的一般方法。

【实验原理】

Protein A 亲和纯化法是制备抗体药物的常用手段。Protein A 全称金黄色葡萄球菌蛋白 A（staphylococal protein A，SPA），是从金黄色葡萄球菌细胞壁分离的一种蛋白质，能特异性地与抗体（主要是 IgG）的 Fc 区域结合。将 protein A 作为配基耦联在琼脂糖介质上，可用于捕捉和结合 IgG。近年来运用基因工程技术对 SPA 序列进行修饰解决了天然 SPA 不耐碱的问题，同时使得 SPA 也耐受蛋白酶，洗脱收集液中 SPA 的脱落更低。

BCA 法测定蛋白质含量的原理是碱性溶液中，蛋白质还原 Cu^{2+}，再与 BCA 反应生成紫色复合物，测定其在 562nm 处的吸收值，并与标准曲线对比，即可计算待测蛋白的浓度。

分析抗体分子的纯度时一般采用 HPLC 法，尽量采用不同原理的色谱柱，除常用的反相色谱柱外还可通过凝胶色谱柱分析聚集体和碎片，通过离子交换色谱柱分析抗体的电荷异质等。目前，毛细管凝胶电泳、毛细管等电聚焦也越来越多地应用到单克隆抗体药物的纯度分析中。SDS - PAGE 是常用的蛋白质纯度和分子量测定方法之一，蛋白质在电场的迁移率与其分子量大小成反比。

采用酶联免疫吸附实验（ELISA）将特异性抗原吸附在固相载体表面，加入纯化的抗体形成抗原抗体复合物，再加入酶标记的二抗，此时固相上的酶量与待测抗体含量成一定比例。最后加入酶的底物生成有色产物，即可根据产物颜色深浅判断抗体的特异性。

【实验材料】

1. 器材

（1）多功能酶标仪

（2）高效液相色谱仪

（3）电泳槽

（4）电泳仪

（5）HitrapTM ProteinA HP 纯化柱

（6）微量移液枪

（7）EP（eppendorf）管　　若干

2. 试剂

（1）PBS 溶液（pH 8.0）：取磷酸二氢钾 0.24g，氯化钠 8g，氯化钾 0.2g，十二水和磷酸氢二钠 3.63g，加 800ml 水溶解，使用浓盐酸调节 pH 至 8.0，定容至 1000ml。

（2）0.1mol/L 甘氨酸溶液（pH 2.5）：取甘氨酸 7.505g 溶于 800ml 水中，使用浓盐酸调节 pH 值至 2.5，定容至 1000ml。

（3）Tris - HCl 缓冲液（pH 9.0）：50ml 0.1mol/L 三羟甲基氨基甲烷（Tris）溶液与 5.7ml 0.1mol/L 盐酸混匀后，加水稀释至 100ml。

（4）20% 乙醇：200ml 乙醇加水定容至 1000ml。

（5）5 × 电泳缓冲液：称取 Tris 15.1g，甘氨酸 94g，SDS 5g，定容至 1000ml。

（6）流动相 A：含 0.1% TFA 的水。

（7）流动相 B：含 0.1% TFA 的乙腈。

（8）PBST：称取磷酸二氢钾（KH_2PO_4）0.2g，磷酸氢二钠（$Na_2HPO_4 \cdot 12H_2O$）3.63g，氯化钠（NaCl）8.0g，氯化钾（KCl）0.2g，吐温 - 20 0.5ml，加水至 1000ml。

【实验方法】

1. 抗 TNFα 抗体的纯化

（1）1000g 离心 10 分钟去除细胞，收集细胞上清液。再次离心，12000g 离心 20 ~ 30 分钟，收集上清液后用 0.45μm 滤膜过滤保存于 4℃ 备用。

（2）装柱　连接 HitrapTM ProteinA HP 纯化柱。

（3）平衡　加入 PBS 溶液平衡柱子。

（4）上样　平衡完毕，将处理好的细胞上清液进行上样。上样结束后，用 PBS 溶液继续平衡亲和柱。

（5）洗脱　基线平衡后，采用 0.1mol/L 甘氨酸溶液（pH 2.5）进行洗脱，当紫外吸收光度仪显示出峰时，收集蛋白到预先含有 Tris - HCl 缓冲液（pH 9.0）的 EP 管中，减少蛋白在酸性条件下保留，避免蛋白聚集。

（6）柱再生　洗脱完成后，用 10 倍柱体积的 0.1mol/L 甘氨酸溶液（pH 2.5）进行洗脱，再用 Tris - HCl 缓冲液（pH 9.0）冲洗柱子，最后用 5 倍柱体积的 20% 乙醇洗涤柱子。

（7）将收集的洗脱蛋白样品用 PBS 透析（pH 8.0），过滤除菌。

2. 抗 TNFα 抗体的分析

（1）BCA 法测蛋白浓度　按试剂盒说明书操作。

（2）还原型 SDS - PAGE 法测定分子量

1）配制 10% SDS 聚丙烯酰胺凝胶，按上样量 10μg/孔制备电泳样品。

2）电泳　浓缩胶 80V 恒压，分离胶 120V 恒压，待溴酚蓝至电泳板底部停止电泳。

3）染色　电泳完毕，取出电泳胶，置染色液中加热 2 分钟，室温轻摇 15 分钟。

4）脱色　胶片转移至脱色液中加热 2 分钟，室温轻摇 10 分钟，至凝胶背景透明。

5）观察目的条带的位置，根据 marker 计算

抗 TNFα 抗体的分子量 =（轻链 + 重链）×2。

（3）纯度测定

1）非还原型 SDS - PAGE 法　方法同（2），上样缓冲液不含巯基乙醇。

2）反相 HPLC　设定流速为 1.0ml/min，柱温为室温，检测波长为 280nm，按上样量 20μg 进行上样。采用梯度洗脱法，1 小时内流动相 B 的浓度从 0% 升至 100%。用 HPLC 系统工作站对结果进行数据处理，采用面积归一化法计算抗体纯度。

（4）抗 TNFα 抗体的特异性测定

1）将 TNFα 用包被缓冲液稀释至 10μg/ml，再按照三倍梯度稀释，舍弃酶联免疫吸附板第 1 列，加入到第 A 行和第 H 行，每孔 100μl。用封板膜贴好，封严，放置于 4℃过夜。第二天，用 PBST 洗涤 3 次。

2）加入 5% 的脱脂牛奶，37℃，封闭 2 小时。

3）将待测抗体用稀释缓冲液稀释至 3000ng/ml 作为起始浓度，再三倍梯度稀释。即起始管内至少 600μl，管编号 S - 1；另取出 10 只离心管，分别加入 400μl 稀释缓冲液，编号 S - 2→S - 11；从 3000ng/ml 待测样品管中（S - 1）吸取 200μl 加入 S - 2 号管，混匀，再从 S - 2 号管中吸取 200μl 加入 S - 3 号管；依次类推稀释至 S - 10 号管。S - 11 号管作为阴性对照。

4）取出封闭完的 ELISA 板，PBST 洗涤 3 遍后，将稀释好的抗体按序号由后至前（S - 11→S - 1）从右至左加入板中，100μl/孔，每个浓度 3 复孔，复孔纵向排列。用封板膜贴好，封严 37℃放置 2h。

5）PBST 洗涤 3 遍后，加入 HRP 标记的抗人 IgG 检测二抗，37℃孵育 2 小时。

6）PBST 洗涤 3 遍后，加入 TMB 底物溶液显色 20 分钟，加入 2mol/L 硫酸终止反应。

7）采用多波长酶标仪在 450nm 和 630nm 处测定吸光度，将读取的吸光值（OD_{450} - OD_{630}）录入 Graphpad 软件计算 EC_{50}。

扫码“练一练”

【思考题】

1. Protein A 纯化抗体的原理是什么？
2. 抗体的鉴定包括哪些指标？和一般重组蛋白质相比，抗体的鉴定有什么特殊指标？

EXPERIMENT 28 Preparation and analysis of anti – TNFα antibody

【Purpose】

1. To master the principle and procedures of protein A affinity purification to prepare antibodies.

2. To learn the general methods of antibody identification.

【Principle】

Affinity purification principle by protein A is a common method of preparing antibody drugs. Staphylococal protein A, isolated from the cell wall of Staphylococcus aureus, can specifically binds to the Fc region of an antibody. Protein A is used to capture and bind IgG when coupled as a ligand on agarose medium. In recent years, gene engineering technology has been used to modify the SPA sequence to solve the problem of alkali tolerance of natural SPA. At the same time, the SPA is also resistant to protease, therefore the SPA less drops off into the elution.

The principle of determination of protein content by BCA is that in alkaline solution, the protein reduces Cu^{2+} into Cu^{+} which reacts with BCA to form purple complex. Determine the absorption value at 562nm and compare with the standard curve can calculate the concentration of the protein.

The HPLC method is usually used to analyze the purity of antibody molecules. The chromatographic columns with different principles should be used as far as possible. Besides RP – HPLC, we can apply gel chromatography to analyze the aggregates and fragments. We can use ion exchange chromatography to measure the charge heterogeneity of antibodies. Recently, capillary electrophoresis and capillary isoelectric focusing have been increasingly applied to analyze the purity of monoclonal antibodies. SDS – PAGE is a common method to determinate protein purity and molecular weight. The mobility of protein in electric field is inversely proportional to its molecular weight.

Enzyme – linked immunosorbent assay (ELISA) was used to identify specificity of the antibody. The specific antigens are adsorbed on the surface of solid – phase carriers. Add purified antibodies to form antigen – antibody complexes, and then add enzyme – labeled antibodies. At this time, the amount of enzyme in solid – phase was proportional to the content of antibodies to be measured. Finally, the substrate of the enzyme was added to produce a colored product, which could be used to determine the specificity of the antibody according to the color of the product.

【Materials】

1. Apparatus

(1) Multi – function microplate reader

(2) High performance liquid chromatography

(3) Electrophoresis tank

(4) Electrophoresis apparatus

(5) HitrapTM proteinA HP purification column

(6) Micropipette

(7) Eppendorf tubes

2. Reagents

(1) PBS solution (pH 8.0)

Take potassium dihydrogen phosphate 0.24g, sodium chloride 8g, potassium chloride 0.2g, dihydrated water and disodium hydrogen phosphate 3.63g, add 800ml water to dissolve, adjust the pH value with concentrated hydrochloric acid to 8.0, to a volume of 1000ml.

(2) 0.1mol/L glycine solution (pH 2.5)

Take 7.505g of glycine dissolved in 800ml of water, adjust the pH to 2.5 with concentrated hydrochloric acid, and dilute to 1000ml.

(3) Tris – HCl buffer (pH 9.0)

50ml of 0.1mol/L Tris solution was mixed with 5.7ml of 0.1mol/L hydrochloric acid and diluted with water to 100ml.

(4) 20% ethanol: 200ml of ethanol was added to 1000ml of water.

(5) 5 × electrophoresis buffer: Weigh 15.1g of Tris, 94g of glycine, 5g of SDS, and dilute to 1L.

(6) buffer A: water with 0.1% TFA.

(7) buffer B: acetonitrile with 0.1% TFA.

(8) PBST: Weigh 0.2g of potassium dihydrogen phosphate (KH_2PO_4), 3.37g of disodium hydrogen phosphate ($Na_2HPO_4 \cdot 12H_2O$), sodium chloride (NaCl) 8.0g, potassium chloride (KCl) 0.2g, Tween – 20 0.5ml, add water to 1000ml.

【Procedures】

1. Purification of anti – TNFα antibody

(1) Remove the cells by centrifuging at 1000g for 10 minutes, collect the supernatant. Centrifuge at 12000 for 20 – 30 minutes, collect the supernatant, filtrate with 0.45μm membrane and store at 4℃.

(2) Packing

Connect the HitrapTM ProteinA HP purification column.

(3) Equilibrium

Add the PBS solution to equilibrate the column.

(4) Loading

After the balance is completed, load the pre – treated supernatant. After loading, continuously equilibrate the affinity column with the PBS solution.

(5) Elution

After baseline equilibration, use 0.1mol/L glycine solution (pH 2.5) for elution. When the UV absorption spectrometer showed a peak, collect the protein into the tubes with Tris – HCl buffer (pH 9.0) to avoid protein aggregation.

(6) Column regeneration

After elution, elute with 10 column volumes of 0.1mol/L glycine solution (pH 2.5), rinse the column with Tris – HCl buffer (pH 9.0), and finally use 5 volumes of 20% ethanol to wash.

(7) Dialyze the collected eluted protein against PBS (pH 8.0), filter with 0.22μm membrane.

2. Analysis of anti – TNFα antibodies

(1) BCA method to measure protein concentration

Follow the kit instructions.

(2) Determination of molecular weight by reduced SDS – PAGE

①Prepare 10% SDS polyacrylamide gel. Load the protein sample 10μg/well.

②Electrophoresis

80V for stacking gel, 120V for resolving gel. Stop electrophoresis till bromophenol blue to the bottom of the electrophoresis plate.

③Dyeing

After electrophoresis is completed, heat the gel in the staining solution for 2 minutes, and shake at room temperature for 15 minutes.

④Decolorization

Transfer the gel to a decolorizing solution, heat for 2 minutes, and shake at room temperature for 10 minutes until the gel background is transparent.

⑤Observe the position of the target band and calculate the molecular weight of the anti – TNFα antibody = (light chain + heavy chain) ×2 according to the markers.

(3) purityanalysis

①Non – reducing SDS – PAGE

The method is the same as 2. 2, and the loading buffer does not contain mercaptoethanol.

②Reversed Phase HPLC

Flow rate: 1. 0ml/min; column temperature: room temperature; detection wavelength: 280nm; sample loading: 20μg. The elution consisted of a linear gradient program from 0% to 100% buffer B over 40min. The results were processed by HPLC system workstation and the area purity normalized to calculate antibody purity.

(4) Specificity of anti – TNFα antibody

①Dilute TNFα to 10μg/ml with a coating buffer, dilute it in a three – fold gradient, discard the first column of the enzyme – linked immunosorbent plate, and add to the A and H rows, 100μl per well. Seal the plate with a sealing film, place it at 4℃ overnight. The next day, wash the plate for 3 times with PBST.

②Add 5% skim milk at 37℃ for 2 hours.

③Dilute the antibody to be tested to 3000ng/ml with a dilution buffer as a starting concentration and dilute three times. That is, at least 600μl in the starting tube, tube number S – 1; take another 10 centrifuge tubes and add 400μl of dilution buffer, number S – 2→S – 11; from 3000ng/ml sample tube to be tested (S – 1). Pipette 200μl into the S – 2 tube, mix well, and then pipette 200μl from the S – 2 tube into the S – 3 tube; and then dilute to the S – 10 tube. Tube No. 11 was used as a negative control.

④Remove the ELISA plate, washthe plate 3 times with PBST, and add the diluted antibody to the plate from the back to the front (S – 11→S – 1), from the right to the front, 100μl/well, each concentration with 3 duplicates. Seal the plate, place it at 37℃ for 2 hours.

⑤Wash the plate with PBST for 3 times, add HRP – labeled anti – human IgG to detect the secondary antibody, and incubate it at 37℃ for 2 hours.

⑥Wash the plate with PBST for 3 times, add TMB substrate solution for 20 minutes, use 2mol/L sulfuric acid to terminate the reaction.

⑦Measure the absorbance at 450nm and 630nm wavelength using a multi - wavelength microplate reader. Calculate the EC_{50} with OD value(OD_{450} - OD_{630}).

【Questions】

1. What is the principle of protein A purified antibody?

2. How to identify antibody? What is the difference between antibodies and common recombinant proteins in identification?

扫码“学一学”

实验二十九　乙肝多肽疫苗的制备与分析

【实验目的】

1. 掌握　离子交换层析法纯化蛋白的技术和双抗体夹心法检测抗原蛋白技术。

2. 了解　利用哺乳动物细胞生产蛋白疫苗的基本原理。

【实验原理】

乙型肝炎是由乙型肝炎病毒（hepatitis B virus，HBV）引起，主要通过血液以及体液传播，具有慢性携带状态的一种传染病。由于目前尚无有效手段治疗乙型肝炎，因此控制乙肝的转播，主要在于控制传染源、切断传播途径及保护易感人群。采用乙肝疫苗来预防，经过多年应用已被证明是一种安全、可靠、有效的方法。

目前，世界上主要有两种基因工程乙肝疫苗：一种是酵母表达的乙肝表面抗原（HBsAg）的主蛋白（S 蛋白）；另一种是由中国仓鼠卵巢细胞（CHO 细胞）表达的乙肝表面抗原 S 蛋白。后一种疫苗通过采用基因重组技术，将乙肝病毒表面抗原 DNA 移入哺乳动物细胞（CHO 细胞）使其具有分泌乙肝表面抗原的能力，通过细胞培养、增殖、分泌出 HBs Ag 到培养液中，收集分泌液，经高速离心、纯化、除菌后，加入佐剂吸附（氢氧化铝）而制成重组乙肝疫苗。疫苗外观有轻微乳白色沉淀。

CHO 细胞属哺乳动物细胞类，它是基因工程表达系统中最高等的宿生细胞，其表达产物更接近于天然产品。同重组乙肝疫苗（酵母）相比，重组乙肝疫苗（CHO 细胞）具有如下特点：CHO 细胞表达的重组 HepB 是高等真核细胞表达产物，其天然结构较低等的真核细胞酵母表达的 HBsAg 蛋白更适合人体吸收，生产过程中不同于酵母使用硫氰酸盐处理，安全性好，十几年大规模使用，未见严重不良反应。且 HBsAg 的 S 蛋白在 CHO 细胞中被糖基化，可提供同样位置的翻译修饰、蛋白复制和大分子装配，因此免疫力持久，抗体维持时间长，具有产品的稳定性。

不同来源的疫苗需选用不同的分离纯化路线，但一般而言，都包括两个基本阶段：初级分离和精制纯化。初级分离阶段的主要任务是分离细胞和培养液，破碎细胞释放产物（如果产物在细胞内），浓缩产物和去除大部分杂质等，这一阶段可选用的分离方法包括细胞破碎技术、离心沉降、盐析和超滤浓缩技术等；精制纯化阶段则选用各种具有高分辨率的技术，以使目的蛋白和少量干扰杂质尽可能分开，达到所需的质量标准，超速离心技术

和各种层析技术成为当前达到此目的的主要方法。CHO 表达系统生产的乙肝疫苗可主要采用两种方法纯化，即疏水层析—超速离心—凝胶过滤法和三步层析法。

【实验材料】

1. 器材

（1）细胞培养箱　　1 台
（2）CO_2 气瓶　　1 个
（3）紫外分光光度计　　1 台
（4）层析柱　　1 个
（5）填料：Butyl－S Sepharose 6 FF，　　1 个
（6）透析袋（100kDa）
（7）BCA 试剂盒

2. 试剂

（1）CHO 重组工程细胞
（2）细胞培养基（DMEM）
（3）小牛血清
（4）缓冲液 A：含 8% 硫酸铵（*W/V*）的 20mM 磷酸钠溶液，pH＝7。
（5）缓冲液 B：20mM 磷酸钠溶液，pH＝7。
（6）异丙醇
（7）琼脂糖凝胶 Butyl－s sepharose 6FF

【实验方法】

1. CHO－28 细胞株种子细胞培养方法

（1）接种细胞。

（2）细胞培养　将细胞悬液按一定比例接种 15L 细胞培养瓶，加生长液摇匀，置于细胞培养箱中 37℃培养。形成单层后，每 2 天换 1 次维持液，收获含 HBsAg 原液，原液留样。

2. HBsAg 蛋白的提取

（1）收集上一步细胞培养液，4℃，8000r/min 离心 30 分钟以去除细胞降解物，在上清中继续加入饱和硫酸铵至终浓度 45%，4℃，8000r/min 30 分钟离心收集沉淀。

（2）柱层析　首先用缓冲液 A 进行柱平衡，用 8% 硫酸铵溶液预溶解沉淀后上柱，含 30%（V/V）异丙醇的缓冲液 B 进行洗脱，紫外分光检测 260/280nm 吸收峰，收集洗脱峰，洗脱液留样。

（3）透析　用 100kDa 分子量的透析袋对收集的洗脱液进行透析除盐，透析液为缓冲液 B，透析液留样。

（4）透析液超滤浓缩，将透析液用截留分子量为 100kDa 的超滤管在 12000r/min 离心 30 分钟。

3. 产品检测

（1）BCA 蛋白含量检测法　参照试剂盒说明书。

（2）ELISA 双抗夹心法检测抗原浓度。

1）包被抗体　用抗体溶液包被 96 孔酶联免疫反应板，每孔 100ng，37℃保温 2 小时或过夜。

2）封闭　适量含10%灭活兔血清和1%BSA的细胞封闭液，每孔加100μl，37℃封闭1小时；用PBS反复离心洗涤3次。

3）加入待测品　将待测品稀释到不同浓度，每孔加入100μl待测样品，37℃保温2小时，用含0.05%Tween－20的PBS离心洗涤3次。

4）加入二抗　加入100μl鼠抗人HBsAg作为二抗，37℃温育1小时，用PBS洗涤6次.

5）加入酶标三抗　加入酶标兔抗鼠－HRP抗体100μl作为三抗，37℃温育1小时，用PBS洗涤6次。

6）显色并测定　加入100μl HRP显色底物（H_2O_2－OPD），反应5～15分钟，用2mol/L H_2SO_4终止反应。在492nm波长测定OD值。根据阳性对照和系列稀释样品测得的OD值，计算出样品的浓度。

（3）15%SDS－PAGE电泳分析样品纯度　电泳方法同前。按常规制备电泳用聚丙烯酰胺凝胶，浓度15%。将细胞培养上清、洗脱液和透析样品处理后，进行15%SDS－PAGE电泳析。电泳2～3小时后，取出凝胶，用0.15%考马斯亮蓝－R250进行染色。过夜后倾出染液，不断脱色，初始时每20分钟换液1次，1小时后，每小时换液1次，直到区带明显，画出电泳区带图谱。

扫码“练一练”

【思考题】

1. 双抗体夹心法检测病毒抗原的原理是什么？
2. 蛋白含量的检测方法有哪些？各有何优缺点。

EXPERIMENT 29　Preparation and analysis of Hepatitis B polypeptide vaccine

【Purpose】

1. To master the technology of purifying protein by ion exchange chromatography and the technique of double antibody sandwich for detecting antigen and protein.

2. To learn the principles of protein vaccine production by mammalian cells.

【Principle】

Hepatitis B is an infectious disease caused by hepatitis B virus(HBV),which is mainly transmitted through blood and body fluids and has a chronic carrier status. Since there is no effective treatment for hepatitis B,the main way to control the spread of hepatitis B is to control the source of infection,cut off the route of transmission,and protect the susceptible population. The use of hepatitis B vaccine for prevention has proved to be a safe,reliable,and effective method after many years of application.

At present,there are two kinds of genetically engineered hepatitis B vaccines in the world,one is the main protein(S protein)of hepatitis B surface antigen(HBsAg)expressed by yeast,the other is hepatitis B surface antigen(S protein)expressed by Chinese hamster ovary cells(CHO cells). The latter vaccines transfer HBsAg DNA into mammalian cells(CHO cells)by gene recombination tech-

nology to make them have the ability to secrete HBsAg. HBsAg is cultured, proliferated, and secreted into the culture medium. The secretion is collected and centrifuged, purified, and sterilized at high speed. Afterwards, the recombinant hepatitis B vaccine was prepared by adding adjuvant to adsorb (aluminium hydroxide). The appearance of the vaccine was milky white precipitation.

CHO cells belong to mammalian cells. They are the highest perennial cells in the gene engineering expression system. Their expression products are closer to natural products. Compared with recombinant hepatitis B vaccine (yeast), recombinant hepatitis B vaccine (CHO cell) has the following characteristics: the recombinant HepB expressed by CHO cell is a product of higher eukaryotic cell expression, and the HBsAg protein expressed by yeast with lower natural structure is more suitable for human absorption. The production process is different from that of yeast treated with thiocyanate. It is safe and has been used on a large scale for more than ten years without serious adverse reactions. Moreover, the S protein of HBs Ag is glycosylated in CHO cells, which can provide translation modification, protein replication, and macromolecule assembly at the same location. Therefore, the immunity of HBsAg is persistent, the antibody maintains for a long time, and the product is stable.

Different isolation and purification routes are needed for vaccines from different sources, but generally speaking, there are two basic stages: primary isolation and refined purification. The main tasks of the primary separation stage are to separate cells and culture media, break up the release products of cells (if the products are in cells), concentrate the products and remove most of the impurities. The methods of separation in this stage include cell breaking technology, centrifugal sedimentation, salting out, and ultrafiltration concentration technology. A high – resolution technology is developed to separate the target protein from a small amount of interfering impurities as far as possible to meet the required quality standards. Ultra – speed centrifugation and various chromatography techniques have become the main methods to achieve this goal. The hepatitis B vaccine produced by CHO expression system can be purified by two methods, both hydrophobic chromatography – ultracentrifugation – gel filtration and the three step chromatography.

【Materials】

1. Apparatus

(1) Cells culture

(2) CO_2

(3) Ultraviolet spectrophotometer

(4) Chromatographic column

(5) Ultrafiltration tube (100kDa)

(6) dialysis bag (100kDa)

(7) BCA kit

2. Reagents

(1) CHO cells

(2) DMEM

(3) Fetal serum

(4) Buffer A: 20mM sodium phosphate solution containing 8% ammonium sulfate (W/V), pH = 7.

(5) Buffer B: 20mM sodium phosphate solution, pH = 7.

(6) Isopropanol

(7) Butyl - s sepharose 6FF

【Procedures】

1. Seed cell culture method of CHO - 28 cell line

(1) Inoculated cells

(2) Cell suspension was inoculated into 15L cell culture flask in a certain proportion, shaken with growth liquid, and cultured at 37℃ in cell culture box. After the formation of monolayer, the maintenance solution was changed once every 2 days, and the original solution containing HBsAg was harvested, and the sample was kept for further study.

2. Extraction of HBsAg protein

(1) The first step of cell culture medium was collected and centrifuged for 30 minutes at 4℃, 8000r/min to remove the degradation products. The final concentration of saturated ammonium sulfate was added to the supernatant until 45%, 4℃ 8000r/min were centrifuged for 30 minutes to collect precipitation.

(2) Column chromatography

First, column equilibrium was carried out with buffer A. After precipitation with 8% ammonium sulfate solution, column B containing 30% (V/V) isopropanol was eluted. The absorption peak of 260/280nm was detected by ultraviolet spectrophotometry. The elution peak was collected and the eluent was retained.

(3) Dialysis

The dialysate collected was desalted by dialysis bag with a molecular weight of 100kDa. The dialysate was buffer B and the dialysate was sampled.

(4) The dialysate was concentrated by ultrafiltration. The dialysate was centrifuged at 12000r/min for 30 minutes with a 100kDa ultrafiltration tube.

3. Product inspection

(1) Detection of BCA Protein Content

Reference kit instructions.

(2) Detection of antigen concentration by ELISA double antibody sandwich method.

①ELISA plates with 96 wells were coated with antibody solution. Each well was 100ng, kept at 37℃ for 2 hours or overnight.

②Blocking

Cell blocking solution containing 10% inactivated rabbit serum and 1% BSA was added to each well for 100ml, blocked for 1 hour at 37℃, and washed with PBS repeatedly by centrifugation for 3 times.

③Adding samples

Dilute the samples to different concentration, add 100ml sample to each well, hold at 37℃ for 2 hours, wash 3 times with PBS centrifuge containing 0.05% Tween - 20.

④Add the second antibody

Add 100ml mouse anti - human HBsAg as the second antibody, incubate at 37℃ for 1 hour, wash with PBS 6 times.

⑤Enzyme - labeled antibodies were added

100ml of rabbit anti - rat - HRP antibody was added as the third antibody, incubated at 37℃ for 1 hour, washed with PBS for 6 times.

⑥Development and determination

Adding 100ml HRP chromogenic substrates (H_2O_2 - OPD), the reaction lasted for 5 - 15 minutes, and terminated with 2mol/L H_2SO_4. OD value was measured at 492nm. According to OD values measured by positive control and series of diluted samples, the concentration of samples was calculated.

(3) Purity Analysis by 15% SDS - PAGE Electrophoresis

Electrophoresis method was the same as before. Polyacrylamide gel was prepared by electrophoresis, with a concentration of 15%. Cell culture supernatants, eluents, and dialysis samples were treated and analyzed by 15% SDS - PAGE electrophoresis. After electrophoresis 2 to 3 hours, the gel was removed and stained with 0. 15% Coomassie brilliant blue - R250. After overnight, the dye solution was poured out and decolorized continuously. At the beginning, the solution was changed every 20 minutes once. After one hour, the solution was changed every hour until the band was obvious. The electrophoretic band pattern was drawn.

【Questions】

1. What is the principle of double antibody sandwich method for detecting viral antigen?

2. What are the detection methods for protein content? Talking about the advantages and disadvantages.

实验三十 抑制 miRNA 功能的反义寡核苷酸制备和活性验证

扫码“学一学”

【实验目的】

1. **掌握** 在哺乳动物细胞中通过反义寡核苷酸抑制 miRNA 的相关实验方法。
2. **熟悉** 抑制 miRNA 功能的反义寡核苷酸设计和制备方法。
3. **了解** 反义药物作用的原理和 miRNA 的定义及作用。

【实验原理】

反义药物，通常是指反义寡核苷酸（antisense oligonucleotides，ASO），主要是利用人工合成或天然存在的 RNA 或 DNA，根据碱基互补配对原则和核酸杂交原理，与目的基因或 mRNA 的特定序列相结合，从基因复制、转录、剪接、转运及翻译等水平上调节靶基因的表达，干扰遗传信息从核酸向蛋白质的传递，从而达到抑制、封闭或破坏靶基因的目的。具体原理是利用 ASO 与 DNA 结合，抑制 DNA 复制和转录，同时 ASO 与 mRNA 前体结合，阻断 RNA 加工、成熟，影响核糖体沿 mRNA 移动，激活 RNase，剪切杂交链中未配对的碱基。在肿瘤研究领域，通过反义寡核苷酸技术，封闭或抑制肿瘤细胞的关键编码基因的表达，从而达到抑制肿瘤细胞的增殖，促进肿瘤细胞凋亡的目的。由于反义寡核苷酸技术的研究深入，使得反义寡核苷酸药物成为治疗肿瘤的潜在新型药物。美国 FDA 已经批准了数

个反义寡核苷酸药物，这包括2018年获批的用于治疗脊髓型肌萎缩症的nusinersen和治疗杜氏肌营养不良症的eteplirsen。

常见的反义药物主要针对DNA和mRNA，随着研究的不断深入，针对非编码RNA的反义药物日渐引起人们关注。MicroRNA（miRNA）是在植物和动物基因组中自然发生、高度保守且短小的非编码RNA分子。它们的长度为17～27个核苷酸，通常通过与互补mRNA序列的3'非翻译区（3'-UTR）结合来调控转录后的mRNA表达，进而导致转录抑制和基因沉默。研究显示，miRNA参与生物体内多种重要的生理过程，并且与多种疾病的发生、发展有重要关系。

【实验材料】

1. 器材

（1）可连接互联网的电脑

（2）离心机

（3）灭菌后锥底离心管（50ml）

（4）免疫荧光和Western印迹法所用设备

（5）层流生物安全柜（Ⅱ级）

（6）体视显微镜

（7）组织培养皿（10cm）

（8）组织培养箱（37℃，5% CO_2），一定湿度

（9）组织培养板（24孔）

2. 试剂

（1）ASO（12.5μmol/L和错配和/或无关对照寡核苷酸）

（2）DharmaFECT 4转染试剂（Dharmacon，目录号T-2004）

（3）DMEM，不含抗生素和血清

（4）PBS，不含钙和镁

（5）胎牛血清（FBS），热灭活

（6）哺乳动物细胞系（如HeLa或NTera2）

（7）青霉素和链霉素

（8）胰蛋白酶-EDTA溶液

【实验方法】

一、抑制miRNA功能的反义寡核苷酸制备

1. 从miR Base（http：//microrna. sanger. ac. uk/sequences）下载目标miRNA序列。

2. 根据Watson-Crick碱基互补配对原则和miRNA序列，设计反义miRNA序列。

3. 在反义miRNA两端加入5个任意碱基（如5'-UCUUA-反义miRNA序列-ACCUU-3'），从而设计ASO序列。

提示：在ASO两端加入任意碱基使其与miRNA靶标更相似，从而可以提高ASO效率；此作用可能通过增强RISC不依赖序列地与ASO结合，从而避免细胞内核酸酶对miRNA中

心靶序列的降解。

4. 用 mFold 默认设置（http：//mfold. bioinfo. rpi. edu/cgi - bin/rna - form1 - 2. 3. cgi）检测 ASO 序列可能的二级结构。如果 ASO 序列可以形成稳定的二级结构，则改变侧链序列并重复步骤 4。

5. 进行 Blast 分析。如果 mRNA 与侧链序列以及靶序列中 13 个以上的碱基互补，则重复步骤 3 至步骤 5。

6. 设计对照寡核苷酸。所有的实验必须包含与实验组 ASO 具有相同侧链序列和相同长度的错配序列（mismatched）对照和/或无关寡核苷酸对照。错配对照是指 ASO 与目标 miRNA 间至少存在 4 个等距离的嘌呤 - 嘌呤错配。无关对照是指来自其他物种非同源的反义引物序列或随意序列组成的 ASO。如步骤 4 和步骤 5 所述，所有对照寡核苷酸应仔细检查可能的二级结构及其与 mRNA 的互补性。通过改变寡核苷酸序列的碱基可以尽量减少二级结构的存在及其与 mRNA 的互补性。

7. 从寡核苷酸合成公司获得 2’ - O - 甲基修饰的 ASO 序列。

提示：许多寡核苷酸合成公司可提供 2’ - O - 甲基修饰的寡核苷酸。2’ - O - 甲基修饰通过增强 ASO 与 miRNA 的亲和力从而提高 ASO 效力，并保护 ASO 不被 RISC 和细胞内核酸酶降解。也可以对 ASO 进行 3’ - 端胆固醇修饰，据报道，3’端胆固醇修饰可以促进 ASO 转运进入细胞。

二、在哺乳动物细胞中通过反义寡核苷酸抑制 miRNA 功能

1. 准备细胞

（1）细胞在 DMEM 培养基（含有 10% FBS、100U/ml 青霉素以及 100μg/ml 链霉素）中培养，含 5% CO_2 的 37℃ 培养箱中培养至 90% 单层。

（2）使用胰蛋白酶 - EDTA 溶液消化细胞使其脱离培养板，然后将细胞重悬于无抗生素、含有 10% FBS 的 DMEM 培养基中。24 孔板每孔接种 500μl 的细胞悬液。

（3）将细胞在含 5% CO_2 的 37℃ 培养箱中孵育过夜，使其达到转染时所需的 30% ~ 40% 单层（约每孔 5×10^4 个细胞）。

提示：最佳细胞密度因细胞系的生长特性而变化。请考虑转染试剂使用说明中的细胞系特异性说明或者通过试验来判断最佳细胞密度。

2. 准备转染试剂　所有实验应该包括一个与试验组 ASO 侧翼序列一样、长度相同的错配和/或无关对照寡核苷酸。所有体积均表示单一试验孔所需的体积。实验至少需要重复 3 次。对于多个孔的转染，需配 10% 的富余试剂以弥补混合过程中溶液的损失。

3. 向一个含 24μl 无抗生素和血清的 DMEM 灭菌试管中加入 1μl（12. 5 pmol）的实验或对照 ASO，用移液管轻轻吹打混匀。

4. 混合液于室温下孵育 5 分钟。

5. 将 ASO 混合液缓缓地加入 DharmaFECT 4 转染试剂混合液中。移液管轻轻吹打，混匀，室温下孵育 20 分钟从而形成 ASO - 脂质复合体。

6. 将步骤 1 中生长介质换成 450μl 无抗生素、含 10% FBS 的 DMEM，加热至 37℃。

7. 将 50μl ASO - 脂质转染介质（来自步骤 5）滴入各个孔中。尽量使其覆盖孔中全部细胞，并轻轻摇动。

8. 将细胞放入含5% CO_2 的37℃培养箱中培养1~2天，若1天后观察到细胞毒性，将转染介质换成新鲜的无抗生素、含10% FBS的DMEM，继续培养。

提示：若转染miRNA特异的ASO和对照ASO后，细胞均发生死亡，则可能是转染试剂或者ASO用量过大引起的；此外，细胞密度可能太低，理想状态下细胞融合度应为30%~40%，每孔约5×10^4个细胞。实验前需要优化转染条件。如果细胞密度太低，需要待细胞密度合适时再进行实验。

扫码"练一练"

若仅转染miRNA特异的ASO后，细胞发生死亡，则可能是抑制性miRNA的靶向mRNA是抗凋亡调节因子，或者miRNA特异性ASO可能含有一个促进天然免疫反应的序列基序。可以尝试换不同化学组成的ASO（如LNA）来进行实验。

9. 通过免疫荧光和Western印迹法检测蛋白质水平，从而分析miRNA靶基因的表达。

EXPERIMENT 30 Preparation and bioactivity evaluation of antisense oligonucleotides inhibiting function of miRNA

【Purpose】

1. To master the relevant experimental methods for inhibiting miRNA by antisense oligonucleotides in mammalian cells.

2. To be familiar with the design and preparation methods of antisense oligonucleotides that inhibit miRNA function.

3. To learn the principles of antisense drugs and the definition and role of miRNAs.

【Principle】

Antisense drugs, usually referred to as antisense oligonucleotides (ASON), are mainly synthetic or naturally occurring RNA or DNA, combined with a specific sequence of the gene or mRNA according to the principle of complementary base pairing and nucleic acid hybridization. The combination of sequences regulates the expression of target genes from the levels of gene duplication, transcription, mRNA splicing, mRNA transport, and mRNA translation, and interferes with the transmission of genetic information from nucleic acids to proteins, thereby achieving the purpose of inhibiting, blocking or destroying target genes. The specific principle is using ASON to bind DNA inhibiting DNA replication and transcription, and ASON binds to mRNA precursors blocking RNA processing and maturation, which affects ribosome migration along mRNA, activates RNase, and cleaves unpaired bases in hybrid chains. In the field of tumor research, antisense oligonucleotide technology is used to block or inhibit the expression of key coding genes of tumor cells, inhibiting the proliferation and promoting the apoptosis of tumor cells. Antisense oligonucleotide drugs have become potential new drugs for treating tumors. Several antisense oligonucleotide drugs have been approved by The US FDA, including nusinersen approved for the treatment of myeloid muscular atrophy and eteplirsen for the treatment of Duchenne muscular dystrophy in 2018.

Common antisense drugs are mainly aimed at DNA and mRNA. With further research, antisense drugs targeting non – coding RNA are attracting researchers' attention. MicroRNAs (miRNAs) are highly conserved and short non – coding RNA molecules found in the genome of plants and animals. They are 17 – 27 nucleotides in length and typically regulate post – transcriptional mRNA expression by binding to the 3′ untranslated region (3′ – UTR) of the complementary mRNA sequence, which in turn leads to transcriptional repression and gene silencing. Studies have shown that miRNAs were involved in many important physiological processes *in vivo* and closely related to the occurrence and development of various diseases.

【Materials】

1. Apparatus

(1) Computer with internet connection

(2) Centrifuge

(3) Sterilized cone bottom centrifuge tubes (50ml)

(4) Equipment for immunofluorescence and Western blotting

(5) Laminar Flow Biosafety Cabinet (Level Ⅱ)

(6) Stereomicroscope

(7) Tissue culture dish (10cm)

(8) Tissue incubator (37℃, 5% CO_2), a certain humidity

(9) Tissue culture plate (24 wells)

2. Reagents

(1) ASO (12.5μmol/L and mismatched and/or unrelated control oligonucleotides)

(2) DharmaFECT 4 Transfection Reagent (Dharmacon, Cat. No. T – 2004)

(3) DMEM, free of antibiotics and serum

(4) PBS, free of calcium and magnesium

(5) Fetal bovine serum (FBS), heat inactivated

(6) Mammalian cell line (eg. HeLa or NTera2)

(7) Penicillin and streptomycin

(8) trypsin – EDTA solution

【Procedures】

1. Preparation of antisense oligonucleotides for inhibiting miRNA function

(1) Download the target miRNA sequence from miR Base (http://microrna.sanger.ac.uk/sequences)

(2) Design antisense miRNA sequences based on miRNA sequences and the principle of complementary base pairing.

(3) Design an ASO sequence by adding 5 arbitrary bases (such as the 5′ – UCUUA – antisensemiRNA sequence – ACCUU – 3′) to the antisense miRNA.

Tips: Adding any base at both ends of the ASO to make it more similar to themiRNA target, which can improve ASO bonding efficiency; RISC does not rely on sequences to integrate with ASO, avoiding the degradation of the miRNA center target sequence by intracellular nuclease.

(4) Use the mFold default setting (http://mfold. bioinfo. rpi. edu/cgi – bin/rna – form1 – 2. 3. cgi) to detect the possible secondary structure of the ASO sequence. If the ASO sequence could form a stable secondary structure, change the side chain sequence and repeat step 4.

(5) Perform a Blast analysis. If the mRNA were complementary to the side chain sequence and 13 or more bases in the target sequence, steps 3 to 5 are repeated.

(6) Design a control oligonucleotide. All experiments must contain a mismatched control and/or an unrelated oligonucleotide control with the same side chain sequence and the same length as the experimental group ASO. The mismatch control means that there are at least four equidistant purine – purine mismatches between the ASO and the target miRNA. An unrelated control refers to an ASO consisting of a non – homologous antisense primer sequence or a random sequence from another species. As described in steps 4 and 5, all control oligonucleotides should be carefully examined for possible secondary structure and their complementarity to mRNA. The presence of secondary structures and their complementarity to mRNA can be minimized by altering the bases of the oligonucleotide sequences.

7. Obtain a 2′ – O – methyl modified ASO sequence from oligonucleotide synthesis company.

Tips: Many oligonucleotide synthesis companies offer 2′ – O – methyl modified oligonucleotides. 2′ – O – methyl modification enhances ASO potency by enhancing the affinity of ASO tomiRNA and protects ASO from degradation by RISC and intracellular nucleases. The 3′ – end cholesterol modification can also be performed on ASO, and it has been reported that the 3′ – end cholesterol modification could promote the transport of ASO into cells.

2. Inhibition of miRNA function by antisense oligonucleotides in mammalian cells

(1) Prepare cells

1) The cells are cultured in DMEM medium (containing 10% FBS, 100U/ml penicillin, and 100μg/ml streptomycin), and culture to a 90% monolayer in a 37℃ incubator containing 5% CO_2.

2) The cells are digested with trypsin – EDTA solution to separate them from the plate, and thenresuspend in DMEM medium containing no antibiotics and containing 10% FBS. 500μl of the cells' suspension are inoculated into each well of a 24 – well plate.

3) The cells are incubated overnight in a 37℃ incubator containing 5% CO_2 to achieve a 30% to 40% monolayer (about 5×10^4 cells per well) required for transfection.

Tips: The optimal cell density varies depending on the growth characteristics of the cell line. Please consider the cell line specificity in the instructions of the transfection reagent or determine the optimal cell density by experiment.

(2) Prepare transfection reagents

All experiments should include a mismatch and/or irrelevant control oligonucleotide of the same length as the test group ASO flanking sequence. All volumes represent the volume required for a single test well. The experiments need to be repeated at least 3 times. For transfection of multiple wells, 10% surplus reagent is required to compensate for the loss of solution during mixing.

(3) Add 1μl (12.5 pmol) of experimental or control ASO to a DMEM – sterilized tube containing 24μl of antibiotic – free and serum, mix gently by pipette.

(4) Incubate the mixture for 5 minutes at room temperature.

(5) Slowly add the ASO mixture to the Dharma FECT 4 transfection reagent mixture. The pipette

was gently pipetted, and the mixture is incubated at room temperature for 20 minutes to form an ASO – lipid complex.

(6) Replace the growth medium in step 1 with 450μl of antibiotic – free DMEM containing 10% FBS and heat to 37℃.

(7) Drop 50μl of ASO – lipid transfection medium (from step 5) into each well. Try to cover all the cells in the well and shake gently.

(8) Incubate the cells in a 37℃ incubator containing 5% CO_2 for 1 to 2 days. If cytotoxicity could be observed after 1 day, replace the transfection medium with fresh antibiotic – free DMEM containing 10% FBS. Continue to culture cells under the same conditions.

Tips: If the cells were killed after transfection of miRNA – specific ASO and control ASO, it may be caused by excessive use of transfection reagent or ASO; the cell density may be too low, ideally, the cell fusion should be 30% – 40%, about 5×10^4 cells per well. Transfection conditions need to be optimized before the experiment. It is necessary to perform the experiment when the cell density is appropriate.

If cells were killed only after transfection of miRNA – specific ASO, it may be that the targeted mRNA of the inhibitory miRNA is an anti – apoptotic regulator, or the miRNA – specific ASO may contain a sequence motif that promotes the innate immune response. Try to use different chemical compositions of ASO (such as LNA).

(9) Analysis of protein expression by immunofluorescence and Western blotting to assess the expression of miRNA target genes.

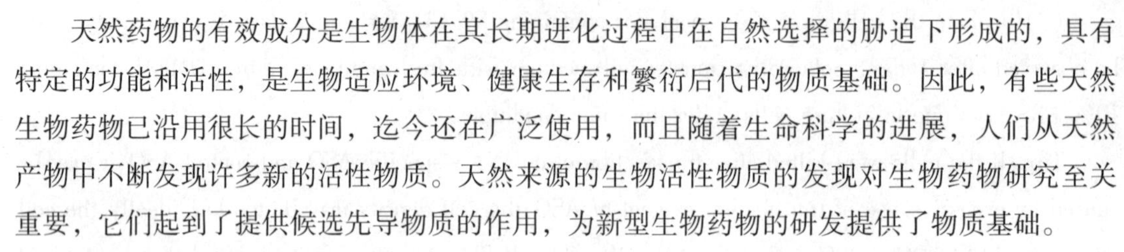

第三部分　设计性实验

扫码“学一学”

实验三十一　天然生物活性物质的发现研究

天然药物的有效成分是生物体在其长期进化过程中在自然选择的胁迫下形成的，具有特定的功能和活性，是生物适应环境、健康生存和繁衍后代的物质基础。因此，有些天然生物药物已沿用很长的时间，迄今还在广泛使用，而且随着生命科学的进展，人们从天然产物中不断发现许多新的活性物质。天然来源的生物活性物质的发现对生物药物研究至关重要，它们起到了提供候选先导物质的作用，为新型生物药物的研发提供了物质基础。

目前，这些活性物质的发现主要集中在以下几个方面。①动物来源的，从动物脏器、血液、分泌物及其代谢物中制备生化活性物质已有很多成功的例子，如从动物胰脏中提取胰岛素、弹性蛋白酶，从猪的肠黏膜中提取肝素等。近年来从鸟类、昆虫类、爬行类、两栖类等动物中寻找具有特殊功能的天然药物，已研究成功的有蛇毒降纤维酶、蛇毒镇痛肽，还发现了多种抗肿瘤蛇毒成分。②人体来源的，如人胎盘因子、人尿及人血中的各种活性物质等。③海洋来源的，如鲨鱼肝再生因子、海洋毒素等。④传统中药来源的，如鹿茸多肽、麝香多肽等。⑤微生物来源的，如香菇多糖等。

其研究思路主要有以下两条。一是以化合物研究为主导的。此类研究首先对所获得的样品应用各种技术进行分离，对分离得到的组分进行活性测定，由测定结果决定是否为新的活性物质。第二种则是以生物活性为导向的。此类研究首先确定样品含某种特殊生物活性的物质，然后以该生物活性为导向进行分离纯化。如研究发现一种美洲巨蜥每年只进食几次，而其血糖却可得以稳定的控制，结果以降糖活性为导向在其唾液中分离得到一种活性多肽，该多肽现在已成功开发为生物药物。

以上两种思路各有利弊。前者应用较普遍但在提取分离和生物活性评价方面结合不够，而且都需要较长的周期和大量的工作，并有盲目性，获得有效活性物质的准确性较低。后者目的明确，实现了分离与活性研究的密切结合，但其活性评价与化学分离难于同步，成为制约研究进度的主要原因，因此建立从分子水平与细胞水平向高通量活性筛选方法格外重要。此外，由于活性评价对样品的需求量较大，给分离工作带来很大负担，在实际工作中还存在许多需要探讨的问题。在具体实践中可以参考实际情况对上述两条思路进行选择。

本设计实验根据所在实验室的具体条件，以天然生物材料（动物来源的、人体来源的、海洋来源的或微生物来源的）为原料，应用生物分离工程技术结合生物活性测定的原理与方法开展生化活性物质（多肽、蛋白质或多糖）的发现研究。

【实验目的】

1. 掌握　多肽、蛋白质或多糖类物质的提取、分离、纯化的一般手段与方法；多肽、蛋白质或多糖类物质纯度检定、分子量检测及含量测定的方法。

2. 学习从资料收集、具体品种工艺设计、制备以及质量控制、数据的收集处理、归纳总结、撰写相关的报告等一个完整的研究过程。

【实验方法】

1. 由教师根据实验室的情况，向学生说明实验要求和分组。

2. 学生分组查阅文献和资料，检索与课题有关的文献和资料，了解实验设计的思路与基本方法，拟定初步实验计划。

3. 学生按分组讨论计划、制定详细实验方案。具体实验方案应包括：所需仪器、试剂；工艺过程流程图和具体实验方案；预期实验的结果；数据结果的处理和统计方法；学时分配。

4. 将实验方案交给指导教师审阅。指导教师对方案认真审阅后，提出修改意见，组织学生讨论，完善方案。

5. 学生修改后的方案经过指导教师批准后，开始准备实验。

6. 实验室的教师要根据学生的方案提前帮助做好准备条件，一些对提高学生能力有比较重要作用的准备工作，让学生在教师指导下完成。实验中如果需要使用的学生不熟悉的仪器设备，应事先指导学生掌握仪器操作方法。

7. 按拟定实验方案进行实验，并详细记录实验结果。

8. 随时分析归纳实验结果，并根据实验情况提出改进的办法（进一步实验的办法）及时调整具体操作。

9. 完成实验论文或科研报告的写作。

【指导原则】

一、蛋白多肽类物质的提取、纯化及分析

（一）多肽、蛋白质类物质的提取、分离、纯化方法

此类方法有多种，一般在分离的初始阶段选择分辨率较低的方法，而在纯化的后期则采用分辨率较高的。针对具体的情况可在以下方法中进行选择。

1. 固相析出分离法　适用于蛋白多肽的初步分离，可将大部分的目的蛋白或是杂质通过极为简单的操作进行选择性获得。具体又分以下几种。

（1）盐析法　该方法是利用各种生物分子在盐溶液中的溶解度的差异，通过向溶液中引入一定量的中性盐，使目的物或杂质沉淀析出，达到纯化之目的。此方法是最早使用的生化分离手段之一，它经济、不需特殊设备而且操作简便安全、应用范围广，较少引起变性。至今仍不失为一种常用的生化分离纯化手段。

（2）有机溶剂沉淀法　该方法利用向溶液中加入一定量的亲水性有机溶剂以降低溶质的溶解度，从而达到使其沉淀析出的目的。其特点为分辨率高于盐析、沉淀剂易于回收，但有导致蛋白变性的风险。

（3）等电点沉淀　该方法利用两性电解质在溶液 pH 处于其 pI 时所带净电荷为零，导致赖以稳定存在的双电层及水化膜削弱或破坏，分子间引力增加从而聚集沉淀析出。其特点是操作极为简单而且不会引入杂质，但由于沉淀不完全，该方法一般不单独使用。

2. 离心技术　分离过程中获得的样品往往为悬浮物，欲进行进一步的分离纯化可考虑采用离心的方法。在离心力场的作用下，加速悬浮液中固体颗粒沉降速度的方法称为离心技术，此技术为生物制药中物质分离纯化的主要手段之一。所采用的离心机按不同的分类

方法可分为多种，但主要还是在速度、容量及是否控温上有所差异。

3. 膜分离技术或过滤技术 过滤技术是生物制药分离纯化中应用最为广泛、最为频繁的技术之一。从原材料的处理直至最终目的物的获得都离不开过滤操作。最原始也是最常见的过滤是指利用固体多孔介质阻留固体颗粒而让液体通过，使固液两相得以分离，随着技术的发展，目前的过滤及膜分离技术可以在分子水平对物质加以区分。

4. 层析技术 根据所采用层析介质的机制不同具体又可分为以下几种。

（1）离子交换层析 该方法利用当溶液的 pH 与蛋白的 pI 存在差异时会使蛋白带一定量的电荷，从而在层析过程中蛋白会结合到层析介质上并且会在一定离子强度下被洗脱下来，利用不同蛋白电荷性质的差异而对其加以区分。在应用此方法时首先对层析柱进行充分平衡，然后上样，再次平衡后以预先设定的盐梯度进行洗脱，对洗脱组分进行分部收集。

（2）分子排阻层析 此方法根据蛋白质的相对分子质量以及蛋白质分子的动力学体积的大小来进行分离的，可应用于蛋白质脱盐和蛋白质分子的分级分离。

（3）疏水层析及反相层析 主要是利用蛋白质疏水性的差异来分离纯化蛋白质。二者的不同在于疏水作用层析通常在水溶液中进行，蛋白在分离过程中仍保持其天然构象，而反相作用层析是在有机相中进行，蛋白经过反相流动相与固定相的作用有时会发生部分变性。

（4）亲和层析 此方法是一种高效的分离纯化手段，不同的蛋白质可以选用不同的特异性亲和配基，如酶和底物、抗原与抗体、糖链和凝集素等。一般是目的蛋白与配基结合而杂蛋白不结合，目的蛋白吸附后再利用快速变换洗脱液和加入竞争剂的方法进行洗脱。由于亲和分离的选择性强，因此在产物纯化中具有较大的潜力。

当几种方法连用时，最好以不同的分离机制为基础，而且经前一种方法处理的样品应能适合于作为后一种方法的料液。如经盐析后得到的样品，不适宜于离子交换层析，但可直接应用于疏水层析。离子交换、疏水及亲和色谱通常可起到蛋白质浓缩的效应，而凝胶过滤色谱常常使样品稀释，在离子交换色谱之后进行疏水层析色谱就很合适，不必经过缓冲液的更换，因为多数蛋白质在高离子强度下与疏水介质结合较强。

亲和层析选择性最强，但不能放在第一步，一方面因为杂质多，易受污染，降低使用寿命；另一方面，体积较大，需用大量的介质，而亲和层析介质一般较贵。因此亲和层析多放在第二步以后。有时为了防止介质中毒，在其前面加一保护柱，通常为不带配基的介质。经过亲和层析后，还可能有脱落的配基存在，而且目的蛋白质在分离和纯化过程中会聚合成二聚体或更高的聚合物，特别是当浓度较高，或含有降解产物时更易形成聚合体，因此最后需经过进一步纯化操作，常使用凝胶过滤色谱，也可用高效液相色谱法，但费用较高。

（二）蛋白多肽类物质纯度检定的方法

大分子类物质没有绝对纯度这一概念，每一种纯度测定方法只是从不同角度对目的物的纯度进行测定。实验室常用的蛋白多肽类物质纯度检定方法有 SDS－PAGE 法、HPLC 法及免疫化学法，所得到的纯度数据分别为电泳纯、色谱纯及免疫纯。

1. SDS－PAGE 法 该方法具体又可分为多种，但其基本原理是一致的。电泳基质存在四个不连续体系即凝胶层、缓冲液离子成分、pH 及电位梯度，从而产生三种物理效应：样品的浓缩效应、凝胶的分子筛效应及一般电泳分离的电荷效应。在电泳过程中不同蛋白

由于 SDS 消除了电荷的影响因素从而仅是根据分子大小表现出不同的迁移率。

2. HPLC 法　按其所采用的不同层析柱可分为多种，但实验室常用的为分子排阻和反相层析，利用 HPLC 可以获得极高的分辨率从而准确地对蛋白纯度进行测定。

3. 免疫化学法　是鉴定蛋白质纯度的有效方法，它根据抗原与抗体反应的特异性，可用已知抗体检查抗原或已知抗原检查抗体。此方法是鉴定蛋白质纯度的特异性方法，但对那些具有相同抗原决定簇的化合物也可能出现同样的反应。

（三）蛋白多肽类物质分子量检测的方法

蛋白多肽的分子量测定方法可分两类，一类是大致测定其表观分子量，而另一类则是具体测定其物理分子量。大致测定蛋白分子量时可使用 SDS－PAGE 或 SEC－HPLC，利用标准蛋白绘制标准曲线，将目的蛋白的迁移率或是保留时间与标准蛋白加以对照即可计算出其分子量。另一类则是利用质谱技术对蛋白的物理分子量进行精确测定，目前此技术发展迅速，现在较为常用的是 MALDI－TOF。

（四）蛋白多肽类物质含量测定的方法

对蛋白多肽进行定量的方法较多，常见的有以下几种。

1. 克氏定氮法　这是测定蛋白质含量的经典方法，其原理是利用蛋白质的含氮量平均在 16%，因此测定蛋白质的含氮量即可计算其含量。

2. 福林－酚试剂法　这是测定蛋白质含量应用最为广泛的一种方法，它利用碱性条件下蛋白质与 Cu^{2+} 形成复合物，还原磷钼酸－磷钨酸试剂形成蓝色化合物，可用比色法测定。其操作简便、灵敏度高，所测定蛋白质的浓度范围为 25～250μg/ml。但此法实际上是蛋白质中酪氨酸及色氨酸与试剂的作用，因此它受到蛋白质氨基酸组成的影响，及不同蛋白质中此两种氨基酸含量的差异使得显色强度有明显差异，此外，酚类等物质对其测定有一定影响，会导致分析的误差。

3. 双缩脲法　在碱性条件下，蛋白质分子中的肽键可与 Cu^{2+} 形成紫红色的络合物，可用比色法定量。此法简单，受蛋白质的氨基酸组成影响较小，但灵敏度小、样品用量大，蛋白质浓度范围为 0.5～10mg/ml。

4. 紫外分光光度法　蛋白质分子中常含有酪氨酸等芳香族氨基酸，在 280nm 处有特征性的紫外吸收，可用于蛋白质定量。此法简单。快速且不损失样品，测定蛋白质的浓度范围在 0.1～0.5mg/ml。如样品中存在有较强紫外吸收的物质时会产生较大误差。蛋白质中含有核酸时可采用下式进行计算：

$$蛋白质的浓度(mg/ml)=1.55A_{280}-0.75A_{260}$$

5. BCA 法　在碱性溶液中，蛋白质将 Cu^{2+} 还原为 Cu^{+} 在于 BCA 试剂生成紫色复合物，于 562nm 处有最大吸收，其强度与蛋白质浓度成正比。此法的优点是单一试剂、终产物稳定，与福林－酚法比较没有干扰物质的影响。尤其是在 Triton－X100、SDS 等表面活性剂中也能够测定。其灵敏度一般在 10～1200mg/ml。

6. Bradford 蛋白分析法　这是一种迅速、可靠的通过染料法测定溶液中蛋白质的方法。利用考马斯亮蓝 G－250 有红蓝两种不同颜色的形式，在一定条件的乙醇及酸性条件下可配成淡红的溶液，当与蛋白质结合后，生成蓝色化合物，反应迅速而灵敏。检测反应化合物在 595nm 的吸光值，可计算出蛋白质的含量。此方法的特点是：快速简便，10 分钟左右便可完成；灵敏度范围一般在 25～200mg/ml，最小可检测至 2.5mg/ml 蛋白质；氨基酸、

肽、EDTA、Tris、糖等无干扰。

二、多糖类物质的提取、纯化及分析

多糖可来自动物、植物和微生物。来源不同，提取分离方法也不同。植物体内含有水解多糖衍生物的酶，必须抑制或破坏酶的作用后，才能制取天然存在形式的多糖。供提取多糖的材料必须新鲜或及时干燥保存，不宜久受高温，以免破坏其原有形式，或因温度升高，使多糖受到内源酶的作用。速冻冷藏是保存提取多糖材料的有效方法，提取方法依照不同种类的多糖的溶解性而定。

（一）多糖的提取

提取多糖时，一般先需进行脱脂，以便多糖释放。方法是将材料粉碎，用甲醇或1∶1乙醇乙醚混合液，加热搅拌1～3小时，也可用石油醚脱脂。动物材料可用丙酮脱脂、脱水处理。

多糖的提取方法主要有以下几种。

1. 难溶于冷水、热水，可溶于稀碱液者　这一类多糖主要是不溶性胶类，如木聚糖、半乳聚糖等。用冷水浸润材料后用0.5mol/NaOH提取，提取液用盐酸中和、浓缩后，加乙醇沉淀得多糖。如在稀碱中仍不易溶出者，可加入硼砂，对甘露聚糖、半乳聚糖等能形成硼酸络合物的多糖，此法可得相当纯的物质。

2. 易溶于温水、难溶于冷水和乙醇者　材料用冷水浸过，用热水提取，必要时可加热至80～90℃搅拌提取，提取液用正丁醇与氯仿混合液除去杂蛋白（或用三氯乙酸除杂蛋白），离心除去杂蛋白后的清液，透析后用乙醇沉淀得多糖。

3. 黏多糖　有些黏多糖可用水或盐溶液直接提取，但因大部分黏多糖与蛋白质结合于细胞中，因此需用酶解法或碱解法使糖—蛋白质间的结合键断裂，促使多糖释放。碱解法可以防止黏多糖分子中硫酸基的水解破坏，也可以同时用酶解法处理组织。提取液中的残留蛋白可以用蛋白质沉淀剂或吸附剂如硫酸铝、藻土等除去。

（二）多糖的纯化

多糖的纯化方法很多，但必须根据目的物的性质及条件选择合适的纯化方法。而且往往用一种方法不易得到理想的结果，因此必要时应考虑合用几种方法。

1. 乙醇沉淀法　是制备黏多糖的最常用手段。乙醇的加入，改变了溶液的极性，导致糖溶解度下降。供乙醇沉淀的多糖溶液，其含多糖的浓度以1%～2%为佳。加完酒精，搅拌数小时，以保证多糖完全沉淀。沉淀物可用无水乙醇、丙酮、乙醚脱水，真空干燥即可得疏松粉末状产品。

2. 分级沉淀法　不同多糖在不同浓度的甲醇、乙醇或丙酮中的溶解度不同，因此可用不同浓度的有机溶剂分级沉淀分子大小不同的黏多糖。在Ca^{2+}、Zn^{2+}等二价金属离子的存在下，采用乙醇分级分离黏多糖可以获得最佳效果。

3. 季铵盐络合法　黏多糖与一些阳离子表面活性剂如十六烷基三甲基溴化铵（CTAB）和十六烷基氯化吡啶（CPC）等能形成季铵盐络合物。这些络合物在低离子强度的水溶液中不溶解，在离子强度大时，这种络合物可以解离、溶解、释放。应用季铵盐沉淀多糖是分级分离复杂黏多糖与从稀溶液中回收黏多糖的最有用方法之一。

4. 离子交换层析法　黏多糖由于具有酸性基团如糖醛酸和各种硫酸基，在溶液中以聚阴离子形式存在，因而可用阴离子交换剂进行交换吸附。常用的阴离子交换剂有D254、Dowexl－X2、ECTEOIA－纤维素、DEAE－C、DEAE－Sephadex A—25和Deacidite FF。吸附时可以使用低盐浓度样液，洗脱时可以逐步提高盐浓度如梯度洗脱或分步阶段洗脱。

5. 凝胶过滤法 可根据多糖分子量大小不同进行分离，常用于多糖分离的凝胶有 Sephadex G 类、Sepharose 6B、Sephacryl S 类等。

（三）多糖含量的测定

1. 蒽酮－硫酸比色法测定糖含量 糖类遇浓硫酸脱水生成糠醛或其衍生物，可与蒽酮试剂结合产生颜色反应，反应后溶液呈蓝绿色，于 620nm 处有最大吸收，吸收值与糖含量成线性关系。

2. 3，5－二硝基水杨酸（DNS）比色法测定还原糖 在碱性溶液中，DNS 与还原糖共热后被还原后棕红色氨基化合物，在一定范围内还原糖的量与反应液的颜色强度成比例关系，利用比色法可测定样品中的含糖量。

3. 苯酚硫酸比色法 苯酚－硫酸试剂可与多糖中的己糖、糖醛酸起显色反应，在 490nm 处有最大吸收，吸收值与糖含量成线性关系。

4. 葡萄糖氧化酶法测定葡萄糖 葡萄糖氧化酶专一氧化 β－葡萄糖，生成葡萄糖酸和过氧化氢，再利用过氧化物酶催化过氧化氢氧化某些物质如邻甲氧苯胺使其从无色转变为有色，通过比色法计算葡萄糖含量。此法专一性高、灵敏度高，适用于测定生成葡萄糖的酶反应。

5. Nelson 法 还原糖将铜试剂还原生成氧化亚酮，在浓硫酸存在下与砷钼酸酸生成蓝色溶液，在 560nm 下的吸收值与糖含量呈线性关系。此方法重复性较好，产物稳定，测定范围为 0. 010 ~0. 18mg。

（四）纯度检查

多糖的纯度只代表某一多糖的相似链长的平均分布，通常所说的多糖纯品也是指具一定分子量范围的均一组分。多糖纯度的鉴定通常有以下几种方法：比旋度法、超离心法、高压电泳法，常压凝胶层析法和高效凝胶渗透色谱法，其中高效凝胶渗透色谱法是目前最常用的也较准确的方法，发展较快，而凝胶过滤法被普遍认为是实验室中最简便可用的方法。

1. 常压凝胶层析法 是根据在凝胶柱上不同分子量的多糖与洗脱体积成一定关系的特性来进行的。凝胶层析的分离过程是在装有多孔物质（交联葡聚糖、多孔硅胶、多孔玻璃等）填料的柱中进行的。选择适宜的凝胶是取得良好分离效果的保证。

2. 高效凝胶渗透色谱法 具有快速，高分辨率和重现性好的优点，因此得到越来越多应用。用于 HPLC 的凝胶柱均为商品柱，可直接使用。选用哪一种性质的填料和用多大的排阻限和渗透限，主要取决于被分离溶质的性质和可能的分子量大小。多糖的检测不采用柱后衍生化方法，而是采用直接检测。最常用的为示差折射检测器，具有中等灵敏度。对于酸性多糖则可以用紫外检测，但多数是采用 RI 和 UV 同时检测。

3. 比旋度法 不同的多糖具有不同的比旋度，在不同浓度的乙醇中具有不同溶解度，如果多糖的水溶液经不同浓度的乙醇沉淀所得的沉淀具有相同比旋度，则该多糖为均一组分。

4. 超离心法 如果多糖在离心力场作用下形成单一区带，说明多糖微粒沉降速度相同，表明其分子的密度、大小和形状相似。

扫码“练一练”

三、多肽、蛋白质或多糖类物质生物活性测定的方法

多肽、蛋白质或多糖类物质的生物活性这一概念极为宽泛，相应的测定方法亦有多种，其具体方法可查阅文献。

Part 3 Designable experiment

EXPERIMENT 31 Discovery of natural bioactive substances

The bioactive components of natural medicine are formed by organisms under the stress of natural selection during long – term evolution. The bioactive components of natural medicine have specific functions and activities, which are the material basis for the organism to adapt to the environment, healthy survival and reproduction. Therefore, some natural biological medicines have been used for a long time and are still widely used. With the development of life sciences, many new bioactive substances are continuously discovered from natural products. The discovery of natural bioactive substances is critical for biopharmaceutical research, and these substances provide candidates for lead materials and provide a material basis for the development of new biopharmaceuticals.

At present, the discovery of bioactive substances is mainly concentrated in the following aspects. There are many successful examples of animal – derived preparation of bioactive substances from animal organs, blood, secretions, and metabolites, such as the extraction of insulin and elastase from the animal's pancreas, the extraction of heparin from the intestinal mucosa of pigs. In recent years, scientists studied natural medicines with special functions from animals such as birds, insects, reptiles and amphibians. The snake venom degrading fibrin and snake venom analgesic peptides have been successfully studied, and various anti – tumor venom components have also been discovered. Bioactive substances from other sources have also been discovered. Human sources such as human placental factors, human urine, and various active substances in human blood. Marine sources, such as shark liver regeneration factors and marine toxins. Traditional Chinese medicine sources, such as pilose antler polypeptides and musk peptides. Microbial sources, such as lentinan.

The main research ideas on natural resources are as follows. One is based on compounds study. The obtained samples are first separated by various techniques, the separated components are subjected to bioactivity measurement, and the measurement results are determined whether the new bioactive substances are used. The second is based on bioactivity evaluation. Such studies first determine that the sample contains substances with specific biological activities and then separate and purify the substances based on the biological activity. For example, scientists found that American monitor lizards only ate several times a year, and the blood sugar could be stably controlled. As a result, an active polypeptide was isolated from its saliva based on hypoglycemic activity. The peptide had been successfully developed into a biological drug.

The above two ideas have their own advantages and disadvantages. The former is more common, but lacks in combination in extraction separation and biological activity evaluation, which both require a long period of time and a large amount of work, and the accuracy of obtaining an effective active substance is low. The purpose of the latter is clear with the close combination of separation and activity research, but its activity evaluation and chemical separationare difficult to synchronize,

which becomes the main reason for restricting the progress of the research. Therefore, it is important to establish a screening method from molecular level and cell level to high - throughput activity. In addition, due to the large demand of the samples for the activity evaluation, it has a great burden on the separation work, and there are still many problems to be discussed in the actual work. In the practice, the above two ideas can be selected with reference to the actual situation.

This design experiment is based on the specific conditions of the laboratory, using natural biological materials (animal - derived, human - derived, marine - derived, or microbial - derived) as raw materials, using biological separation engineering techniques combined with the principle of biological activity measurement and methods for the discovery of bioactive substances (polypeptides, proteins, or polysaccharides).

【Purpose】

1. Master the general methods for extraction, separation, and purification of peptides, proteins or polysaccharides.

2. Master methods for purity determination of peptides, proteins or polysaccharides.

3. Master methods for detecting molecular weight of polypeptides, proteins or polysaccharides.

4. Master methods for determining the content of polypeptides, proteins or polysaccharides.

5. Learn a complete research process from data collection, process design, preparation, and quality control, data collection and processing, summary and writing related reports.

【Procedures】

1. Teachers explain the experiment requirements and groupings according to the situation of the laboratory.

2. Students read the literature and materials, search the literature and materials related to the subject, understand the ideas and basic methods of the experimental design, and design a preliminary experimental plan in groups.

3. Students discuss the plan and develop detailed experimental plans in groups. Specific experimental protocols should include required instruments, reagents, process flow chart and specific experimental plan, the expected result of the experiment, processing and statistical methods of results; school time allocation.

4. Submit the experimental plan to the teacher for review. After carefully review the plan and propose revisions, teachers organize students to discuss and improve the plan.

5. After the students' revised plans are approved by the teachers, the experiment begins.

6. The laboratory teachers should prepare the related materials in advance, according to the students' plan, and some preparatory work that plays an important role in improving the students' ability should be complete by students under the guidance of the teacher. In the experiment, if students need to use the equipment that is unfamiliar to the students, teacher should instruct students to master the operation method of the instrument in advance.

7. Experiments are carried out according to the proposed experimental protocol, and the experimental results are recorded in detail.

8. Analyze the results of the experiment at all time, and propose an improved method (further experimental method) according to the experimental situation to adjust the specific operation in time.

9. Finish the experimental reports.

【Principle】

1. Extraction, purification, and analysis of peptides and proteins

(1) Methods for extraction, separation, and purification of peptides and proteins

There are a variety of methods, generally selecting a lower resolution method in the initial stage of separation, and a higher resolution in the latter stage of purification. For the specific situation, you can choose from the following methods.

1) Solid phase precipitation separation

The method is suitable for the preliminary separation of protein and polypeptides, and most of the target proteins or impurities can be selectively obtained by an extremely simple operation. Specifically, it is divided into the following categories.

① Salting out

The method utilizes the difference in solubility of various biomolecules in a salt solution, and introduces a certain amount of a neutral salt into the solution to precipitate a target substance or an impurity, thereby achieving the purpose of purification. This method is one of the earliest biochemical separation methods. It has the advantages of economy, no special equipment, simple and safe operation, wide application range and less degeneration. It is still a commonly used biochemical separation and purification method.

② Organic solvent precipitation

The method utilizes a certain amount of a hydrophilic organic solvent added to the solution to lower the solubility of the solute, thereby achieving the purpose of precipitation. It is characterized by a higher resolution than salting out and precipitant that is easy to recycle, but has a risk of denaturation of the protein.

③ Isoelectric precipitation

The method utilizes an ampholyte with no charge when the pH of the sample's solution is at its pI, resulting in weakening or destruction of the electric double layer and the hydration film which are stably present, and the intermolecular attraction is increased to accumulate and precipitate. It is characterized by extremely simple operation and no impurities. But since the precipitation is incomplete, the method is generally not used alone.

2) Centrifugation

The samples obtained during the separation process are often suspended, and centrifugation maybe considered for further separation and purification. Under the action of the centrifugal force field, the method of accelerating the sedimentation rate of solid particles in the suspension is called centrifugation technology, and this technology is one of the main methods for separation and purification of substances in biopharmaceuticals. The centrifuges can be divided into various types according to different classification methods, but mainly differ in speed, capacity and temperature control.

3) Membrane separation technology or filtration technology

Filtration technology is one of the most widely used and most frequently used technologies for biopharmaceutical separation and purification. Filtration operation is indispensable from the processing of the raw materials to the final acquisition of the target. The most primitive and most common filtration refers to the use of solid porous media to retain solid particles and allow liquid to pass, so that the solid – liquid two phases can be separated. With the development of technology, the current filtration and membrane separation technologies can separate substances at the molecular level.

4) Chromatography

According to the different mechanisms of the chromatographic medium, chromatography can be divided into the following types.

①ion exchange chromatography

The method utilizes the difference in charge of the protein when the pH of the solution differs from the pI of the protein. During chromatography, the protein will bind to the chromatographic medium and will elute the components at a certain ionic strength, utilizing the different charge properties of different proteins to separate them. When applying this method, the column is first thoroughly equilibrated, then loaded, re – equilibrated, and eluted with a predetermined salt gradient to collect the eluted fraction.

②Molecular exclusion chromatography

This method is based on the relative molecular weight of the protein and the kinetic volume of the protein molecule, and can be applied to protein desalination and fractionation of protein molecules.

③Hydrophobic chromatography and reversed phase chromatography

The main purpose is to utilize the difference in hydrophobicity of proteins to separate and purify proteins. The difference between hydrophobic chromatography and reversed phase chromatography is that the hydrophobic interaction chromatography is usually carried out in an aqueous solution, the protein retaining its natural conformation during the separation process, and the reversed phase chromatography is carried out in the organic phase, partial denaturation sometimes occurring when the protein is subjected to reversed phase mobile phase and stationary phase.

④Affinity chromatography

This method is an efficient means of separation and purification. Different proteins can usedifferent specific affinity ligands, such as enzymes and substrates, antigens and antibodies, sugar chains and lectins. Generally, the target protein is bound to the ligand and the impurity protein is not. The target protein is eluted by rapidly changing the eluate and adding a competitor. Due to the strong selectivity of affinity separation, there is greater potential for product purification.

When several methods are used in combination, it is preferable to use different separation mechanisms and high compatibility. Samples obtained after salting out are not suitable for ion exchange chromatography, but can be directly applied to hydrophobic chromatography. Ion exchange, hydrophobic, and affinity chromatography usually increase the protein concentration, while gel filtration chromatography often dilutes the sample. Hydrophobic chromatography after ion exchange chromatography is suitable, without buffer replacement, because most proteins bind strongly to hydrophobic media at high ionic strength.

Affinity chromatography has the highest selectivity, but it cannot be placed in the first step. Firstly, it could be easily contaminated by the impurities and the service life would be shorten accordingly. Secondly, it is bulky and requires a large amount of medium. The media is generally more expensive. Therefore, the affinity chromatography is placed second or after the second step. Sometimes, to prevent media poisoning, a guard column is placed in front of the column, usually a medium without a ligand. After affinity chromatography, there may be exfoliated ligands, and the target protein will polymerize intodimers or higher polymers during separation and purification, especially when the concentration is high, or when it contains degradation products. Therefore, the final purification process often uses gel filtration chromatography. High performance liquid chromatography can also be used, but the cost is higher.

(2) Methods for determining the purity of protein polypeptides

Macromolecules do not have the concept of absolute purity. Each method of purity determination only measures the purity of the target from different angles. The purity methods of protein or peptides commonly used in laboratories are SDS – PAGE, HPLC, and immunochemistry. The purity data obtained are electrophoresis pure, chromatographic pure, andimmunopure.

①SDS – PAGE method

There are four discontinuous systems in the electrophoresis matrix, namely the gel layer, the ion component, the pH, and the potential gradients, which produce three physical effects: the concentration effect of the sample, the molecular sieve effect of the gel, and the charge effect of general electrophoretic separation. Different proteins exhibit different mobility just depending on molecular size during electrophoresis because SDS eliminates the influence of charge.

②HPLC method

According to the different columns, it can be divided into many types, but what the laboratory commonly used are molecular exclusion and reversed – phase chromatography. HPLC can rely on the high resolution to accurately determine the protein purity.

③Immunochemistry

Immunological technique is an effective method to identify protein purity. Depending on the specific reaction between the antigen and the antibody, the antibody can be examined by a known antibody or an antigen. This method is a specific method for identifying the purity of a protein, but the same reaction may occur for compounds having the same antigenic determinant.

(3) Methods for detecting molecular weight of protein polypeptides

The methods for determining the molecular weight of a polypeptide can be divided into two categories, one is to roughly determine its apparent molecular weight, and the other is to specifically determine its physical molecular weight. SDS – PAGE or SEC – HPLC which using a standard protein to draw a standard curve, is used for determining the apparent molecular weight of the sample by comparing the mobility or retention time of the target protein with the standard proteins. The other is to use mass spectrometry to accurately determine the physical molecular weight of proteins, and MALDI – TOF is now more commonly used.

(4) Methods for determining the content of protein or polypeptides

There are many methods for quantifying the content of protein or peptides, and the following are common.

①Kjeldahl method

This is a classical method for determining the protein content. The principle is that the nitrogen content of the protein is 16% on average, so the content of the protein can be determined by the nitrogen content.

②Forint – phenol reagent method

This is the most widely used method for determining protein content. Proteins can form complex with Cu^{2+} under alkaline conditions which reduce the phosphomolybdic acid – phosphoric acid to a blue compound. It is easy to operate and has high sensitivity. The concentration of the protein measured ranges from 25 to 250μg/ml. However, this method is the action of tyrosine and tryptophan in the protein with the reagent, so the amino acids composition of the protein will affect the results. The difference of the content of the two amino acids in different proteins makes the color intensity significantly different. Substances such as phenols have a certain influence on their determination, which may lead to errors in analysis.

③Biuret method

Under alkaline conditions, peptide bonds in protein can form a purplish red complex with Cu^{2+}, which can be quantified by colorimetry. This method is simple, and is less affected by the amino acid composition of the protein, but the sensitivity is low, the amount of the sample is large, and the protein concentration ranges from 0.5 to 10mg/ml.

④Ultraviolet spectrophotometry

Proteins often contain aromatic amino acids such as tyrosine, which having characteristic ultraviolet absorption at 280nm. This method is simple, quick and without loss of sample. The concentration of the protein ranges from 0.1 to 0.5mg/ml. If there is a substance with strong ultraviolet absorption in the sample, a large error will occur. When a protein contains a nucleic acid, it can be calculated by the following formula:

$$\text{Protein concentration (mg/ml)} = 1.55A_{280} - 0.75A_{260}$$

⑤BCA method

In an alkaline solution, the protein reduces Cu^{2+} to Cu^{+} and then form a purple complex with BCA reagent which producing the maximum absorption at 562nm, and its intensity is proportional to the protein concentration. The advantage of this method is that the single reagent and the final product are stable, and there is no interference effect compared with the forin – phenol method. In particular, it can be measured in surfactants such as Triton – X100 and SDS. Its sensitivity is generally between 10 and 1200μg/ml.

⑥Bradford Protein Analysis

This is a fast and reliable method to determine proteins in solution by dye method. Coomassie Brilliant Blue G – 250 is available in two different colors, red and blue. Under certain conditions of ethanol and acidic conditions, it can be formulated into a reddish solution. When combined with protein, the reddish will turn to blue quickly and sensitively. The absorbance of the reaction compound at 595nm was measured to calculate the protein content. The method is fast and simple, can be completed in about 10 minutes. Sensitivity range is generally 25 – 200μg/ml, the minimum detectable

limit is 2. 5μg/ml protein. Amino acids, peptides, EDTA, Tris, sugar, etc. have no interference in this method.

2. The extraction, purification, and analysis of polysaccharides

Polysaccharides can be derived from animals, plants, and microorganisms. The extraction and separation methods are different depending on the source. Because there exist enzymes hydrolyzing polysaccharide derivative in plants, we can obtain polysaccharides only after inhibiting or destroying those enzymes. The material for extracting the polysaccharide must be stored fresh or in a timely manner, and should not be subjected to high temperature for a long time, so as not to damage its original form. Quick freezing is an effective method for preserving the extraction of polysaccharide materials. The extraction method also depends on the solubility properties of different kinds of polysaccharides.

(1) Extraction of polysaccharides

First, it is generally necessary to defat to release the polysaccharide. The method: pulverize the material, then heat and stir for 1 to 3 hours in a mixture of methanol or 1 : 1 ethanol ether. The mixture can be placed by petroleum. Animal materials can be degreased and dehydrated with acetone.

The following methods are mainly used for extracting polysaccharides.

①Hard to dissolve in cold water, hot water, but soluble in dilute lye

This type of polysaccharides is mainly insoluble gums such as xylan, galactan, etc. After infiltrating the material with cold water, it was processed with 0. 5mol/L NaOH, and the extract was neutralized with hydrochloric acid, concentrated, and then precipitated with ethanol to obtain a polysaccharide. If it is not easily dissolved in a dilute alkali, borax may be added to form a boric acid complex such as mannan or galactan, and this method may obtain a relatively pure substance.

②Soluble in warm water, but difficult to be dissolved in cold water and ethanol

The material is immersed in cold water, extracted with hot water (heated to 80 – 90℃ if necessary). The extract is mixed with n – butanol and chloroform or trichloroacetic acid to remove the bound protein. The supernatant after centrifugation is precipitated with ethanol to obtain polysaccharides after dialyzing.

③Mucopolysaccharides

Some mucopolysaccharides can be directly extracted with water or a salt solution, but most of mucopolysaccharides are bound to the protein in the cell, so it is necessary to use an enzymatic or alkaline solution to break the bond to release polysaccharides. Alkaline and enzymatic hydrolysis can prevent the hydrolysis of the sulfate group in the mucopolysaccharide molecule. The residual protein in the extract can be removed with a protein precipitant or an adsorbent such as aluminum sulfate, algae or the like.

(2) Purification of polysaccharides

There are many purification methods for polysaccharides, but it is necessary to select a suitable purification method according to the nature and conditions of the target. Moreover, sometimes we need to combine different methods.

①Ethanol precipitation

Ethanol changes the polarity of the solution and results in the solubility decrease of sugar. The best concentration of polysaccharide solution is 1% to 2%. Stir for a few hours to ensure complete precipitation of the polysaccharide after the addition of alcohol. The precipitate can be dehydrated with anhydrous ethanol, acetone and diethyl ether, and a loose powder could be obtained after vacuum drying.

②Fractional precipitation method

The solubility of different polysaccharides in different concentrations of methanol, ethanol or acetone is different. Therefore, different concentrations of organic solvents can be used to fractionally precipitate mucopolysaccharides with different molecular sizes especially in the presence of divalent metal ions such as Ca^{2+} and Zn^{2+}.

③Quaternary Ammonium Salt Complexes

Mucopolysaccharides can form Quaternary Ammonium Salt Complexes (QASC) with some cationic surfactants such as CTAB and CPC. These complexes could not be dissolved in aqueous solutions of low ionic strength, and when the ionic strength became high, the complex can be dissociated, dissolved, and released. QASC is one of the most useful methods for fractionating complex mucopolysaccharides and recovering mucopolysaccharides from dilute solutions.

④Ion exchange chromatography Mucopolysaccharides are present in the form of polyanions in solution due to their acidic groups such as uronic acid and various sulfate groups, and thus can be exchange – adsorbed with an anion exchanger. Commonly used anion exchangers include D254, Dowexl – X2, ECTEOIA – cellulose, DEAE – C, DEAE – Sephadex A – 25 and Deacidite FF. Low salt concentration samples can be used for adsorption. When eluting, the salt concentration can be gradually increased, such as gradient elution or stepwise elution.

⑤Gel filtration method

This method can separate polysaccharides according to their molecular weights. The gels commonly used for polysaccharide separation include series of Sephadex G, Sepharose 6B, and Sephacryl S.

(3) Determination of polysaccharide content

①Determination of sugar content by anthrone – sulfuric acid colorimetry.

The sugar is dehydrated by concentrated sulfuric acid to form furfural or its derivative, which can produce color reaction with anthrone with the maximum absorption at 620nm.

②Determination of reducing sugars by 3,5 – dinitrosalicylic acid (DNS) colorimetry.

In an alkaline solution, The DNS is reduced to a brown – red amino compound after co – heating with the reducing sugar. The amount of the reducing sugar is proportional to the color intensity of the reaction liquid in a certain range which can be colorimetric.

③Phenol sulfate colorimetry

The phenol – sulfuric acid can react with the hexose and uronic acid in the polysaccharide having a maximum color absorption at 490nm, and the absorption value is linear with the sugar content.

④Determination of glucose by glucose oxidase method.

Glucose oxidase specifically oxidizes β – glucose to produce gluconic acid and hydrogen peroxide, and then uses peroxidase to catalyze the oxidation of certain substances such as o – aniline to change from colorless to colored. This method is highly specific and sensitive and is suitable for measuring the enzyme reaction that produces glucose.

⑤Nelson method

The reducing sugar reduces the copper to formoxymethylene oxide with the absorption value at 560nm and the absorption is linear with the sugar content. The method has good repeatability and the product is stable, and the measurement range is from 0. 010 to 0. 18mg.

(4) Purity inspection

The purity of the polysaccharide only represents the average distribution of the similar chain length of a certain polysaccharide, and the so - called pure polysaccharide also refers to a uniform component having a certain molecular weight range. There exist many ways to determine the purity of polysaccharides. Among them, high performance gel permeation chromatography is the most commonly used method. But the gel filtration method is generally considered to be the easiest and most available method in the laboratory.

①Atmospheric pressure gel chromatography

Gel chromatography is carried out based on the characteristics of the polysaccharides of different molecular weights on the gel column in relation to the elution volume. The separation process of the gel chromatography is carried out in a column packed with a porous substance (crosslinked dextran, porous silica gel, porous glass, etc). Choosing the right gel is a guarantee of good separation.

②High performance gel permeation chromatography

HPLC has the advantages of fastness, high resolution, and good reproducibility, and thus has been increasingly used. The gel columns for HPLC are commercial columns and can be used directly. The type of the medium, exclusion limits and permeation limits choosing depend primarily on the nature of the separated solutes and the possible molecular weight. The detection of the polysaccharides does not use post - column derivatization, but direct detection. The most commonly used method is a differential refractive detector with moderate sensitivity. For acidic polysaccharides, UV detection can be used, but most of them are detected simultaneously by RI and UV.

③Specific rotation method

Different polysaccharides have different specific rotations and different solubilities in different concentrations of ethanol. If the aqueous solution of the polysaccharide is precipitated with different concentrations of ethanol, and if the precipitate has the same specific rotation, the polysaccharide could4) be considered as a homogeneous component.

④Ultracentrifugation

If the polysaccharide forms a single zone under the action of the centrifugal force field, which the sedimentation speed of the polysaccharide particles is the same, indicating that the density, size, and shape of the molecules are similar.

3. Methods for determining biological activity of polypeptides, proteins or polysaccharides

The concept of biological activity of polypeptides, proteins or polysaccharides is extremely broad, and there are many corresponding methods for determination. The specific methods can be found in the literatures and references.

实验三十二 重组蛋白类药物的表达、制备及分析

扫码“学一学”

广义基因工程是重组 DNA 技术的产业化设计与应用，包括上游技术和下游技术两大部分。上游技术指的是基因重组、克隆和表达的设计与构建（即狭义基因工程）；而下游技术则涉及基因工程菌或细胞的大规模培养以及基因产物的分离纯化及分析过程，最终制造出重组蛋白质乃至自然界没有的新型蛋白质和新物种，为人类提供服务。

狭义基因工程是将一个含目的基因的 DNA 片段经体外操作与载体连接，并转入宿主细胞，而表达产生外源蛋白质的过程。用于基因克隆的载体有质粒、噬菌体、考斯质粒（cosmid）、人造染色体载体等，针对不同的载体，可以选择合适的宿主细胞，常用的宿主细胞表达系统分别为原核细胞表达系统和真核细胞表达系统。真核细胞表达系统主要有酵母表达体系和动植物细胞表达体系，动物细胞表达体系主要以 CHO、NS0、SP2/0、HEK293 表达的重组蛋白质为主，是生物制药最理想的表达系统，但是其培养基成本昂贵，培养环境要求高，工业化生产难度较大；而原核细胞表达系统研究最多的是大肠埃希菌体系，其成本低，周期短，但是纯化工艺相对要求严格。

选择合适的表达载体或宿主，从构建、表达、制备、分析定性到生产都存在着成功和失败两种可能，本实验即是学习如何通过充分的调研、合理的设计及实验方法，利用基因工程技术制备具有生物活性的重组蛋白类药物。

【实验目的】

1. 掌握 重组工程菌的培养和重组蛋白表达方法；重组工程菌及重组蛋白的鉴定方法；重组蛋白纯度及分子量的鉴定方法；重组蛋白含量的测定方法；电泳的基本操作和对目的条带观察分析。

2. 学习从资料收集、目的基因获得、重组体构建、工程菌的克隆与鉴定、目的蛋白的表达、纯化制备以及质量控制、数据的收集处理、归纳总结、撰写相关的报告等一个完整的研究过程。

【实验方法】

1. 由教师根据实验室的情况，向学生说明实验要求和分组。

2. 学生分组查阅文献和资料，检索与课题有关的文献和资料，了解实验设计的思路与基本方法，拟定初步实验计划。

3. 学生按分组讨论计划、制定详细实验方案。具体实验方案应包括：所需仪器、试剂；工艺过程流程图和具体实验方案；预期实验的结果；数据结果的处理和统计方法；学时分配。

4. 将实验方案交给指导教师审阅。指导教师对方案认真审阅后，提出修改意见，组织学生讨论，完善方案。

5. 学生修改后的方案经过指导教师批准后，开始准备实验。

6. 实验室的教师要根据学生的方案提前帮助做好条件准备，一些对提高学生能力有比较重要作用的准备工作，让学生在教师指导下完成。实验中如果需要使用的学生不熟悉的仪器设备，应事先指导学生掌握仪器操作方法。

7. 拟定实验方案进行实验，并详细记录实验结果。

8. 随时分析归纳实验结果，并根据实验情况提出改进的办法（进一步实验的办法）及时调整具体操作。

9. 完成实验论文或科研报告的写作。

【指导原则】

DNA 重组药物制造主要包括以下步骤：获得目的基因；将目的基因和载体连接，构建 DNA 重组体；将 DNA 重组体转入宿主菌构建工程菌；工程菌的发酵；外源基因表达产物的分离纯化；产品的检验等。

1. 目的基因的获得

（1）反转录法　以目的基因转录成的信使 RNA（mRNA）为模板，反转录成互补的单链 DNA（cDNA），然后在酶的作用下合成双链 DNA，从而获取所需的基因。

（2）PCR 法或反转录 PCR（RT - PCR）

典型 PCR 反应包括：模板变性　94℃以上（1 ~ 2 分钟）

退火　50 ~ 55℃（1 ~ 2 分钟）

延伸　72℃（1 ~ 2 分钟）

在高温聚合酶作用下，以 DNA 单链为模板，由引物起始从 5′→3′延伸，可合成 2kb（经多次循环），错误率一般为 0.25%。

（3）化学合成法　60 ~ 100bp 长度为宜。

2. DNA 重组体的构建　根据宿主菌表达系统不同，选择基因载体（Vector）。

（1）*E. coli* 质粒：非表达型（PBR 322），表达型（PUC），实用型（PBV 220，PET 系统）；λ 噬菌体：非表达型（λgt10），表达型（λgt11）。

目的基因 < 10kb 选用质粒为基因载体，目的基因 > 10kb 选用 λ 噬菌体 DNA 为基因载体。

（2）芽孢杆菌 PUB110、pE194 和 pC194。

（3）链霉菌 PIJ101、PSG5。

cDNA 与载体连接方法：同聚尾连接法；人工接头连接法。

3. 重组体的表达系统

（1）原核表达系统　①*E. coli*；②芽孢杆菌 *Bacillus*；③链霉菌 *Streptomyces*。

优点：易大量生产，成本低，周期短。

缺点：多为胞内表达、提取困难，易生成包涵体、含起始密码 Met（AUG），有内毒素毒性。

（2）真核表达系统

1）酵母表达，毕赤酵母（*pichia psatoris*）受甲醇诱导。

优点：易培养无毒性，易高密度发酵（100g/L），高表达 12 ~ 14g 蛋白/L，成本低，产物可糖基化，有分泌表达。

2）动物细胞　哺乳动物细胞——CHO 细胞，昆虫细胞——家蚕细胞。

4. 工程菌的发酵　选育高效表达工程菌，优化发酵条件（培养基组成、接种量、温度、溶氧、pH、诱导作用、发酵动力学）。

5. 表达产物的纯化

（1）产物的表达形式　根据外源基因表达产物在宿主细胞中的定位，可将表达方式分为分泌型表达和胞内表达。

1）外源蛋白的分泌表达是通过将外源基因融合到编码信号肽序列的下游来实现的。将外源基因接在信号肽之后，表达产物在信号肽的引导下跨膜分泌出胞外，同时在宿主细胞膜上存在特异的信号肽酶，它识别并切掉信号肽，从而释放出有生物活性的外源基因表达产物。

2）如果表达产物前没有信号肽序列，它可以可溶形式或不溶形式（包涵体 inclusion body）存在于细胞中。在工业生产中常用大肠埃希菌作为宿主菌来生产目的蛋白，在大肠埃希菌中当外源蛋白的表达量达到20%以上时，它们一般就会以包涵体的形式存在。所谓包涵体是指由于表达部位的低电势及外源蛋白分子的特殊结构如 Cys 含量较高、低电荷、无糖基化等，使得外源蛋白与其周围的杂蛋白、核酸等形成的不溶性聚合体。

（2）表达蛋白的提取

1）如果蛋白质以胞内可溶表达形式存在，则收集菌体后破壁，离心取上清液，然后用亲合层析或离子交换法进行纯化。分泌型表达产物的发酵液的体积很大，但浓度较低，因此必须在纯化前富集或浓缩，通常可用吸附、沉淀或超滤的方法来进行富集或浓缩。因宿主细胞内存在各种蛋白水解酶，破壁后和产物一同释放到细胞上清液中，在纯化过程中还常采取适当的保护措施，如低温、加入保护剂、尽量缩短纯化工艺时间等措施来防止产物的降解和破坏。

2）如果产物以不溶的包涵体形式存在，则可通过离心的方法将包涵体与可溶性杂质分离，常用5000～10000g离心使包涵体沉淀下来，可避免胞内酶的降解破坏，同时包涵体中目的蛋白质的纯度较高，可达20%～80%。但是，此表达形式最大的缺点是包涵体中的蛋白质是无活性形式，必须经变性复性过程重新折叠，常用的方法是以促溶剂（如尿素、盐酸胍、SDS）溶解，然后在适当条件下（pH、离子强度与稀释）复性。

3）另外，表达产物还可存在于大肠埃希细胞周质中，这是介于细胞内可溶性表达和分泌表达之间的一种形式，它可以避开细胞内可溶性蛋白和培养基中蛋白类杂质，在一定程度上有利于分离纯化。大肠埃希菌经低浓度溶菌酶处理后，可采用渗透冲击的方法来获得周质蛋白。由于周质中仅有为数不多的几种分泌蛋白，同时又无蛋白水解酶的污染，因此通常能够回收到高质量的产物。但其缺点是渗透冲击的方法破壁不完全，往往产物的收率较低。

（3）重组蛋白的复性　所谓复性是指变性的包涵体蛋白在适当的条件下折叠成有活性的蛋白质的过程，是利用包涵体获得外源蛋白最关键也是最复杂的一步。主要有两种方法。

1）将溶液稀释，导致变性剂的浓度降低，促使蛋白质复性。此法很简单，只需加入大量的水或缓冲液；缺点是增大了后处理的加工体积，降低了蛋白质的浓度。

2）用透析、超滤或电渗析法除去变性剂。有时包涵体中的蛋白质含有两个以上的二硫键，其中有可能发生错误连接。为此，在复原之前需用还原剂打断—S—S—键，使其变成—SH，复性后再加入氧化剂使两个—SH形成正确的二硫键。常用的还原剂有二硫苏糖醇（1～50mmol/L）、β-巯基乙醇（0.5～50mmol/L）、还原型谷胱甘肽（1～50mmol/L）。常用的氧化剂有谷胱甘肽、空气（在碱性条件下）、半胱氨酸。

复性过程是一个十分复杂的过程，迄今为止人们还没有完全了解它的反应机制，在具体操作中要不断地摸索最适条件，对不同的表达产物包涵体的复性条件不同，需通过实验来确定，而且复性的难易和蛋白质的种类及结构有很大关系。

（4）表达蛋白纯化　基因工程产物常需采用层析来进行精制以达到药用标准。在选择层析类型和条件时要综合考虑蛋白质的性质，如蛋白质的等电点和表面电荷的分布，蛋白质是两性分子，其带电性质随 pH 值的变化而变化。

一般来说，等电点处于极端位置（pI <5 或 pI >8）的基因工程蛋白质应该首选离子交换层析方法进行分离，这样很容易就可以除去大部分的杂质，但在应用时要注意考虑目的蛋白质的稳定性。

亲和层析是一种高效的分离纯化手段，不同的蛋白质可以选用不同的特异性亲和配基，如酶和底物、抗原与抗体、糖链和凝集素等。一般是目的蛋白与配基结合而杂蛋白不结合，目的蛋白吸附后再利用快速变换洗脱液和加入竞争剂的方法进行洗脱。由于亲和分离的选择性强，因此在产物纯化中具有较大的潜力。

疏水作用层析和反相作用层析利用蛋白质疏水性的差异来分离纯化蛋白质。二者的不同在于疏水作用层析通常在水溶液中进行，蛋白在分离过程中仍保持其天然构象，而反相作用层析是在有机相中进行，蛋白经过反相流动相与固定相的作用有时会发生部分变性。

凝胶排阻层析根据蛋白质的相对分子质量以及蛋白质分子的动力学体积的大小来进行分离的，它可应用于蛋白质脱盐和蛋白质分子的分级分离。

考虑到工业生产成本，一般早期尽可能采用高效的分离手段，如通常先用非特异、低分辨的操作单元（沉淀、超滤和吸附等），以尽快缩小样品体积，提高产物浓度，去除最主要的杂质（包括非蛋白类杂质）；然后采用高分辨率的操作单元（如具有高选择性的离子交换色谱和亲和色谱）；而将凝胶排阻色谱这类分离规模小、分离速度慢的操作单元放在最后，以提高分离效果。

当几种方法连用时，最好以不同的分离机制为基础，而且经前一种方法处理的样品应能适合于作为后一种方法的料液。如经盐析后得到的样品，不适宜于离子交换层析，但可直接应用于疏水层析。离子交换、疏水及亲和色谱通常可起到蛋白质浓缩的效应，而凝胶过滤色谱常常使样品稀释，在离子交换色谱之后进行疏水层析色谱就很合适，不必经过缓冲液的更换，因为多数蛋白质在高离子强度下与疏水介质结合较强。亲和层析选择性最强，但不能放在第一步，一方面因为杂质多，易受污染，降低使用寿命；另一方面，体积较大，需用大量的介质，而亲和层析介质一般较贵。因此亲和层析多放在第二步以后。有时为了防止介质中毒，在其前面加一保护柱，通常为不带配基的介质。经过亲和层析后，还可能有脱落的配基存在，而且目的蛋白质在分离和纯化过程中会聚合成二聚体或更高的聚合物，特别是当浓度较高，或含有降解产物时更易形成聚合体，因此最后需经过进一步纯化操作，常使用凝胶过滤色谱，也可用高效液相色谱法，但费用较高。

6. 重组药物的质量控制　重组药物与其他传统方法生产的药品有许多不同之处，它利用活细胞作为表达系统，并具有复杂的分子结构。它的生产涉及生物材料和生物学过程，如发酵、细胞培养、分离纯化目的产物，这些过程有其固有的易变性。

重组药物的质量控制包括原材料、培养过程、纯化工艺过程和最终产品的质量控制。

（1）原材料质量控制　往往采用细胞学、表型鉴定、抗生素抗性检测、限制性内切酶图谱测定、序列分析与稳定性监控等方法。需明确目的基因的来源、克隆经过，提供表达载体的名称、结构、遗传特性及其各组成部分（如复制子、启动子）的来源与功能，构建中所用位点的酶切图谱，抗生素抗性标志物等；应提供宿主细胞的名称、来源、传代历史、检定结果及其生物学特性等；还需阐明载体引入宿主细胞的方法及载体在宿主细胞内的状态，如是否整合到染色体内及在其中的拷贝数，并证明宿主细胞与载体结合后的遗传稳定性；提供插入基因与表达载体两侧端控制区内的核苷酸序列，详细叙述在生产过程中，启动与控制克隆基因在宿主细胞中表达的方法及水平等。

（2）培养过程质量控制　要求种子克隆纯而且稳定，在培养过程中工程菌不应出现突变或质粒丢失现象。原始种子批须确证克隆基因 DNA 序列，详细叙述种子批来源、方式、保存及预计使用期，保存与复苏时宿主载体表达系统的稳定性。对菌种最高允许的传代次数、持续培养时间等也必须做详细说明。

（3）最终产品的质量控制　主要包括产品的鉴别、纯度、活性、安全性、稳定性和一致性。目前有许多方法可用于对重组技术所获蛋白质药物产品进行全面鉴定，如用各种电泳技术分析、高效液相色谱分析、肽图分析、氨基酸成分分析、部分氨基酸序列分析及免疫学分析的方法等；对其纯度测定通常采用的方法有还原性及非还原性 SDS - PAGE、等电点聚焦、各种 HPLC、毛细管电泳（CE）等，需有两种以上不同机制的分析方法相互佐证，以便对目的蛋白质的含量进行综合评价。

扫码“练一练”

EXPERIMENT 32　Expression, preparation, and analysis of recombinant protein drugs

Genetic engineering can be defined as the industrialized design and application of recombinant DNA technology, including upstream and downstream technologies. Upstream technology refers to the design and construction of gene recombination, cloning, and expression (i. e. narrow sense genetic engineering). The downstream technology involves the large - scale cultivation of genetically engineered bacteria or cells and the process of isolation, purification, and analysis of gene products, resulted in the production of recombinant protcins and new species of proteins that are not available in nature to provide services to mankind.

In a narrow sense, genetic engineering is a process in which a DNA fragment containing the target gene is manipulated *in vitro* to connect with the vector and transferred into the host cell to produce exogenous proteins. Vectors used for gene cloning include plasmid, bacteriophage, cosmid, artificial chromosome vector and so on. For different vectors, suitable host cells can be selected. Commonly used host cells belong to prokaryotic cell expression system and eukaryotic cell expression system respectively. Eukaryotic cell expression system mainly includes yeast expression system, animal and plant cell expression system. Animal cell expression system is mainly composed of CHO, NS0, SP2/0, HEK293. It is the most ideal expression system for biopharmaceuticals. However, the cost of the medium is high, the cultivation environment is demanding, and the industrialized production is difficult. *Escherichia coli* is the most widely studied prokaryotic expression system, which has low cost and short cycle, but the purification process is relatively strict.

After choosing the appropriate expression vector or host, there is a possibility of failure in construction, expression, preparation, qualitative analysis, and production. This experiment is to learn how to prepare recombinant protein with biological activity through adequate research, reasonable design, and experimental methods, using genetic engineering technology.

【Purpose】

1. To master the methods of culture and expression of recombinant engineering bacteria.

2. To master the identification methods of recombinant engineering bacteria and recombinant proteins.

3. To master the methods of determining purity and molecular weight of recombinant protein.

4. To master the method of measuring recombinant protein concentration.

5. To grasp the basic operation of electrophoresis and observe and analyze the target bands.

6. To learn a complete research process with data collection, target gene acquisition, recombinant construction, cloning, and identification of Engineering bacteria, expression of target protein, purification and preparation, quality control, data collection and processing, induction and summary, and related reports.

【Procedures】

1. Teachers should explain the experiment requirements and divide student into different groups.

2. Students are grouped to consult the literature and information, search the literature andinformation related to the subject, understand the ideas and basic methods of experimental design, and draw up the preliminary experimental plan.

3. Students discuss plans in groups and make detailed experimental plans. The specific experimental scheme should include Instruments and reagents, process flow chart and specific experimental scheme, expected experimental results, data processing and statistical methods, time allocation.

4. Deliver the experimental plan to thetutor for review. After carefully reviewing, the instructor puts forward suggestions for revision, organizes students to discuss and improves the scheme.

5. Prepare the experiment after the revised procedure are approved by the tutor.

6. Teachers in the laboratory should helpstudents prepare the reagents and equipment according to the student's plan. Some preparations which play an important role in improving students' abilities should be completed under the guidance of teachers. In the experiment, if the students are not familiar with the instrument and equipment, they should be instructed to master the operation method of the instrument in advance.

7. Draw up the experimental scheme for experiment and record the experimental results in detail.

8. Analyze and summarize the experimental results at any time and put forward the improved method(the method of further experiment) according to the experimental situation to adjust the specific operation in time.

9. Complete writing scientific research reports.

【Principle】

DNA recombinant drug manufacturing mainly includes the following steps: obtaining target genes, connecting target genes and vectors to construct DNA recombinants, transferring DNA recombinants into host bacteria to construct engineering bacteria, fermentation of engineering bacteria, iso-

lation and purification of exogenous gene expression products, inspection of products and so on.

1. Acquisition of target genes

(1) Reverse transcription

Using the mRNA from the target gene as a template, it is reversely transcribed into complementary single – stranded DNA(cDNA), which is then synthesized by enzymes to obtain the required genes.

(2) PCR or RT – PCR

Typical PCR reactions include: Templatedenaturation above 94℃(1 – 2 minutes)
Annealing 50 – 55℃(1 – 2 minutes)
Extension 72℃(1 – 2 minutes)

Under the action of high temperature polymerase, 2kb can be synthesized from 5′ to 3′, starting from the primer, using DNA single strand as template. The error rate is generally 0.25%.

(3) Chemical synthesis 60 – 100bp is suitable.

2. Construction of DNA recombinant

Select gene vectors according to different expression systems of host bacteria.

(1) *E. coli* plasmid, non – expressive (PBR 322), expressive (PUC), practical (PBV 220), PET system; λ phage DNA as vectors, non – expressive λgt10, and expressive λgt11.

Target gene < 10KB, select plasmid as gene vector.

Target gene > 10KB, selection λphage DNA as gene vector.

(2) *Bacillus*: PUB110 pE194 and pC194.

(3) *Streptomyces*: PIJ101 and PSG5.

Connection of cDNA and carrier: homo – tail connection method; artificial joint connection method.

3. Recombinant expression system

(1) Prokaryotic expression system ①*E. coli*; ②*Bacillus*; ③*Streptomyces*.

Advantages: easy mass production, low cost, short cycle.

Disadvantages: Most are intracellular expression, difficult extraction, easy to generate inclusion bodies, containing the initial code Met(AUG), and endotoxin toxicity.

(2) Eukaryotic expression system

①Yeast expression Pichia psatoris induced by methanol

Advantages: easy to cultivate non – toxic, easy to high – density fermentation(100g/L), high expression(12 – 14g protein/L), low cost, products can be glycosylated, secretory expression.

②Animal cells expression Mammalian Cell CHO, Insect Cell – Silkworm Cell

4. Fermentation of Engineering Bacteria

Breeding of highly expressive engineering bacteria and optimizing fermentation conditions(medium composition, inoculation volume, temperature, dissolved oxygen, pH, induction, fermentation kinetics).

5. Purification of expressed products

(1) Expression of product

According to the localization of exogenous gene expression products in host cells, the expression patterns can be divided into secretory expression and intracellular expression.

①The secretory expression of exogenous proteins is achieved by fusing exogenous genes into the downstream of the signal peptide sequence. After the exogenous gene is connected to the signal peptide, the expressed product is secreted out of the cell through the membrane under the guidance of the signal peptide. At the same time, there is a specific signal peptidase on the host cell membrane. It recognizes and cuts off the signal peptide, thus releasing the biologically active exogenous gene expression product.

②If there is no signal peptide sequence before the expression product, it can exist in cells in soluble form or insoluble form (including inclusion body). *Escherichia coli* is commonly used as host bacteria in industrial production to produce target proteins. When the expression of exogenous proteins in *Escherichia coli* is more than 20%, they generally exist in the form of inclusion bodies. The so-called inclusion body refers to the insoluble polymer formed by the low potential of the expression site and the special structure of exogenous protein molecule, such as high Cys content, low charge, no glycosylation, etc., which makes the exogenous protein and its surrounding impurities, nucleic acids, and so on.

(2) Extraction of expressed protein

①If the protein is soluble intracellular, collect the cells and break down the cell wall. Collect the supernatant after centrifuging, and then purify the protein by affinity chromatography or ion exchange method. For secretory expression, the volume of fermentation broth is large, but the concentration of products is very low. Therefore, it must be enriched or concentrated before purification. Usually, it can be enriched or concentrated by adsorption, precipitation, or ultrafiltration. Because there are various proteolytic enzymes in host cells, which are released into cell supernatant after wall breaking, appropriate protective measures such as low temperature, adding protective agents and shortening the time of purification process are often taken to prevent the degradation and destruction of products.

②If the product exists in the form of insoluble inclusion body, the inclusion body and soluble impurities can be separated by centrifugation. The inclusion body can be precipitated by centrifugation of 5000 - 10000g, which can avoid the degradation and destruction of intracellular enzymes. At the same time, the purity of the target protein in the inclusion body is high, up to 20% - 80%. But the biggest disadvantage of this expression is that the protein in inclusion body is inactive and must be refolded through denaturation refolding process. The commonly used method is to dissolve the protein in solvents (such as urea, guanidine hydrochloride, SDS) and then refolding under appropriate conditions (pH, ionic strength and dilution).

③In addition, the expressed product can also exist in the periplasm of *E. coli*, which is a form between soluble expression and secretory expression in cells. It can avoid the protein impurities in the cell soluble protein and the medium, and to a certain extent, it is conducive to isolation and purification. The periplasmic protein of *E. coli* can be obtained by osmotic impingement after low concentration lysozyme treatment. Because there are only a few secreted proteins in periplasm and there are no proteolytic enzymes, high quality products can usually be recovered. But its disadvantage is that the method of penetration impact is incomplete, and the yield of products is often low.

(3) Renaturation of recombinant protein

Renaturation refers to the process by which denatured body proteins fold into active proteins under appropriate conditions. It is the most critical and complex step to obtain foreign proteins from inclusion bodies. There are two main ways.

①One is to dilute the solution, resulting in the reduction of denaturant concentration and protein refolding. This method is simple, only need to add a large amount of water or buffer, but the disadvantage is to increase the processing volume of the post – treatment, reduce the concentration of protein.

②The other is to remove denaturants by dialysis, ultrafiltration or electrodialysis. Sometimes proteins in inclusion bodies contain more than two disulfide bonds, which may lead to incorrect connections. For this reason, the reductant should be used to break the—S—S—bond before restoration, and then the oxidants should be added to make the two—SH form the correct disulfide bond. Commonly used reducing agents are dithiothreitol (1 – 50mmol/L), beta – mercaptoethanol (0.5 – 50mmol/L) and reduced glutathione (1 – 50mmol/L). Commonly used oxidants are glutathione, air (in alkaline conditions), cysteine.

The refolding process is a very complex process. So far, the reaction mechanism has not been fully understood. In the specific operation, the optimum conditions should be constantly explored. The refolding conditions of inclusions of different expressed products are different, which need to be determined by experiments. Moreover, the difficulty of refolding is closely related to the type and structure of proteins.

(4) Purification of expressed protein

Genetically engineered products often need to be refined by chromatography to meet medicinal standards. When choosing chromatographic types and conditions, the properties of proteins, such as the distribution of isoelectric point and surface charge of proteins, should be considered comprehensively. Proteins are amphoteric molecules, and their charged properties change with the change of pH value.

Generally speaking, ion exchange chromatography should be the first choice for the separation of genetically engineered proteins whose isoelectric point is at extreme position ($pI < 5$ or $pI > 8$), so that most of the impurities can be easily removed, but the stability of the target protein should be considered in application.

Affinity chromatography is an efficient method for separation and purification. Different proteins can choose different specific affinity ligands, such as enzymes and substrates, antigens and antibodies, sugar chains and lectins. In general, the target protein binds to the ligand while the protein impurity does not. After the target protein is adsorbed, it is eluted by rapid transformation eluent and adding competitors. Because of the high selectivity of affinity separation, it has great potential in product purification.

Hydrophobic chromatography and reverse – phase chromatography utilize differences in hydrophobicity of proteins to isolate and purify proteins. The difference between them is that hydrophobic interaction chromatography is usually carried out in aqueous solution. Protein still maintains its natural conformation in the process of separation. Reverse phase interaction chromatography is carried out in organic phase. Protein sometimes undergoes partial denaturation through the interaction of re-

verse flow phase and stationary phase.

Gel exclusion chromatography separates proteins based on the relative molecular weight of proteins and the kinetic volume of protein molecules. It can be used for protein desalting and protein fractionation.

Considering the cost of industrial production, the most efficient separation methods are usually used in the early stage, such as non – specific, low – resolution operating units (precipitation, ultrafiltration and adsorption) to reduce the sample volume as soon as possible, increase the product concentration, remove the most important impurities (including non – protein impurities); and then high – resolution operation units are adopted (such as highly selective ion exchange chromatography and affinity chromatography). The gel exclusion chromatography, which is a small separation unit with low separation speed, is placed at the end to improve the separation effect.

When several methods are used together, it is better to consider different separation mechanisms. And the sample treated by the former method should be suitable for the feed liquid used as the latter method. For example, samples obtained after salting out are not suitable for ion exchange chromatography but can be directly applied to hydrophobic chromatography. Ion exchange, hydrophobic and affinity chromatography usually play an important role in protein concentration. Gel filtration chromatography often dilute the sample. After ion exchange chromatography, hydrophobic chromatography is very suitable, without the need for buffer replacement, because most proteins bind strongly to hydrophobic media at high ionic strength. Affinity chromatography has the highest selectivity, but it cannot be put in the first step. On the one hand, it is easy to be polluted because of many impurities, which reduces the service life; on the other hand, it needs a large number of affinity chromatography medium, which is generally very expensive. Therefore, affinity chromatography is mostly after the second step. Sometimes, in order to prevent the poisoning of the medium, a protective column is added in front of it, usually a medium without ligands. After affinity chromatography, there may also be exfoliated ligands, and the target proteins will polymerize into dimers or higher polymers during separation and purification, especially when the concentration is high or containing degradation products. Therefore, further purification operation is needed. Gel filtration chromatography is often used, and high – performance liquid chromatography can be used, but the cost is high.

6. Quality Control of Recombinant Drugs

Recombinant drugs differ from other traditional drugs in many ways. They use living cells as expression systems and have complex molecular structures. Its production involves biological materials and biological processes, such as fermentation, cell culture, separation and purification of target products, which have inherent variability.

The quality control of recombinant drugs includes the quality control of raw materials, culture process, purification process, and final products.

(1) Quality Control of Raw Materials

Cytology, phenotypic identification, antibiotic resistance detection, restriction endonuclease mapping, sequence analysis, and stability monitoring are often used. It is necessary to clarify the origin and cloning process of the target gene, to provide the name, structure, genetic characteristics of the expression vector and the source and function of its components (such as replicators and promot-

ers), to construct the digestion map of the loci used in the construction, antibiotic resistance markers, etc. The name, origin, passage history, verification results and biological characteristics of host cells should be provided. The method of introducing vector into host cells and the status of vector in host cells should also be clarified, such as whether it integrates into chromosomes and the number of copies, and the genetic stability of host cells after binding with vectors should be proved. The nucleotide sequences inserted in the control region of both sides of the gene and expression vector are provided. The methods and levels of initiating and controlling the expression of cloned genes in host cells during production are described in detail.

(2) Quality control of culture process

Seed cloning is required to be pure and stable. Mutation or plasmid loss should not occur in engineering bacteria during culture. The DNA sequence of the cloned gene must be confirmed in the original seed batch, and the source, mode, preservation, and expected life of the seed batch, as well as the stability of the host vector expression system during storage and recovery, should be described in detail. The maximum allowable number of passages and the duration of continuous culture of the strain must also be explained in detail.

(3) Quality Control of Final Products

It mainly includes product identification, purity, activity, safety, stability, and consistency. At present, many methods can be used for comprehensive identification of protein drug products obtained by recombinant technology, such as electrophoretic analysis, high performance liquid chromatography, peptide map analysis, amino acid composition analysis, partial amino acid sequence analysis, and immunological analysis. Methods for purity determination include reductive and non - reductive SDS - PAGE, isoelectric focusing, various kinds of HPLC, capillary electrophoresis (CE) and so on. Two or more different mechanistic analysis methods are needed to comprehensively evaluate the purity of the target protein.

实验三十三 生物药物中试放大研究

扫码“学一学”

中试放大是由小试转入工业化生产的过渡性研究工作，对小试工艺能否成功地进入规模化生产至关重要，其研究围绕着如何提高收率、改进操作、提高质量、形成批量生产等方面进行。一个工艺研究项目的最终目的是能在生产上采用，因此，当实验室研究工作进行到一定阶段，就应考虑中试放大，以验证实验室工艺路线的可行性以及在实验室阶段尚未发现的问题。

通过中试研究要达到三个基本要求：①建立稳定的制造规程，为正式生产提供多种必需的工艺条件与参数；②为临床前研究与临床实验的评价提供足量的合格产品；③拟定符合 GMP 要求的制造规程与检定规程。

本中试放大研究可根据学校的条件，以天然生物药物的制备工艺或基因工程药物的制备工艺为研究对象。

【实验目的】

1. 验证实验室工艺路线的可行性。

2. 研究工艺参数，制定工艺规程和检定规程，为正式生产提供工艺参数，保证能在以后生产中应用。

3. 掌握中试放大应具备的条件，并学习中试放大的方法。

4. 掌握中试放大研究的内容，并学习撰写中试放大后的总结报告。

【计划安排】

1. 由教师根据本校的条件，给学生说明实验要求和分组。

2. 学生分组查阅参考文献，讨论和制定详细实验方案，包括所需仪器设备、型号、试剂、方法、工艺过程、数据结果处理等。方案中应该有明确的学时分配。

3. 将实验方案交给指导教师审阅。指导教师对方案认真审阅后，提出修改意见，返还给学生修改、完善方案。

4. 学生修改后的方案经过指导教师批准后，开始进行中试实验。

5. 中试实验室的教师要根据学生的方案提前帮助做好条件准备。其中一些对提高学生能力有比较重要作用的准备工作，组织学生在教师指导下完成。

6. 教师随时在现场指导。介绍中试实验中要使用的学生不熟悉的仪器设备及方法。

7. 要求学生随时观察实验现象、做好实验数据记录，随时分析归纳实验结果，并根据实验情况提出改进的办法（进一步实验的办法）及时调整具体操作。

8. 每组独立撰写中试实验报告，以及实验成功与失败的体会、实验注意事项等。

【指导原则】

一、进入中试应具备的条件

实验进行到什么阶段才能进入中试尚难制定一个标准，但至少下列一些内容在进入中试前应该基本具备：①收率稳定，质量可靠；②操作条件已经确定，产品、中间体及原料的分析方法已经确定；③生物材料的资源（包括菌种，细胞株等）已确定并已系统鉴定；④进行过物料平衡，“三废”问题已有初步的处理方法；⑤提出中试规模及所需原辅料的规格和数量；⑥提出安全生产的要求。

根据上述要求，在考察工艺条件的研究阶段中，必须注意和解决下列问题。

1. 原辅材料规格的过渡实验 在小试后，一般采用的原辅材料（如原料、试剂、溶剂、纯化载体等）规格较高，目的是为了排除原料中所含杂质的不良影响，从而保证实验结果的准确性。但是当工艺路线确定之后，在进一步考察工艺条件时，应尽量改用大规模生产时容易得到的原辅材料。为此，应考察某些工业规格的原辅材料所含杂质对反应收率和产品质量的影响，制定原辅材料质量标准，规定各种杂质的允许限度。

2. 设备选型与材料质量实验 在小试阶段，大部分实验是在小型玻璃仪器中进行，但在工业生产中，物料要接触到各种设备材料，如微生物发酵罐、细胞培养罐、固定化生物反应器、多种层析材料以及产品后处理的过滤浓缩、结晶、干燥设备等。有时某种材质对某一反应有极大影响，甚至使整个反应无法进行，因此在中试时，要对设备材料的质量及设备的选型进行实验，为工业化生产提供数据。

3. 反应条件限度实验 可以找到最适宜的工艺条件（如培养基种类、反应温度、压力、pH 等），一般应有一个许可范围。有些反应对工艺条件要求很严，超过一定限度后，

就会造成重大损失，如使生物活性丧失或超过设备能力，造成事故。在这种情况下，应进行工艺条件的限度实验，以全面掌握反应规律。

4. 原辅材料，中间体及产品质量分析方法研究　在生物制药工艺研究中，有许多原辅材料，尤其是中间体和新产品均无现成分析方法，因此必须研究它们的鉴定方法，以便制定简便易行、准确可靠的检验方法。

5. 下游工艺的研究　上游工艺固然十分重要，如基因克隆、细胞融合、微生物培养等，是生物药物生产的源泉，但下游工艺包括产品的提取、分离、纯化、母液处理、溶剂回收等生化工程操作也必须认真对待，因为这是产品的收率、质量及经济效益好坏的关键所在，因此必须研究尽量简化的下游工艺操作，采用新工艺、新技术和新设备，以提高劳动生产率，降低成本。

二、中试放大方法

中试放大的方法有经验放大法，相似放大法和数学模型放大法。经验放大法主要是凭借经验通过逐级放大（实验装置、中间装置、中型装置和大型装置）来摸索反应器的特征。制药工艺研究中试放大主要采用经验放大法，这是化工科研中采用的主要方法。

三、中试放大内容

为中试放大研究总结内容主要有：①工艺路线与各步反应方法的最后确定；②设备材质与型号的选择；③反应器的规模选择及反应搅拌器型式与搅拌速度的考查。④生产反应条件的研究；⑤工艺流程与操作方法的确定；⑥物料衡算；⑦安全生产与“三废”防治措施研究；⑧原辅材料，中间体的物理性质和化工常数的测定；⑨原辅材料，中间体质量标准的制定；⑩消耗定额，原料成本，操作工时与生产周期等的计算。

扫码“练一练”

中试放大完成后，根据中试总结报告与生产任务等可进行车间设计，制定定型设备选购计划及非标设备的设计、制造，然后按照施工图进行车间的厂房建设和设备安装。在全部生产设备和辅助设备安装完成后，即可制定生产工艺规程，交付试生产。

EXPERIMENT 33　Study on pilot scale－up of biopharmaceuticals

The pilot scale is a transitional research work from small trials to industrial production. It is crucial for the small test process to successfully enter large－scale production. The research focuses on how to improve yield, improve operation, improve quality, and form industrial production batches. The ultimate goal of a process research project is to be able to be used in production. Therefore, when the laboratory research work is carried out to a certain stage, the pilot scale should be considered to verify the feasibility of the laboratory process route and the undetected process at the laboratory stage.

Through the pilot study, three basic requirements must be met: ①establish a stable manufacturing procedure, provide a variety of necessary process conditions and parameters for formal production; ②provide sufficient qualification for the evaluation of preclinical and clinical trials; ③Develop manufacturing procedures and verification procedures that meet GMP requirements.

This pilot scale study can be based on the conditions of the school, the preparation process of natural biopharmaceuticals or the preparation process of genetic engineering drugs.

【Purpose】

1. To verify the feasibility of laboratory technological routes

2. To study the process parameters, formulate process rules, and verification rules, providing process parameters for formal production, and ensure its application in future production.

3. To master the conditions of pilot scale – up and learn the methods of pilot scale – up.

4. To master the content of pilot scale – up, and learn to write a summary report after pilot scale – up study.

【Procedures】

1. The teachers are required to explain the experimental requirements and group according to the conditions of the school.

2. Students review the references, discuss, and develop detailed protocols, including required equipment, models, reagents, methods, processes, and data results processing. There should be a clear allocation of hours in the program.

3. Submit the protocol to the instructor for review. After the instructor carefully reviews the plan, he or she proposes a revised opinion and returns it to the student to modify and improve the plan.

4. After the student's revised plan is approved by the instructor, the pilot test will begin.

5. The teachers in the pilot laboratory should prepare for the conditions in advance according to the student's plan. Some of them are preparatory work that playa more important role in improving students' ability, and organize students to complete under the guidance of teachers.

6. The teacher is on hand to guide. Introduce the equipment and methods that students are not familiar with in the pilot test.

7. Students are required to observe the experimental phenomena at any time, do a good job of recording experimental data, analyze and summarize the experimental results at any time, andpropose improved methods (further experimental methods) according to the experimental conditions to adjust the specific operations in a timely manner.

8. Each group independently writes a pilot test report, as well as the experience of the success and failure of the experiment, and the precautions for the experiment.

【Requirements】

1. The conditions for entering the pilot test

It is difficult to establish a standard when the experiment is going to enter the pilot test, but at least the following contents should be basically available before entering the pilot test: ①The yield is stable and the quality is reliable. ②Operating conditions, analytical methods for products, intermediates and raw materials have been determined. ③ Resources of biomaterials (including strains, cell lines, etc.) have been identified and systematically identified. ④After material balance, the problem

of "three wastes" has been preliminarily dealt with. ⑤The scale of pilot test and the specifications and quantities of raw materials and accessories required are put forward. ⑥Requirements for safe production.

According to the above requirements, in the research stage of examining the process conditions, the following problems must be addressed and solved.

(1) Transition experiment on specifications of raw and auxiliary materials

After the small test, the raw and auxiliary materials (such as raw materials, reagents, solvents, purification carriers, etc.) generally used are of high specification, and the purpose is to eliminate the adverse effects of impurities contained in the raw materials, thereby ensuring the accuracy of the experimental results. However, when the process route is determined, further examination of the process conditions is needed, the raw and auxiliary materials that are easily available in mass production should be used as much as possible. To this end, the impact of impurities contained in raw and auxiliary materials of certain industrial specifications on thc rcaction yield and product quality should be examined, the quality standards of raw and auxiliary materials should be established, and the allowable limits of various impurities should be specified.

(2) Equipment Selection and Material Quality experiments

In the small test phase, most of the experiments were carried out in small glass instruments, but in industrial production, the materials were exposed to various equipment materials, such as microbial fermentation tanks, cell culture tanks, immobilized bioreactors, various chromatograms and filtration, crystallization, drying equipment, etc. of materials and product aftertreatment. Sometimes a material has a great influence on a certain reaction, and even the entire reaction cannot be carried out. Therefore, in the pilot test, the quality of the equipment and the selection of the equipment should be tested to provide data for industrial production.

(3) Reaction Conditions Limit Expcriments

The reaction conditions limit experiment can find the most suitable process conditions (such as the type of medium, reaction temperature, pressure, pH, etc.), generally there should be a permit range. Some reactions are very strict with the process conditions. If they exceed a certain limit, they will cause significant losses, such as loss of biological activity, or exceeding the capacity of the equipment, resulting in an accident. In this case, the limit experiment of the process conditions should be carried out to fully grasp the reaction law.

(4) Study on Analytical Method of Raw and Auxiliary Materials, Intermediates and Product Quality

In the biopharmaceutical process research, there are many raw and auxiliary materials, especially intermediates and new products, and there is no ready – made analysis method. Therefore, their identification methods must be studied in order to develop a simple, accurate, and reliable test method.

(5) Study on Downstream Technology

Upstream processes are important, such as gene cloning, cell fusion, microbial culture, etc., which are the source of biopharmaceutical production, but downstream processes including product extraction, separation, purification, mother liquor treatment, solvent recovery, and other biochemical engineering operations must also be taken seriously. Because this is the key to the yield, quality and

economic benefits of the product, it is necessary to study the downstream process operations as simple as possible, using new processes, new technologies and new equipment to increase labor productivity and reduce costs.

2. Methods

The method of pilot scale amplification has empirical amplification method, similar amplification method and mathematical model amplification method. The empirical amplification method is mainly based on experience to explore the characteristics of the reactor by step – by – step amplification (experimental devices, intermediate devices, medium – sized devices, and large – scale devices). The pilot scale of pharmaceutical process research mainly uses empirical amplification, which is the main method used in chemical research.

3. Contents

The summary of the research for the pilot test is: ①Final determination of process route and step reaction method. ②Selection of material and model of equipment. ③Selection of reactor scale and investigation of reaction stirrer type and stirring speed. ④Study on production reaction conditions. ⑤Determination of process flow and operation method. ⑥Material balance. ⑦Study on safety production and prevention measures of "three wastes". ⑧Determination of physical properties and chemical constants of raw and auxiliary materials and intermediates. ⑨Formulation of quality standards for raw and auxiliary materials and intermediates. ⑩Calculations of consumption quota, raw material cost, operating hours and production cycle, etc.

After the completion of the pilot scale, according to the pilot report and production tasks, workshop design, non – standard equipment design and manufacturing, plant construction and equipment installationshould be recorded. After the installation of all production equipment and auxiliary equipment is completed, the production process rules can be formulated and the trial production can be delivered.

参考文献

1. 周长林．微生物实验与指导［M］．3 版．北京：中国医药科技出版社，2015.

2. 张怡轩．生物药物分析［M］．2 版．北京：中国医药科技出版社，2015.

3. 谭树华．分子生物学实验与指导［M］．北京：中国医药科技出版社，2003.

4. 吴梧桐主编．生物制药工艺学［M］．4 版．北京：中国医药科技出版社，2015.

5. 俞俊棠，唐孝宣，邬邢彦．新编生物制药工艺学［M］．北京：化学工业出版社，2003.

6. 陈钧辉，陶力，李俊等．生物化学实验［M］．3 版．北京：科学出版社 2003.

7. 韦平和主编．生物化学实验与指导［M］．北京：中国医药科技出版社，2003.

8. 司徒镇强，吴军正．细胞培养［M］．北京：世界图书出版公司，1996.

9. 陈秀林．CM－纤维素固定化脂肪酶的研究［J］．海峡药学，2005，17（1），35.

10. 肖丽霞，史永旭，姜涌明．以壳聚糖多空诸位亲和吸附剂载体分离纯化猪凝血酶的研究［J］．食品研究，2000，（21）31.

11. 上官棣华，赵睿，刘国诠．一种分离抗凝血酶Ⅲ的高效亲和色谱填料的制备及表征［J］．高等学校化学学报，2002，23（2），203.

12. 张宏，谭竹钧．壳聚糖微珠作为亲和层析载体的制备及其用于分离牛凝血酶的研究［J］．内蒙古大学学报，2003，34（1），37.

13. 李良铸，李明桦．最新生化药物制备技术［M］．北京：中国医药科技出版社，2000.

14. 汪家政，范明主编．蛋白质技术手册［M］．北京：科学出版社，2000.

15. D. R. 马歇尔编著．蛋白质纯化与鉴定实验指南［M］．北京：科学出版社，2002.

16. 吴敬，赖龙生，刘景晶，吴梧桐．重组 L－门冬酰胺酶的提取纯化工艺［J］．中国医药工业杂志，2000，31（2）：50.

17. 刘景晶，李晶，吴梧桐．大肠肝菌 L－门冬酰胺酶Ⅱ高效表达［J］．中国药科大学学报，1996，27（11）：696.

18. 赖龙生，吴敬，翟源等．L－天冬酰胺酶Ⅱ基因工程菌的培养［J］．药物生物技术，1999，6（4）：140.

19. 曹荣月，申克宇，王学军，凤姣，刘景晶．重组 L－门冬酰胺酶Ⅱ工程菌的表达和 PEG 的化学修饰［J］．生物学杂志，2005，22（3）：8.

20. 吴敬，吴梧桐，赖龙生，刘景晶．重组 L－门冬酰胺酶Ⅱ的中试工艺及其性质的研究［J］．中国药学杂志，2000，35（40）：268.

21. J. 萨姆布鲁，E. F. 弗里奇，T. 曼尼阿蒂斯．分子克隆实验指南［M］．2 版．北京：科学出版社，1998.

22. 崔莉，谭树华，章良，吴梧桐．重组水蛭素Ⅲ的分离纯化与鉴定［J］．中国药科大学学报，2005，36（1）：78.

23. 崔莉，谭树华，吴梧桐．重组水蛭素Ⅲ的工程菌的培养及高密度发酵［J］．药物生物技术，2004，11（1）：22.

24. TanShuhua，WuWutong，LiuJingjing，et al. Effivient expression and secretion of recombinant hirudinⅢ in E. coli using the LasparaginaseⅡ singal scquence［J］. Protein Expression and Purification，2002，25（3）：430.

25. 李永明．实用分子生物学方法手册［M］. 北京：科学出版社，2001.

26. 谭树华，吴梧桐，刘景晶．在大肠杆菌中分泌表达水蛭素基因的初步研究［J］. 药物生物技术，1998，5（4）：197.

27. 陈桂良，陈钢．重组水蛭素的表达纯化及活性检测方法进展［J］. 药物生物技术，2004，11（1）：52.

28. 梁国栋．最新分子生物学实验技术［M］. 北京：科学出版社，2001.

29. 谭树华，吴梧桐，刘景晶．水蛭素（HV3）基因的合成、克隆与鉴定［J］. 药物生物技术，1998，5（1）：18.

30. 李良铸，李明桦．最新生化药物制备技术［M］. 北京：中国医药科技出版社，2000.

31. 方成，何执中．快速蛋白液相色谱法纯化威廉环毛蚓纤溶酶及其活性研究［J］. 药物生物技术，2004，11（1）：45.

32. 李菁华．蚓激酶研究与应用进展［J］. 锦州医学院学报，2004，25（5）：58.

33. 师治贤，王俊德．生物大分子的液相色谱分离和制备［M］. 北京：科学出版社，1996.

34. 张玉彬，王旻，吴梧桐，等．基因工程菌 CTB2 的固定化及其制备 L－苯丙氨酸的研究［J］. 中国药科大学学报，1996，27（7）：445.

35. 史燕东，吴梧桐，张玉彬，等．表达芳香族氨基酸转氨酶基因工程菌的细胞固定化初步研究［J］. 中国药科大学学报，1994，25（5）：316.

36. 赵景联．胰岛素生产技术［J］. 轻化工实用技术，1992（4）：42.

37. 徐英权，安帮弢．胰岛素大规模生产研究［J］. 黑龙江医药，1995，8（4）：233.

38. 吴世斌，陆海波．胰岛素粗品生产中盐析工艺的改良［J］. 中国生化药物志，2001，22（2）：88.

39. 李湛君，杨昭鹏，徐康森．重组人胰岛素效价测定中生物测定与理化测定的相关性验证［J］. 药物分析杂志，1998，18（4）：241.

40. 吕炜锋．酶法制取 L－丙氨酸工艺中试研究［J］. 药物生物技术，1996，3（2）：94.

41. 杨安钢，毛积芳，药立波．生物化学与分子生物学实验技术［M］. 北京：高等教育出版社，2001.

42. 齐香君．现代生物制药工艺学［M］. 北京：化学工业出版社，2004.

43. 陈建龙，祁建城，曹仪植，等．固定化酶研究进展［J］. 化学与生物工程，2007，23（3）：7.

44. 罗贵民，曹淑桂，张今．酶工程［M］. 北京：化学工业出版社，2002.

45. 于萍，赵先亮，张利宁．间接抗体夹心酶联免疫吸附法测定人红细胞生成素（EPO）体外活性［J］. 齐鲁药事，2005，24：150.

46. 徐志利．重组人促红细胞生成素的制备及临床应用［J］. 药学进展，1993，7

(2): 65.

47. Zhou W, Bi J, Janson JC, et al. Ion – exchange chromatography of hepatitis B virus surface antigen from a recombinant Chinese hamster ovary cell line [J]. J Chromatogr A. 2005 Nov 18; 1095 (1 –2): 119.

48. 张国强, 赵铠. 从乙肝疫苗的制备看疫苗纯化技术的发展 [J]. 微生物学杂志, 2006, 26 (5): 59.

49. Dermot Walls, Sinéad T. Loughran. Protein Chromatography [M]. Humana Press. 2017.

50. Tong Y, Fang X, Tian H, et al. De novo generation of specific humanIgGs by in vitro immunization using autologous proteins containing immunogenic p – nitrophenylalanine [J]. MAbs, 2019, 11 (2): 401.

51. Da S E M S J, Otávio A V C, Ennes I, et al. Simple immunoaffinity method to purify recombinant hepatitis B surface antigen secreted by transfected mammalian cells. [J]. Journal of Chromatography B, 2003, 787 (2): 303.

52. 郭敏, 李育敏, 费嘉, 张洹. 以 microRNA –21 为靶标的反义寡核苷酸对人白血病 K562 细胞的抑制作用 [J]. 中国病理生理杂志, 2009, 25 (6): 1127.

53. YoshiokaKotaro, Kunieda Taiki, Asami Yutaro, et al. Highly efficient silencing of microRNA by heteroduplex oligonucleotides [J]. Nucleic acids research, 2019, 47 (14).

54. YamakawaKeiko, Nakano – Narusawa Yuko, Hashimoto Nozomi, Yokohira Masanao, Matsuda Yoko. Development and Clinical Trials of Nucleic Acid Medicines for Pancreatic Cancer Treatment [J]. International journal of molecular sciences, 2019, 20 (17).

55. 谭云. 干扰素的功能和制备方法 [M]. 生物学教学, 2010, 35 (1): 65.

56. 郭兆斌, 韩玲, 刘亮亮, 李君兰. 猪胰脏中胰酶的提取工艺优化研究 [J]. 食品科学, 2019, 30 (22): 162.

57. 郭君, 杨孝辉. 正交实验优化猪胰脏中胰岛素的提取工艺 [J]. 广州化工, 2011, 39 (11): 93 –95.